RATIONAL QUEUEING

RATIONAL QUEUEING

Refael Hassin

Tel Aviv University, Israel

CRC Press
Taylor & Francis Group
Boca Raton   London   New York

CRC Press is an imprint of the
Taylor & Francis Group, an **informa** business

A CHAPMAN & HALL BOOK

CRC Press
Taylor & Francis Group
6000 Broken Sound Parkway NW, Suite 300
Boca Raton, FL 33487-2742

First issued in paperback 2020

© 2016 by Taylor & Francis Group, LLC
CRC Press is an imprint of Taylor & Francis Group, an Informa business

No claim to original U.S. Government works

ISBN-13: 978-1-4987-4527-7 (hbk)
ISBN-13: 978-0-367-65855-7 (pbk)

Visit the Taylor & Francis Web site at
http://www.taylorandfrancis.com

and the CRC Press Web site at
http://www.crcpress.com

Contents

Preface

The idea for writing a survey on rational queueing arose in the summer of 2010 when I realized how unfamiliar I was with the latest research in spite of the fact that a book I co-authored on this subject was published in 2003.

Research on strategic behavior in queueing systems is carried out across a wide variety of disciplines. My intention in writing this book is first and foremost to make a contribution to the research community in this area which can benefit from a comprehensive survey providing a single overall inclusive account.

As often happens, once underway it readily became apparent to me that this endeavor would require much more time and effort than originally anticipated. As I began studying the existing literature, more and more I realized how dynamically this topic is evolving. Nearly half of the surveyed papers included here were published or distributed in 2011 or later, that is to say, after I started working on this book!

My own research on strategic behavior in queues focuses on equilibrium behavior and social optimality. Other topics covered were new to me when I started writing, and I benefited significantly from the cooperation and important feedback of many of those whose work is summarized here.

Special thanks go to Antonis Economou, Amihai Glazer, and Nina Hassin for reading the complete manuscript and providing numerous comments, and to Carl Sperber, my language editor, for his excellent editing. This research was supported by the Israel Science Foundation (grant No. 1015/11).

About the author

Refael Hassin is a prominent figure in the area of rational queueing, a first-class scholar with an international and very highly esteemed reputation. His research has mostly focused on discrete optimization, but after writing his first paper on strategic queueing in 1983, he became interested in the economics of queues. He is well recognized both for his research contributions and for his excellent previous book on the topic. He has received four grants from the Israel Science Foundation to investigate strategic queueing systems.

Chapter 1

Introduction

1.1 Rational queueing

Rational queueing pertains to queueing systems with interactions among agents rationally acting to maximize goals defined by primitive data.

Queueing theory research can be classified as:

- performance analysis,

- optimal design and control,[1]

- analysis of rational (strategic) behavior.

These categories can also be defined according to the number of *decision makers* (**DMs**): In the first, performance analysis, there are no **DMs**. In the second, system optimization, there is a single **DM** (e.g., a central planner), whereas in the third there are at least two **DMs** whose decisions interact (e.g., profit-maximizing servers and customers, or a service system owner and a server operating under contract).

In other words, rational queueing deals with *games* played by *agents*, or *players*, who operate in a queueing system according to an economic choice model. The solution is, in most cases, an *equilibrium*, referring to a state where players cannot improve their payoffs by unilateral changes in behavior. The only comprehensive reference that pertains to strategic queueing is the book by Hassin and Haviv [1] (2003).

The fundamental questions asked in this line of research concern the existence, uniqueness, and characterization of the equilibrium. Another important research goal is to compare the equilibrium to the best aggregate outcome that could be expected for the participating agents; that is to say, the *socially optimal* or *system-optimal* (**SO**) solution. In particular, there is interest in whether the equilibrium solution is **SO**, and if not, then how can the **DM** induce the **SO** solution, i.e., *coordinate* or *regulate* the system.

---

[1]According to Stidham [594] (2009) *design* corresponds to *static* optimization of parameters such as arrival and service rates, *control* corresponds to *dynamic* optimization in response to changes in the state of the system. The book [594] provides a comprehensive treatment of optimal design of queueing systems.

In many cases, rational queueing insights are special cases of general principles. For example, the inefficiency of the equilibrium in Naor's [497] model follows from negative externalities associated with joining the queue, as in non-queueing congested systems.[2] However, the queue structure, particularly under Markovian assumptions, often enables explicit solutions and definitive results. Such outcomes are not possible in a more general model of congested systems. Moreover, the queue structure often enables unique non-price control through manipulations of the queue regime. These properties make rational queueing differ in many respects from more general economic theory.

1.2   Scope

The starting point of this survey is the aforementioned book [1] and we focus on literature not covered there. With few exceptions, papers treated there do not appear in the bibliography here and the reader is referred to the appropriate parts of that book for complementary information.[3] The scope of this survey is somewhat extended relative to [1] and includes models with implicit strategic behavior and bounded rationality.

Our main focus will be on *stochastic* queueing models. Deterministic service systems are not covered in our survey unless the model strongly relates to stochastic queues like, for example, when it serves to approximate a stochastic counterpart.

Our focus is on *queueing games*, which are games played in queueing systems. Typically, the players (or agents) are customers, system owners, and social planners. For the most part, this survey refers to models that start with primitive data (such as service values and waiting costs) and with the players' objective functions. We do include, however, models where players' behavior is guided by exogenous rules. An example of this would be when aggregate demand for service is given by a function of the price of service and the waiting-time distribution. The reason being that such a function can often be implied by rational behavior (see §1.5).

Other types of models may result from implicit strategic behavior. For example, models assuming a state-dependent arrival rate resulting from *probabilistic joining*, where a customer observing a queue of size n joins with probability $p(n)$, or balks otherwise. The joining probabilities $p(n)$ are usually exogenous,[4] monotone nonincreasing, and induces a state-dependent joining

---

[2]See, for example, G. Hardin, "The tragedy of the commons," *Science* **162** (1968) 1243-1248.

[3]Papers listed in the bibliography already treated in [1] are marked with [* * *].

[4]See [397, 188] and [1] §2.5 for models where the joining probability is endogenous. The model with uniformly distributed waiting cost rates described there has recently been rediscovered in [268].

behavior which can be explained by heterogeneity in service valuations or waiting-costs. The focus of many papers that assume probabilistic joining is not strategic, and we only include those that have an underlying decision model, or give a strategic interpretation to probabilistic joining behavior.

A second class of models for which the same arguments apply deals with reneging from a queue. In many cases reneging is assumed to occur at a fixed rate. We concentrate on papers that include economic decision making models that yield the reneging behavior. Similar comments apply with regard to retrial models.

It is a challenge to impose boundaries on topics included in a survey. In many cases it is difficult to decide whether a model even deals with a queueing system. For example, systems like $M/M/s/s$ and $M/M/\infty$ are considered queueing systems though they have no queues and waiting costs may be irrelevant. The performance measure in $M/M/s/s$ is typically the blocking probability rather than the waiting time. We will frequently describe here models dealing with general non-queueing delay functions, especially when they lead to interesting queueing insights.

A different approach for classifying strategic queueing models is given in a short exposition by Altman [42] (2005), who classifies models according to the question they ask: to queue or not to queue; when to queue; where to queue; and how much to queue. We also refer the reader to the relevant surveys by Afèche [10] (2006), in particular §2, and Haviv [320] (2011).

1.3 Mode of description

Although there are exceptions to every rule, we generally adopt the following guidelines:

- **Emphasis:**

  - This survey emphasizes *qualitative* aspects. For each paper we describe the main assumptions and qualitative results.

  - We avoid mentioning common assumptions (such as independence of stochastic processes, or obvious nonnegativity of parameters) and technical assumptions made in the interest of facilitating the presentation.

  - We do mention, however, assumptions that restrict generality of insights obtained.

  - We often define constants implicitly. For example, a statement that the demand function is $\lambda = a - bp$ where p denotes price implicitly defines the constants a and b.

- **Classification** Most papers in this survey belong to subjects in more than a single chapter, and we try to place each paper according to its main theme. However, the assigning of subjects is often nearly arbitrary due to imprecise definitions or overlapping of topics. For example, supply chains and networks of complementary services share many features and in some cases assignment to either of these subjects is reasonable. Similarly, models of competition are closely related to those on routing to parallel servers as in both cases customers select their preferred servers. In some cases a paper's description is divided into two parts, each in a different chapter. In other cases it is placed in one chapter, with a comment in another chapter that mentions its relevance. However, obvious relations are not mentioned. For example, almost all of the papers contain customer decisions, but the papers placed in the customer decisions chapter are only those whose focus is on customer decisions.

- **Solution:** The main solution concept is some type of an equilibrium. In most cases it is a Nash equilibrium or variation (like Wardrop, subgame perfect, or Bayesian). We often do not explicitly refer to the type of equilibrium, simply calling it an equilibrium.

- **Applications:** Unless the model applies mainly for a certain type of application we do not describe the application that motivated the research, but rather emphasize the generic model. Our terminology is mostly queueing theoretical. Interested readers will be able to find models closest to their particular subjects of research or application and use our short summary to decide whether the paper is relevant for their needs.[5] Yet when the model makes little sense outside of its motivating context, we will use the application's terms.

- **Order:** Summaries within a section are arranged chronologically according to date of publication.

- **Cross references:** Many sections start with a list of related papers described in other parts of this survey. These lists are not comprehensive and the readers are invited to use the subject index to locate additional relevant references.

- **Notation:** There is no standard notation for most parameters and variables associated with queueing models. With few exceptions, we follow the notation used in the surveyed paper. However, since there is almost a universal use of λ for the arrival rate and μ for the service rate, we use these notations without redefining them. We also often use ρ for the utilization factor, when it equals λ/μ.

---

[5]A similar approach is adopted in [594]. See page x of [594] for justification.

1.4   Basic models and assumptions

The first mathematical model of a queueing system with rational customers was formulated by Naor [497] (1969) (see [1] §2). Naor's queue is *observable*, meaning customers know the queue length when deciding whether to *join* the queue or to *balk*. The model assumes an M/M/1 server, homogeneous customers valuing service at R and incurring waiting costs at rate C per unit time while waiting and during service.

The first *unobservable* models, where customers do not know the queue length when deciding whether to join or balk were formulated by Littlechild [448] (1974) under the assumption of customer heterogeneity, and by Edelson and Hildebrand (E&H) [221] (1975) under the assumption of customer homogeneity. The E&H model is the unobservable version of Naor's model (see [1] §3).[6]

Generally, it is assumed that the cost associated with joining a queue is a two-variable function, $C(p, w)$, where p is a monetary fee, and w measures the sojourn time. A common assumption is that $C(p, w)$ is a function of $p + c(w)$ which is the *full price* (or, *full cost*) incurred by the customer. A customer would join the queue if the service value is at least the expected full price. Similarly, when choosing among different servers offering the same service, customers prefer servers with a smaller full price (see §1.5).

Naor's model assumes customers to be *risk neutral*, and this assumption is common to most of the literature. We make this assumption everywhere unless otherwise explicitly stated. Thus, it is understood that customers maximize expected net benefit. In Naor's model, the net benefit is (w.l.o.g.) 0 for balking, and $R - p - Cw$ for joining, where w is the expected waiting time conditioned on available information. It follows that in Naor's observable case there is a pure equilibrium strategy defined by a *threshold* or *critical* value that distinguishes between queue lengths that prescribe joining and those which prescribe balking. In the E&H unobservable case, either all customers join in equilibrium, or none do, or the equilibrium strategy is mixed and the joining rate is such that customers are indifferent between joining and balking.

The common optimality criterion is the long-run average return for the server or for the system. In some cases discounted return is optimized, but the qualitative conclusions are often the same, and unless the difference is important we will not specify this assumption.

---

[6]The **SO** arrival rate in this model has already been computed by **Hillier [341] (1964)**, see [594] §1.2.1.

1.5   Demand

Waiting is the main feature of a queueing system, and in most cases waiting is costly. Most models assume waiting cost to be borne by the customers. In other models the firm bears the waiting costs, known as *holding costs*. Common performance measures that affect customer welfare are: *service value* – utility obtained upon service completion; *waiting time* – time a customer spends in the system from arrival to service completion; *cost* – nominal payments associated with service; *loss probability* – the probability a customer request is not fulfilled (either being rejected or following the customer's reneging decision); *throughput* – average rate of served requests; and *fill rate* – the probability of immediately obtaining service without queueing. Customer net utility can be a function of several of these performance measures. It is common to distinguish relevant measures by saying that a customer is *price sensitive, delay sensitive*, etc.[7]

A commonly used approach is the Littlechild-Mendelson demand model (see [1] §3.3, §4.4.3, and §8.1). Suppose that the potential arrival rate to the system is Λ, and customers have heterogeneous service valuations. The *value function* $V(\lambda)$ is the total value generated when the fraction λ/Λ of the customers with the highest service values is served. This function is increasing and concave. Alternatively, suppose that customers are homogeneous but the quality of service deteriorates as more customers are served. Then $V(\lambda)$ is the system value rate of serving λ customers per unit time.[8] Given the expected cost x associated with joining the system and waiting to be served, customers who value service above x will join. Hence a demand function $\lambda(x)$ can be derived from the relation $V'(\lambda) = x$. For example, $V(\lambda) = a\lambda - b\lambda^2$ with $a, b > 0$ yields the linear demand function $\lambda(x) = (a - x)/2b$. In many cases, x is the full price composed of a nominal fee and waiting costs.

1.5.1   Substitution effects

An important part of the strategic queueing literature deals with non-monopolistic markets (see [1] §7). Consider a market with two firms providing the same service (or, *substitutable* services) at prices $p_1 < p_2$ and delays $W_1 > W_2$. Suppose that the waiting-cost rate C is distributed within the population according to a cdf F, and the service value R is the same at both service

---

[7]It is common to assume that customers are sensitive to quality of service, but the meaning of this term is not unique and it may refer to expected sojourn time, rate of service, or value of service.

[8]This interpretation applies to [420], where the value from serving demand $\Gamma_j = \sum_i \lambda_{ij}$ depends on the aggregate demand Γ_j, but not on its components (λ_{ij}) that belong to different users.

providers. Then, the customers choosing firm 1 are those with C value such that $p_1 + CW_1 < \min\{R, p_2 + CW_2\}$, those with $p_2 + CW_2 < \min\{R, p_1 + CW_1\}$ will choose firm 2, and the rest will balk. Let Λ be the *market size*, i.e., the potential demand rate, then the demand functions are:

$$\lambda_1 = \Lambda F\left(\min\left(\frac{p_2 - p_1}{W_1 - W_2}, \frac{R - p_1}{W_1}\right)\right)$$

$$\lambda_2 = \Lambda F\left(\frac{R - p_2}{W_2}\right) - \Lambda F\left(\frac{p_2 - p_1}{W_1 - W_2}\right).$$

(1.1)

The effects that competitors' prices and delays have on a firm's demand are commonly referred to as *substitution effects*, and reflects situations where an increase in price or delay might lead users to instead use a substitute.

A commonly used alternative is the linear demand function which also incorporates substitution effects:

$$\lambda_i = a_i - b_i p_i + \beta_i p_j - c_i W_i + \gamma_i W_j \quad j = 3 - i \ \ i = 1, 2. \quad (1.2)$$

The β constants are the *substitutability coefficients*.

Note that the demand function (1.2) allows customers to choose firm 2 even when firm 1 dominates in both dimensions of price and delay, that is, $p_1 < p_2$ and $W_1 < W_2$. Justification for this would be explained through other variables that are not explicit in the model, such as location and quality, and the market structure is one of *monopolistic competition*. In contrast, under the full-price approach only firms with the minimum full price receive positive demand.

1.5.2 Attraction models

In most strategic queueing models each customer chooses the option that maximizes his utility. A different approach is represented by *attraction demand models*. Here, there is a function that determines the probability of choosing an option, with better options being chosen more often. In general, if the *attractiveness* of option i is v_i, then the probability it is chosen is $a_i(v_i)/\sum a_j(v_j)$. For example, v_i can be the value of obtaining service at firm i and it can depend on such parameters as the price and quality (often measured in terms of delay) of this firm. Some common attractiveness functions are (see [243]):

- **Multinomial logit (MNL):** $a_i(v) = \exp(-\alpha_i v)$, with $\alpha_i > 0$.

- **Cobb-Douglas:** $a_i(v) = v^{-\alpha_i}$, with $\alpha_i > 1$.

- **Linear:** $a_i(v) = \alpha_i - \beta_i v$ with $\alpha_i, \beta_i > 0$.

The attraction model is supported by a derivation corresponding to a set of axioms about the market behavior.[9] An alternative approach leading to an attraction model[10] derives the logit choice by assuming players make errors in choosing their strategy. The resulting equilibrium is called *Quantal Response Equilibrium* (QRE), and with the logit function it becomes a *logit quantal response equilibrium* (LQRE). It is common to apply QRE when analyzing laboratory experiments. Some authors (e.g., [332, 351, 442]), explicitly refer to logit choice as irrational and incorporate a parameter that expresses the *degree of irrationality*.

1.5.3 Hotelling-type models

Hotelling-type models assume customers reside in a *linear city*. Each customer considers the distance between his location and the server's location and the congestion at the server while deciding whether to join a queue. The location of the server is often a decision variable. In a more general context, the server's location may indicate the relative positioning of a product on a one-dimensional parameter space, and customers express preference for a product positioned closer to their ideal types.

1.6 Information

The information of decision makers when making strategic decisions is a crucial part of the model and the effect of improved information is a major subject of research. Some popular information levels are:

- **Observable workload:** Customer service durations are known upon arrival and decision makers know the system workload, which is the total remaining service time for all customers in the system. When customers are risk neutral, the outcome is similar when queue length and expected residual (or outstanding) service time is observable (for example [520]).

- **Observable queue:** Full real-time information on the system state of the system is available. It is assumed, however, that the system workload is not observable, unless otherwise stated.

---

[9]Mainly, if the attraction of a brand changes by a given amount, the market share of the other brands is affected equally. See D.E. Bell, R.L. Keeney, and J.D.C. Little, 'A market share theorem," *Journal of Marketing Research* **12**(2) (1975) 136-141, and P.S.H. Leeflang, D.R. Wittink, M. Wedel, and P.A. Naert, *Building Models for Marketing Decisions* Kluwer (2000) §9.4.

[10]See, R.D. McKelvey and T.R. Palfrey, "Quantal response equilibria for normal form game," *Games and Economic Behavior* **10**(1) (1995) 6-38.

- **Unobservable queue:** Real-time information on the state is not available.

- **Almost unobservable queue:** Decision makers have information on the state of the server, such as being repaired, idle, etc., but not on queue length.

- **Almost-observable queue:** Decision makers know the queue length but not the state of the server.

The almost-observable and almost-unobservable terms were coined in [115] and can be classified as **partially observable**. There are many other models of partial information. For example: (i) customers know whether they are in service but cannot observe the queue length; (ii) queue length is observable but no information is available on residual service time of the current service (when service is not exponential) or arrival history (when the arrival process is not Poisson), or queue composition (for example, in a priority queue); (iii) there are two (or more) queues but only one is observable [50, 242, 304], or only the total number of customers is observable [182]; (iv) customers know whether the server is busy and whether the queue is empty [324, 573]; (v) the number of customers joining the queue since the arrival of the currently served customer is known, but customers also renege from the queue and this process is not observable [671, 414]; (vi) in a multi-queue system, customers observe their own queue but not the other queues [74]. Fung [239] (2001) describes other interesting settings of partial information related to multi-step service processes. For example, a potential customer may see the occupancy of a restaurant or the number of people waiting to be seated, but not their orders nor the queues at the kitchen. Similarly, in a queueing network one may observe the total number of customers in the network but not their locations.

It is common to emphasize the extreme cases of information by referring to the observable and unobservable models as *fully observable* and *full unobservable.*

Limited communication constraints may also result in partial information. In [640] it is assumed that reports on the realization of a random parameter on some interval are subject to a limit on the size of the message space. Consequently the interval is partitioned and only the subinterval containing the realized value is announced. The compartmental model of [218] can also be interpreted as an example of limited communication.

Environmental uncertainty arises when decision makers lack information concerning system *parameters* like service rate, waiting costs, or quality of service.

- Terminology regarding queue information is inconsistent, and terms like "full information" or "partial information" have different meanings in different papers. For example, in [284], the term *no information* is used to describe the unobservable case, *partial information* when queue length is observable, and *full information* for an observable workload.[11]

- Since most of the literature is on unobservable models we present most of the summaries of observable models in a single chapter. Unobservable models are classified according to secondary features, and we often do not explicitly state that the models pertain to unobservable queues.

- In some papers, the delay costs are charged to the server and customers are assumed to be price sensitive but not delay sensitive. In such cases it is unimportant whether the queue observable because customers are indifferent to queue-length information.

- Dynamic, state-dependent control provides queue-length information. In particular, if different prices are charged at different states a rational customer can deduce the state from the price, thereby indirectly making the queue length observable. In other cases, there is one price when congestion is low and another price when congestion is high which provides customers with partial information on the system state.

1.7   Social optimality

In equilibrium, every agent applies an *individually optimal* (**IO**) strategy given the strategies of the other agents (i.e., the strategy profile). It is common, therefore, to refer to the equilibrium as the solution obtained under **IO** behavior.

The socially optimal (**SO**) solution itself is not of central interest in this survey, and our interest in it primarily is as a benchmark for comparison with the equilibrium solution. For example, we are interested in whether it is possible to apply incentive mechanisms that induce an equilibrium that is also **SO**. In such a case it is common to say the mechanism *coordinates* the system.

Often the **SO** solution is denoted as *centralized*, meaning it is obtained when everybody obeys a *central planner* whose goal is to maximize the system's benefit. This is in contrast to *decentralized* equilibrium.

A recent surge of research quantifies system *inefficiency* (or suboptimality) with the main focus being on worst case analysis. This type of research has been applied in the context of queueing systems as early as 1985 [592]. The term *Price of Anarchy* (**PoA**) has been popularized by this research and

---

[11]Some authors refer to *visible* and *invisible* queues. However, the term "invisible queue" forms the impression customers are not aware of the existence of a queue, like for example when the queue is part of a multi-step service process [239]. In contrast, "unobservable" clearly indicates customers cannot observe the queue though they are aware of its existence.

has been extensively used in routing models with various latency functions. See, for example, Part III of the book of Nisan, Roughgarden, Tardos, and Vazirani [503] (2007), and in particular the survey included by Vöcking. Our interest here is mainly in results that explicitly use latency functions related to queueing systems.

The Price of Anarchy is the ratio of the **SO** value of a performance measure to the worst case value of this measure under equilibrium.[12] The *price of stability* is similarly defined, except it considers the best rather than the worst equilibrium, and is more appropriate when the preferred equilibrium can be induced (see, for example, [229]).

A more conservative measure than **PoA** is the *coincident cost degradation* [381, 382], which measures the degree of Pareto inferiority and provides a ratio such that the equilibrium payoffs of *each customer* can be increased by at least this value. These measures are also used to compare equilibrium solutions in two systems (see §8.6). This is appropriate when the interest is in the frontier of individual welfare values. When all customers have the same expected utility under the equilibrium and optimal solutions, the coincident cost degradation value is identical to the **PoA**.[13]

The definition of the **SO** strategy may differ according to context, and that choice will affect the way equilibrium inefficiency is measured. We mention two examples:

- In §4.1.1 we describe models where customers independently decide when to arrive to the system. A strategy is a density function over an interval. The **SO** strategy computed in [310] gives the social-welfare maximizing density function. Another possible definition would be the optimal appointment system in which customers are given deterministic arrival times [327].

- Suppose homogeneous customers choose from (or are routed to) a finite set of unobservable queues. Symmetric equilibrium defines the probability of a customer selecting each queue. The **SO** solution may also define such routing probabilities ([85],[1] §3.7, [594] §1.5), but a better solution is presented in [61] where routing is controlled by a *broker* with memory in a round-robin strategy alternating equally among the servers (§5.4).

---

[12]When costs are considered, **PoA** is the ratio between the worst equilibrium value and the **SO** value so that always **PoA** $\geqslant 1$.

[13]These theoretical worst-case measures do not necessarily reflect real-life practice. For example, **Qiu, Yang, Zhang, and Shenker [538] (2006)** demonstrate through simulations that in a realistic Internet-like environment and under various queueing latency functions the equilibrium routing latency is close to optimal.

1.8   Useful concepts

1.8.1   Externalities

Externalities in a queueing system are naturally generated when a customer joins the queue, and are generally expected to be negative. However, joining may also be associated with positive externalities, like for example when long queues induce faster service, or in models with (positive) *network effects* (see §5.2).

Other actions by customers can also generate positive externalities. For example, when some of the customers of a multi-queue system inspect several queues before joining the shortest one the variation in the queue lengths decreases, which may increase the expected utility for other customers.

Interesting aspects of negative externalities in simple queueing systems are discussed by Haviv and Ritov [329] (1998). In these systems, the externality is the difference between the actual total waiting time of the others when a given customer is present and when he is not. It can be computed as the difference between the expected waiting time of a customer who attains the lowest possible preemptive priority and the expected waiting time in the system. For example, in an M/M/1 queue this gives $\frac{1}{\mu(1-\rho)^2} - \frac{1}{\mu-\lambda} = \frac{\rho}{\mu(1-\rho)^2}$.[14] A disadvantage of this measure is that the externalities exceed the total waiting time. The authors overcome this drawback by defining an alternative measure, the (expected) *tangible externality*, defined as the (expected) total queueing time added to the others *while the customer is in service*.

1.8.2   Avoid the crowd or follow the crowd

A central distinction in queues with a single decision type is whether customers are more encouraged to follow a given type of action when others do it with increasing tendency.

The importance of this distinction and its consequences on *symmetric* equilibrium solutions in queueing systems with homogeneous customers was recognized by Hassin and Haviv [307] (1997), where the following terms were coined. When an individual best response to a threshold strategy adopted by the rest of the population is a nondecreasing function of the threshold, this behavior is called *follow the crowd*, or FTC; the opposite behavior is called *avoid the crowd*, or ATC.[15] One may think that ATC is a natural characteristic of queueing models, thus it would come as a surprise that a large number of nat-

---

[14]See also [322].

[15]ATC and FTC are closely related to the game theoretical concepts *strategic substitutes* and *strategic complements*. These terms are usually used in contexts of two-player games. When studying rational queueing, we often consider many players, assume that all but one use the same strategy, and observe its effect on the particular player. Technically, however, the situations are similar.

ural queueing models exhibit FTC behavior. Several of these are mentioned in [1].

Queueing systems often enjoy *scale economics*, i.e., the system's cost per unit of served demand decreases with served demand. This expresses the advantage provided by increased demand to the overall performance of the system. In contrast, FTC expresses advantages to an individual's welfare brought about when others join him in the same behavior.

A simple example of ATC is the unobservable E&H queue. A strategy is defined by a joining probability p. The expected queue length increases in p and therefore motivation to join decreases with p. Therefore, this is an ATC situation. Similarly simple is the example analyzed by Hassin and Haviv in [1] §4.2. An M/M/1 queue has two absolute priority classes. Buying high priority comes at a fixed cost and the strategy is defined by the probability of buying priority. The authors prove that customer motivation to buy priority increases with this probability and therefore this is an FTC case.

ATC leads to a unique symmetric equilibrium whereas FTC typically leads to multiple symmetric equilibria. This difference is also explained in the introduction of [1].

ATC and FTC are defined for situations with a one-dimensional state variable. In many interesting models the system state is two dimensional, say (i, j), and customer strategy is characterized by a *switching curve*. Thus for every given i the values of j for which a certain action is prescribed are those below a threshold n_i. Typically these thresholds are monotone in i. Here we can often observe ATC or FTC behavior. There are such ATC models ([306, 389], for example) that exhibit uniqueness of the symmetric equilibrium, but no general theory is known.

ATC and FTC behaviors can also be observed when the decisions being made are multidimensional. For instance, when a customer chooses from three options and ATC is observed for any two of them when the probability of selecting the third is fixed. This is a *pairwise-ATC* situation [313].

Though a rigorous proof that a model is ATC or FTC may turn out to be quite involved, these properties can often be intuitively expected. Our approach will be to supply this intuition and refer the reader to the original paper for the analytical proof.

One would be wrong in thinking that ATC is associated with negative externalities while FTC is associated with positive externalities. ATC and FTC refer to effects on behavior, whereas externalities refer to effects on welfare. For example, joining a priority queue is associated with negative externalities but the behavior is FTC (see [1] §4.1-2).[16]

When we deal with *asymmetric* equilibria, ATC behavior does not guarantee uniqueness of the solution. Suppose for instance that there are two players and each selects a number in $[0, 1]$. If the best response to a choice x by one

---

[16]Joining in Naor's model is associated with negative externalities, but the model is neither ATC nor FTC since a customer decision to join or balk is independent of the strategies adopted by the others.

of them is $1 - x$, then this is an ATC situation where *every* $x \in [0,1]$ is an equilibrium. The best response to x is $1 - x$ and the best response to $1 - x$ is indeed x. However, there is just one symmetric equilibrium, at $1/2$. A similar situation is described in [446], see §5.4.

1.8.3 The $c\mu$- and $Gc\mu$-rules

Optimal priority strategies in unobservable queues with convex delay costs were considered by Haji and Newell [296] (1971) and generalized by Van Mieghem (see [1] §4.6) and Mandelbaum and Stolyar [466] (2004). The generalizations show that a variation of the well-known $c\mu$-rule is close to optimal: the priority index at any give time is proportional to the product of service rate times marginal cost of delay.

1.8.4 Asymmetric information

In a model with heterogeneous customers there is natural interest in segmenting the market and exploiting customer attributes to increase social welfare or profits. For example, customers can be discriminated by price or by priority. The only restriction is that customers' expected utility must be nonnegative so customers agree to participate in the game. Such restrictions are referred to as *individual rationality* (**IR**) constraints (or participation constraints). Often, however, when dealing with strategic customers the model assumes *asymmetric information*, i.e., each customer has private information that is not available to the queue manager, like service valuation, time cost, or service requirement. In equilibrium, each customer selects at most one contract offered by the server such that the expected utility associated with this contract satisfies two types of constraints: The **IR** constraints which assure that the utility is nonnegative, and *incentive-compatibility* (**IC**) constraints which assure that it is not less than any of the expected utilities from the alternative options. When the contract satisfies these requirements, discrimination by customer types is not necessary and customers voluntarily choose the option intended for them to use.

The *revelation principle* is very useful in solving models such as these. Suppose there are n customer types. The revelation principle states that if there exists a menu of contract options that induces an equilibrium with a certain payoff, then a payoff at least as good can always be induced by a menu of $k \leqslant n$ options and an equilibrium in which all customers of a given type select the same option.

Supply-chain literature often considers long-term contracts which motivate agents to behave according to the contract designer's goal. In some cases the models assume symmetric information (e.g., [120, 657]) but contracts are tied to output rather than to an observable effort. Other models assume that effort is not directly observable and only output is contractible. We note however that asymmetric information is not typical in a long-term relation. The following quotation from [533] describes in a setting where the supplier's service rate is not directly observable by the retailer, "By observing the timing of the agent's output (jumps in the inventory level), the principal can draw inference about his choice of production rate." Thus meaning that in the long run the service rate, even the complete service distribution, can be inferred and becomes contractible. However, the authors prefer to link payments to directly observed variables: "by making payments to the agent contingent on the inventory process, the principal can create incentives for the agent to control the production rate in the manner she desires."

1.8.5   Heavy traffic

Halfin and Whitt [297] (1981) consider a sequence of $M/M/s$ queues and show that as demand rate Λ grows, the probability of positive queueing delay converges to a number in $(0,1)$ iff s grows with Λ according to a *square-root staffing rule* $s = R + \beta\sqrt{R} + o(\sqrt{R})$, where $R = \Lambda/\beta$ and β is a positive constant.

The following classification is defined by Borst, Mandelbaum, and Reiman [99] (2004), depending on how monopolistic firms respond to an increase in market size: (i) *efficiency-driven* (**ED**) where almost all customers experience some delay before starting service; (ii) *quality-driven* (**QD**) where almost all customers start service immediately; (iii) *quality- and efficiency-driven* (**QED**) where a nontrivial fraction of customers receive service immediately. The last is also known as the Halfin-Whitt regime.

In some models, heavy traffic is induced by the optimal solution rather than being an assumption of the model. See [65, 66, 422, 528, 531, 532].

1.8.6   Achievable region

In many multiclass queueing systems the steady-state performance measures, in particular the expected system time, satisfy conservation laws such that total performance over all classes is invariant under work-conserving queue regimes. The *performance space*, or *achievable region*, is then a polyhedron that can be characterized by its extreme points. A class of scheduling policies is *complete* if any point in the achievable region can be realized with a policy from this class. Relevant results relating to this subject are surveyed in [312] and Appendix A of [594].

Coffman and Mitrani [170] (1980) characterize the achievable region of $M/M/1$ queues with n classes and preemption allowed. The extreme points of the achievable region correspond to the preemptive priority disciplines associated with the $n!$ permutations of the classes. Every other point in the achievable region can be obtained by randomizing over absolute priority disciplines. Similar results have been shown for more general models, for example, in Federgruen and Groenevelt [233] (1988), and Shanthikumar and Yao [568]

(1992). This characterization enables us to compute optimal service schedules (priority disciplines) which maximize system-wide performance objectives.

A drawback of this approach, especially when customers act strategically, is that it requires randomization among absolute-priority regimes in a way that will not affect customer behavior. The straightforward way would be to randomly choose (with the appropriate probabilities) an absolute-priority regime yet conceal it from the customers. Another idea, offered by Coffman and Mitrani, is to randomize at the beginning of every busy period and adopt the resulting discipline to the end of the busy period. Hassin, Puerto, and Fernández [312] (2009) use *discriminatory processor sharing* (DPS). Under this regime there exist weights $x_i \in (0, 1)$, $\sum x_i = 1$, representing *relative priorities*;[17] if n_i i-customers are present in the system, a j-customer receives a fraction $p_j = x_j / \sum n_i x_i$ of the service capacity. If service is exponential, this is equivalent to having a lottery upon service completion giving this service to any given j-customer with probability p_j.[18] Every point in the relative interior of the performance space is achievable by a suitable choice of relative priorities, and therefore system optimization may be achieved without the need to conceal the details of the priority rule. Sinha, Rangaraj, and Hemachandra [580] (2010) optimize over the achievable region through *delay-dependent dynamic priorities*. See §6.6.2.

1.8.7   Double-ended queues and MTS servers

When firms hold an inventory of finished goods it is common to say they *make to stock* (MTS). Otherwise, if customers must wait for their requests to be processed, the firms *make to order* (MTO). The latter terminology does not imply that the firm produces customized products but simply that it holds no stock. An MTO firm can be modeled as a queueing system where customers wait for their orders to be completed. In some cases the firm maintains a queue at no-stock periods and makes to stock when the queue is empty. This situation can be described by a *double-ended queue* where positive queue lengths refer to customers, as under MTO, and negative queue lengths refer to the number of units held in stock. A common policy in MTS systems is to produce as long as the queue length is below a *base-stock* level b. When stock level reaches b, production stops until a new customer arrives. When the queue length is negative an arriving customer immediately obtains a unit and the queue length increases by one.[19]

---

[17]When these weights are equal for all customer classes we get the *egalitarian processor sharing* (EPS) regime.

[18]See Remark 1.1 in [309].

[19]The MTO policy is simply the case where $b = 0$.

1.8.8 Bounded rationality

We focus on models that assume rational decisions. Often rational decisions require simple calculations, for example estimating the expected waiting costs in a first-come first-served observable Markovian queue. However, in other cases the rational decision model assumes the individual can solve complicated problems, like estimating the equilibrium solution when customers use private signals about the quality of service at different servers.[20] It is quite unlikely that customers can make such calculations in practice and many papers assume, explicitly or implicitly, that customers follow simpler guidelines.[21] We include these models and point out where the model deviates from the rationality assumption. Some of these cases can be explained by a *bounded rationality* assumption – individuals are unable to perform the necessary calculations.

The following example illustrates this point. Reneging from an unobservable queue means that each customer is associated with a reneging time, and a customer will abandon the queue if service does not start (or end) within this time. It is commonplace to assume that the reneging time is an exponential random variable. Whitt [660] explains the assumptions that support such behavior as follows:

> We assume that each customer is willing to wait a fixed time before beginning service, called the delay threshold. However, different customers may have different delay thresholds ... so that we assume that successive customers are willing to wait random times that are independent and identically distributed ... With probability β, a customer is unwilling to wait any amount of time, and so will balk.

This description relates to customer heterogeneity, but ignores customer's strategic considerations. Rational behavior can lead to reneging for several reasons, for example, a service distribution with decreasing failure rate, decreasing value of service, increasing waiting-cost rates, and improved estimates of system parameters obtained by observing the queue.

A rational customer is expected to use his knowledge on system parameters and customer strategies when deciding on his own strategy.

---

[20] In another example, Zohar, Mandelbaum, and Shimkin [719] remark that rational customers are expected to base their reneging decision on the entire delay distribution, however, "a typical customer can hardly be expected to form a clear estimate of the entire waiting-time distribution based on limited experience."

[21] **Lu, Musalem, Olivares, and Schilkrut [459] (2013)** describe an empirical study on the effect of queue length and number of servers on queue joining behavior in a deli section of a supermarket. While the required calculation here is quite simple, customers adopted a very simple approximation and primarily relied on the queue length as a measure of congestion, ignoring the number of servers. Therefore, pooling queues may encourage balking. **Conte, Scarsini, and Sürücü [173] (2015)** conduct a laboratory experiment where individuals selecting among servers with different price, service rate, and queue length. They find that the higher the time pressure the greater the observed deviation from rational behavior.

General literature on bounded rationality is surveyed in [351].

1.8.9   Delay guarantees and quote sensitivity

It is common to identify quality of service with a delay-related measure. Service obtained through queueing is related to the class of *experience goods* – products which must be used for their quality to be observed. In an observable queue a customer knows queue length, whereas waiting time remains a random variable. In unobservable queues even queue length is a random variable.

In the presence of asymmetric information the firm may have better ability to estimate the delay measure than do its customers. The firm usually has more accurate knowledge than its customers on the system state and parameters. The firm may find that revealing such information to customers serves its goals, or it may prefer to conceal this information.

Firms often exploit customer naivety by strategically choosing a policy of lead-time guarantees to maximize profits. In particular, the joining strategy of naive customers may depend on *promised delivery time* (PDT) rather than on the actual waiting-time distribution. Rational customers may suspect that the announced delay is biased due to the firm's desire to attract potential customers.

We reserve the term PDT to delay announcements made when there are penalties imposed on the firm for late and, possibly also early, delivery. These penalties motivate the firm to quote true values and increase its credibility. A different approach assumes the firm compensates its customers, fully or partially, for lateness, and this will affect their strategy. The compensation mechanism serves as a signal of the firm's desire and ability to fulfill its commitments. Still, the PDT is a consequence of the firm's optimization and it need not be a reliable measure.

Rational customers use the quoted information to form their own estimates on the statistical properties of the delay distribution.[22] Many models assume, however, that customers take delay announcements at face value even when there is no guarantee of reliability.[23] This behavior can be interpreted as a form of bounded rationality.

When capacity is a decision variable, the firm sets the smallest capacity that fulfills the commitment. This means that capacity and lead time are interchangeable decision variables.

Delay guarantees can take various forms, the most common are (i) a definite reliable guarantee accompanied with the option of the server outsourcing (or "expediting") the work or purchasing the demanded item at a price that

---

[22]See [31] for a model of customers using the server's announcements in a rational way.

[23]**Keskinocak and Tayur [399] (2004)** describe over 100 papers on due-date management (DDM) in service systems. They comment that "Most of the research on DDM ignores the impact of the quoted due dates on the customers' decision to place an order." The only exceptions are six papers (described there in §6.1) which allow due date dependent customer decisions, and nine papers (in §6.2) which allow due date and price-dependent decisions.

exceeds its own production cost, (ii) a guarantee on expected lead time, (iii) a *delay standard* (or, delay guarantee) D and a *reliability level* α such that the *delivery time reliability constraint* $\Pr(W < D) \geqslant \alpha$, is satisfied. The reliability level α is also referred to as lead-time reliability guarantee, or service level.

Distinction between the last two forms is often not essential. For example, the sojourn time of an M/M/1 queue is distributed $\exp(\mu - \lambda)$, and the α-fractile of the distribution is $W_\alpha = \frac{-\ln(1-\alpha)}{\mu-\lambda}$, which is a scaled version of the expected sojourn time $\frac{1}{\mu-\lambda}$. Therefore, the particular choice of α by the firm should not affect customer behavior, but when it does have an effect we attribute it to bounded rationality (see §11.3.3).

Information supplied by a server may either be static like for example the expected sojourn time, or may be dynamically updated according to the current state of the system. When the PDT is state-dependent rational customers can deduce the state of the system from the quoted delay guarantee, but under a bounded rationality assumption they ignore this information.

Kopalle and Lehmann [408] (2006). investigate a multi-period non-queueing model of experience goods. Similar ideas could be used to study the dynamics of expectations and endogenize the penalties associated with inaccurate PDTs in strategic queueing models.

1.9   Terminology conventions

Indices and summation: When dealing with a finite number of agents or agent types, they are often indexed. We refer to "server i," or to an "i-customer," or to demand rate λ_i without explicitly describing the range, for example, $i = 1, \ldots, n$. Similarly, when summing indexed quantities with no risk of ambiguity we omit the range of summation and write, for example, $\sum \lambda_i$ for $\sum_{i=1}^{n} \lambda_i$.

Markovian queues: This term is used when the distributions of all random variables related to time, such as arrivals, service, reneging, and vacations are assumed to be exponential.

Perfectly correlated service valuations and waiting costs: The most common types of customer heterogeneity are waiting costs and service valuations. In some cases the model is simplified by assuming perfect correlation between these parameters, i.e., $C = f(V)$, where C is the waiting-cost rate and V is the service value. Some models assume these parameters are *affinely related*, i.e., $C = a + b \cdot V$, or *linearly related*, when $a = 0$ and without loss of generality, $b = 1$.

Atomic customers: Customers are said to be *atomic* if each has a non-negligible demand. Otherwise, when there is a continuum of users, each with an infinitesimal amount of demand, they are *non-atomic*. It is common to refer to jobs submitted by an atomic customer as a cooperative class, and the resulting model is said to involve *class* or *sectoral* optimization.

Queue regimes: Unless otherwise mentioned, the queue discipline is assumed to be first-come first-served (FCFS). The notation LCFS (LCFS-PR) is used to denote the last-come first-served (preemptive-resume) queue. The discipline of *serving in random order* is denoted SIRO. Egalitarian processor sharing is denoted EPS.

Abbreviations: In addition to queue regimes, we use the following abbreviations:
ATC - avoid the crowd
E&H - Edelson and Hildebrand's unobservable queue
FTC - follow the crowd
IO - individually optimal
IR - individual rationality (or, individually rational)
IC - incentive compatibility (or, incentive compatible)
MTO - make to order
MTS - make to stock
PDT - promised delivery time
SO - socially optimal (system optimal)
cdf - cumulative distribution function
pdf - probability distribution function

Terminology: Different papers, especially when appearing in journals from different areas, use different terminology to describe the same entity. We use these terms interchangeably. Here are some examples:
server = firm/provider/agent/system manager
arrival rate = market size/demand
effective arrival rate = throughput
traffic intensity = load/utilization
capacity = service rate/processing rate/production rate
delay = waiting time/lead time/sojourn time/response time/
     system time/order time fulfillment time/completion time/flow time/
     responsiveness/latency/end-to-end delay
waiting cost = delay cost/delay sensitivity/time cost
waiting-cost rate = impatience factor/parameter
service value = willingness to pay/reservation price
socially optimal = welfare maximizing/efficient/centralized
flexible customers = non-dedicated customers/switchable customers
self-financing = break-even budget/balanced budget
reneging = abandonment/defection/cancellation

1.10 Plan

Our starting point, Chapter 2, deals with observable queues. The main reason for this choice is that whether a queue can be observed by customers seems to be the first and most important assumption that distinguishes a model. These models assume state-dependent behavior which is typical of queues and distinguishes them from other congested systems. Most other models assume unobservable queues and the assumption is so common that it is often not explicitly mentioned in the paper.

Clearly, whether a queue is observable is a matter of information available to customers. Chapter 3 concentrates on other types of information in queueing systems but also considers comparisons between observable and unobservable variations and incentives for information disclosure.

The three common objectives in a queueing system are maximization of individual utilities, of social welfare, or of profits. Chapters 4 through 7 contain relevant models according to this classification. Individual non-cooperative behavior is the subject of Chapter 4, social optimization is the subject of Chapter 5. Profit maximization is divided into two parts, monopolistic environments are considered in Chapter 6, competitive markets are the subject of Chapter 7. Chapter 8 deals with queueing networks, from simple parallel servers to general network structures. Planned vacations are common in strategic queueing and forced vacations (such as breakdowns) strongly affect system behavior. Chapter 9 is devoted to these models. Supply chains are the subject of Chapter 10. In these models agents have different goals but they all profit when the system operates efficiently and are ready to enter profitable contractual agreements. Our last chapter relaxes the assumption of fully rational agents and allows bounded rationality.

Chapter 2

Observable queues

A queue is *observable* if customers know its length either by direct observation it or by indirect inference. For example, queue length can be inferred from state-dependent (dynamic) price quotes of the server. Otherwise the queue is *unobservable*. The rational queueing literature mostly deals with unobservable queues, and this fact is often taken for granted without being explicitly mentioned. The main reason for focusing on unobservable queues is, probably, that they are easier to analyze than observable queues. Yet, observable queues carry with them unique interesting features.

2.1 Extensions and variations of Naor's model

2.1.1 Extensions

Li [439] (1992) considers an MTS generalization of Naor's model, where a firm can also produce to stock. The firm incurs a cost c per unit of production and a unit holding cost rate of h, and charges an exogenously fixed price p per unit of sale. When the stock level is positive, an arriving demand is immediately satisfied. Otherwise, customers join iff the queue is shorter than a threshold value.

The author derives explicit formulas for the joining threshold and the base-stock level and shows that:

- The optimal base-stock level is higher if customers are less patient, value the product less, or have more attractive alternatives.

Brooms [111] (2005) considers a Markovian single-server FCFS system with state-dependent service rate: the service rate increases in the queue length. The number of customers in the system is bounded by its buffer size B. The arrival rate consists of two independent streams. One stream consisting of customers who enter the queue if its length is less than B [1] and the other consisting of strategic customers who join only if the expected waiting time in the system is less than a threshold θ (and the number of customers already in the system is less than B).

Since the service rate increases in the number of customers, a customer gains more from joining when the tendency of others to join increases. The author rigorously proves the intuitive property that this is an FTC model (Lemma 5), and hence multiplicity of equilibria is possible. The paper ends with a model of dynamic learning that leads to a pure stable equilibrium.

Chen and Kulkarni [139, 140] (2006, 2007) extend Naor's model to two customer classes. Class 1 customers obtain preemptive resume priority over class 2 customers. Both individual and social objectives incorporate discounting. The subsystem consisting of class 1 customers provides a solution to Naor's single-class model with discounting. In [139] the authors assume the service value is obtained upon service completion, as in Chen and Frank (2001) (see [1] §2.7, where the threshold with $p = 0$ coincides with (1.1) in [139]). In [140] the service value is received when the customer joins the queue, and this assumption leads to a different solution.

The authors consider individual, class, and social optimization. They find that:

- More 1-customers, but fewer 2-customers are accepted by the class-optimal policy than by the **SO** policy. The 1-customers result is clearly expected because of the negative externalities imposed by 1-customers. The 2-customers result is explained as follows: Since the class-optimal policy admits more 1-customers, the effect a 2-customer has *on other 2-customers* will be higher. Therefore, the class-optimal policy admits less 2-customers.

---

[1]This type of customer is not an essential part of the model, but it simplifies the analysis assuring that the recurrent set of states is not a function of the strategy used by the strategic customers. This way there is no need to refer to issues as dealt with in [308].

- More 1-customers join in equilibrium than under the class-optimal (and hence also the **SO**) policy. However, the relation of the **IO** joining of 2-customers and the other two policies can be arbitrary.

Sun and Li [603] (2012) consider the joining decisions of customers with the nonlinear waiting cost functions $c \cdot t^m$, $m = 2, 3$. They derive equilibrium thresholds and provide numerical examples where the order of the equilibrium, social, and profit-maximizing thresholds is as in Naor's linear cost model.

Ziani, Rahmoune, and Radjef [715] (2015) solve a variation of Naor's model with pairs of customers arriving. When a pair arrives they independently decide whether or not to join and are randomly ordered if both join. In the observable case the authors compute the unique (mixed) symmetric threshold strategy.[2] In the unobservable case the authors compute the unique symmetric joining probability. They present an example in which the average customer joining rate is larger in the unobservable case except when service value is small.[3]

Wang, Zhang, and Zhang [648] (2014) consider Naor's model with a quadratic utility function $U(W) = R - CE(W) - A \cdot C^2 \text{VAR}(W)$, where W is waiting time. The authors fully characterize of the equilibrium, **SO**, and profit-maximizing solutions in both observable and unobservable cases. Qualitative results when customers are risk averse are similar to those obtained by Naor and E&H under risk neutrality, and the authors point out interesting properties when customers are risk seeking:

- In the **Observable case:** When A is very negative even social welfare benefits from high congestion and the **SO** threshold may be infinite.

- In the **Unobservable case:** Risk seeking induces FTC behavior. When A is sufficiently negative multiplicity of equilibria (two stable solutions and one unstable solution) is possible. Moreover, the joining probability is not necessarily smaller under social optimization than in equilibrium.

Shone, Knight, Harper, Williams, and Minty [575] (2015) extend Naor's insight that the state space induced by the **SO** policy is contained in the state space induced by **IO** behavior in equilibrium. They prove this result for a model with N FCFS M/M/C_i facilities with heterogeneous waiting-cost rates and service values. The result also holds under customer heterogeneity.

Shi and Lian [571] (2016) consider double-ended Markovian queues motivated by taxi service. Taxi drivers are not strategic in this model and the

---

[2]They also describe the pure asymmetric strategies.

[3]The authors conclude that the queue owner generally prefers to suppress queue-length information. Note that arguments similar to those presented in the introduction to §3 indicate social welfare is higher when the queue is observable.

maximum number of taxis allowed in the station is N. Customers incur linear waiting costs while taxi operators incur linear waiting costs and a fixed cost per trip. The value of a ride and price per ride are exogenous.

The authors compute **IO** and **SO** customer joining thresholds in an observable version and joining probabilities in an unobservable version and discuss ways to increase social welfare by controlling N or by the use of taxation and subsidies.

2.1.2   Non-exponential service distribution

Some non-Markovian extensions of Naor's results are described in [1] §2.10. For example, customer strategy is of the (pure) threshold type also in the observable G/M/s queue. The outcome is similar when the workload or the time elapsed since the beginning of the current service is observable in a G/G/1 queue. The M/D/1 system under this assumption has been analyzed by **Adler and Naor** [7] **(1969)**. Also see [684] for Erlang service distribution with dynamic pricing, [350] for Erlang service with multiple vacations, and [469] for a related model with bulk service and general inter-service time distribution, [696] for an extension of the M/G/1 customer decision model assuming a generally distributed setup time when a customer arrives to an empty system, and [637] for the **SO** threshold in the GI/M/c extension of Naor's model.

Altman and Hassin [49] **(2002)** considered an M/G/1 queue where queue length, but not the expected residual service time, is observable. They demonstrate that, since the queue length provides information about the expected residual service time an equilibrium threshold strategy does not necessarily exist. This left open the question of existence and uniqueness of the equilibrium in such systems and conditions for the existence of an equilibrium threshold strategy.

A partial answer to these questions was given by **Haviv and Kerner** [324] **(2007)**. They assumed an M/G/1 system where an arriving customer can observe whether the server is idle (in which case the customer enters service) or busy, and if busy, also whether the queue is empty or not. A symmetric equilibrium is characterized by a pair (p, q) where p is the probability of joining when arriving to an empty queue and q is the probability of joining when arriving to a nonempty queue. The determination of p and q is done recursively.[4] It is observed that the value of q is not relevant to a customer arriving to an empty queue. Therefore, the equilibrium value(s) of p can be computed first. Then, p is used to compute the equilibrium q. Some interesting features of the solution are:

- Depending on the service distribution, the expected utility from joining

---

[4]Similar to [215].

an empty queue can decrease with p (ATC behavior) and a unique equilibrium p exists, or it can increase with p (FTC behavior) and multiple equilibria usually exist. It is also possible that none of these cases will occur.

- In contrast, ATC always prevails with respect to selecting q; the higher the value of q selected by others, the higher the expected queue length given that it is positive, and therefore a customer's tendency to join a nonempty queue is smaller. Hence, for any given p value there is a unique equilibrium q value.

- Examples show that equilibria with $q > p$ exist, that is to say, the tendency to join a nonempty queue may be greater than that of joining an empty queue. This result is in line with [49].

Kerner [397] (2011) completes the analysis of the M/G/1 queue where queue length is observable but not the residual service. Depending on the service distribution, one may obtain different types of behavior, including ATC with a unique equilibrium, or FTC with multiple equilibria. Assuming the probability of joining an empty system is 1, an equilibrium is characterized by a vector $(p_1, p_2, \ldots, p_n, \ldots)$ giving the joining probability when an arriving customer observes n customers in the system. As in [324], the equilibrium probabilities are solved recursively starting with p_1.

The author proves the following results:

- An equilibrium always exists, but it need not be unique or characterized by a threshold strategy.

- When the service random variable X has *decreasing mean residual life*, i.e., $E(X - t | X > t)$ decreases with t, there is a unique equilibrium strategy, and it is a threshold strategy.

- When X has *increasing mean residual life*, i.e., $E(X - t | X > t)$ increases in t, all equilibrium strategies are of the threshold type, and at least one is pure.

2.1.3   Asymptotic analysis and price of anarchy

Gilboa-Freedman, Hassin, and Kerner [254] (2014) quantify the extent of the inefficiency observed by Naor [497], that customer equilibrium behavior in a queueing system is not **SO**. Naor's model has two parameters, namely the traffic intensity $\rho = \lambda/\mu$ and the normalized service value $\nu = R\mu/C$, where R is the service value (or reward) and C is the waiting-cost rate.

The authors investigate the behavior of **PoA** as a function of these two parameters, and the main results characterize **PoA**(ρ), which is the supremum of the **PoA** over all possible ν-values for a given value of ρ:

- For most real-life applications, i.e., when $\rho < 0.98$, $\mathbf{PoA}(\rho) \leqslant 1.5$.

- When $\rho \leqslant 1$, $\mathbf{PoA}(\rho) \leqslant 2$.

- $\mathbf{PoA}(\rho)$ is unbounded for any $\rho > 1$.

Borgs, Chayes, Doroudi, Harchol-Balter, and Xu [98] (2014) investigate properties of the **SO** threshold, n^*, in Naor's model. They find a new representation of n^* using the *Lambert W function* and use it to show that the welfare rate under n^* is closely approximated by $\mu R - C(n^* + 1)$. That is to say, *the maximal welfare rate is approximately what would be obtained if queue length is constantly kept at its maximal length* (and therefore the server is continuously busy).

The authors also find that when $R \to \infty$, n^* behaves as $(1 - \rho)R\mu/C$ if $\rho < 1$, as $\sqrt{2R\mu/C}$ if $\rho = 1$, and as $\log_\rho(R\mu/C)$ if $\rho > 1$.

Hassin and Koshman [311] (2014) compare the maximum profits obtained in Naor's model and the profits obtained when the price is $R - C/\mu$, so customers join iff the server is free (i.e., no queue is maintained). They show that the latter policy always guarantees at least half of the maximal profit, and in most cases the loss is quite small. The ratio of 0.5 is asymptotically obtained when $\rho = 1$ and $\nu = \frac{R\mu}{C} \to \infty$.[5]

2.1.4 Reneging

See §11.4 for bounded rationality models with reneging decisions.

Cripps and Thomas [176] (2014) consider a discrete-time model of an observable single-server queue with homogeneous customers who maximize discounted payoff. The server is *bad* (i.e., not functioning) with a given probability and *good* otherwise.[6] In every period there is a single arrival and the number of customers a good server can serve is geometrically distributed. The value associated with balking is smaller than the benefit a customer receives upon service completion. The customer's goal is to maximize his discounted payoff.

An arriving customer first faces a join-or-balk decision. Once joining, the customer continues to observe the queue. Termination of service proves the server is good. Otherwise, the best-informed customer is the one at the head of the queue. If this customer reneges then all other customers in the queue should follow.[7] The decision is therefore how long should a customer who

---

[5]The significance of $\rho = 1$ in Naor's model is interesting and unexpected as all costs and profits are continuous functions for $0 < \rho < \infty$. See also [254] where it is shown that the **PoA** in Naor's model is finite iff $\rho \leqslant 1$.

[6]Note the similarity to the model of Mandelbaum and Shimkin (2000) (see [1] §5.2.3) where calling customers do not know whether their call has been accepted or rejected, and the authors determine the equilibrium reneging strategy.

[7]See [519, 473] for a model where customers ignore this information.

arrived to an empty queue and has not seen a service completion wait before recognizing that the probability of a good server becomes sufficiently small to justify reneging.

The main result is a characterization of a symmetric equilibrium policy of the following type: Balk if queue length exceeds a threshold; if queue length is positive but below threshold join and stay until served or renege when the first in the queue reneges. If the queue is empty, then join and apply a (mixed) threshold reneging time strategy.

Debo, Hassin, and Veeraraghavan [183] (2012) consider a model similar to that of [176] but with the emphasis on the signal a customer obtained when arriving at an empty system. The server is either good or bad with a known probability. The value of good (bad) service is positive (negative). When service ends, its quality is observed by the queueing customers and all renege if it is bad. In the base model there is no waiting cost, and thus the only decision is whether to join an empty system.

The authors compute the equilibrium and **SO** probabilities of joining an empty system and show that due to positive externalities associated with joining an empty system, the equilibrium joining probability is smaller than the respective **SO** probability. The authors also discuss how queue discipline mechanisms, like LCFS and SIRO, can be used to improve social welfare.

Afèche and Sarhangian [19] (2015) study pricing under rational reneging for an observable M/M/1 queue with two predetermined preemptive priority classes. Customers maximize expected service reward minus a linear delay cost minus service fees. The special feature of this model is that low-priority customers can renege when a high-priority customer arrives, and the authors focus on their behavior. Therefore, the model is simplified by assuming zero service fee for high-priority customers and thus they join as in Naor's model and have no incentive to renege.

The authors show the equilibrium join/balk/renege strategy of low-priority customers has a threshold structure that depends both on queue composition and pricing structure. The authors consider pricing as a means to control the behavior of low-priority customers in the system. A distinguishing feature of the model is that in the presence of reneging, charging the customer upon entering the system or upon service completion leads to different outcomes. Welfare maximization requires charging only a service fee and no entrance fee (i.e., for order placement or joining the queue). In contrast, revenue maximization requires charging not only a service fee, but also an entrance fee. Moreover, charging only an entrance fee may generate more or less revenue than charging only a service fee.

2.1.5 Feedback queues

Brooms and Collins [112] (2013) generalize classic results of Naor (1969) and Yechiali (1971) ([1] §2) to Bernoulli feedback queues. Consider a

GI/M/1 queue where a customer who completes service is fed back to the end
of the queue with probability p. Arriving customers observe queue length and
decide whether to join or to balk. This extension of Naor's basic model adds to
it an interesting property, namely that a customer's decision also depends on
the actions taken by future arrivals in the sense that these customers have the
potential to be ahead of him once he is fed back to the end of the queue. Hence,
the equilibrium strategy here is no longer a dominant strategy. The authors
do not explicitly mention the customer parameter of service value, but instead
define a penalty on balking. They add a crucial assumption that is not needed
in Naor's original model: *Customers who join the queue may not renege during
their sojourn.*[8] They characterize the equilibrium (possibly mixed) threshold
and socially optimal (pure) threshold and prove several structural properties
that lead to the following conclusions (which conform with the intuition that
this is an ATC system and joining is associated with negative externalities):

- For the M/M/1 case there exists a unique symmetric equilibrium, with
  a (possibly mixed) threshold strategy.

- The same property holds in the GI/M/1 case if the strategies are restricted such that the joining probability, given a queue length n, is
  non-increasing in n (but no counterexample is known when this restriction is lifted).

- In the M/M/1 case the **SO** threshold is determined by the same formula
  as Naor's with the exception that $R\mu/C$ is multiplied by $(1 - p)$.

- The **SO** threshold is smaller or equal to the equilibrium threshold.

2.1.6   Priorities

Models of strategic priority selection in observable queues are surveyed
in [1] §4.1. An arriving customer observes the number of customers at each
priority level and decides which priority to buy. The regime within each class is
usually FCFS. Hassin and Haviv [307] (1997) observe that the model leads to
an FTC behavior and multiplicity of (pure and mixed) threshold equilibria.[9]

Altmann, Daanen, Oliver, and Suárez [52] (2002) present numerical analysis of priority purchase in a system with two (preemptive) priority
classes and Poisson arrivals of homogeneous customers. The model assumes
an EPS regime within a priority class. The authors observe that the equilibrium joining strategy is the same as when the service order within each
priority class is FCFS: for thresholds n_1, n_2, customers join the low-priority

---

[8]Without this assumption, a customer whose job is fed back faces exactly the same
decision as a new arrival.

[9]It seems that [307] is the first to consider mixed threshold equilibria in the context of
a queueing system.

queue if it has at most n_1 customers; otherwise they join the high-priority queue if it has fewer then n_2 customers, and they balk if there are already n_1 low-priority customers and n_2 high-priority customers. The authors offer an algorithm for computing pure equilibrium strategies; however, they do not discuss mixed strategies nor multiple solutions. The theoretical results obtained are compared to experimental data. An instance with heterogeneous waiting-cost rates is also analyzed.

Printezis and Burnetas [535] (2008) assume an arrival process which includes a class of time-sensitive customers given an offer to buy priority options. This class is small and decisions of these customers do not affect the system's steady-state distribution. It is further assumed that a customer in that class is expected to arrive to the system N times in sufficiently spaced intervals so that the system states remain independent. At the time of arrival, each customer observes the queue length and decides whether to exercise an option or to join the end of the queue and wait his turn. Their decision variables are the number of options to buy and when to exercise an option. The option price is determined by a profit-maximizing queue manager.

The authors solve this model and mention the natural extension allowing for the class of customers able to buy priority options to be significantly large.

Engel and Hassin [227] (2016) consider an M/M/1 system with two first-come first-served queues, an observable *system queue* (SQ) and an unobservable *virtual queue* (VQ). An arriving customer who finds the server busy decides which queue to join based on the SQ's observed length and the conditional VQ expected waiting time.[10] Customers in the SQ have non-preemptive priority over those in the VQ, but waiting in the SQ is more costly than waiting in the VQ.

The authors derive the system steady-state behavior under a (mixed) customers' threshold joining strategy and show that, similar to other priority systems, the system demonstrates FTC behavior and multiplicity of equilibria.

2.1.7 Competition

Christ and Avi-Itzhak [168] (2002) consider capacity competition in a Markovian model with two competing servers sharing a common queue. Service price is exogenous and servers compete by setting service rates. However, customers cannot observe these rates and therefore do not discriminate between servers. When both servers are idle, an arriving customer randomly selects one, and when only one is idle the new arrival joins that one. If both

---

[10]This situation resembles Hassin's (1996) "gas stations" model (see [1] §7.6) where customers choose between an observable queue and an unobservable queue. The main difference is that the gas stations model assumes separate servers while here there is a commons server for the two queues.

servers are busy, an arriving customer joins the queue with an exogenous probability that decreases with queue length, but is *independent of the service rates*. Operating cost is increasing and convex in μ, and revenue from serving a customer is independent of the customer's sojourn time. The authors prove the existence of a unique symmetric equilibrium in this game.

Avi-Itzhak, Golany, and Rothblum [75] (2006) explore globally optimal service rates in the model of [168], that is, solutions that maximize the aggregate profit of the two servers. They prove that the solution has symmetric service rates which are smaller than those obtained under competition, and that the system can be coordinated by imposing a penalty proportional to the service rate.

Deck, Kimbrough, and Mongrain [189] (2014) consider competition between two firms that set state-independent (static) prices. One firm operates two single-server queues and the other operates only one such queue. Customers have heterogeneous waiting-cost rates and identical deterministic service requirements. The queues are observable; the choice of one of the three queues or balking is straightforward.

The authors concentrate on price competition between the two firms. They consider several instances of this setting and compare numerically derived equilibrium prices with behavior observed in laboratory experiments. The main research goal is to study the effect of allowing the two-queues firm to charge different prices. In this case, the firm can reduce the price in one queue to extract greater market share while increasing the price in the other to exploit impatient customers. Allowing for price discrimination results in a lower average price and increased consumer welfare, but, surprisingly, reduces profitability of both firms.

2.2   The dual approach

IO behavior is often easier to compute than its **SO** counterpart, as already reflected in Naor's work. Hassin [302] (1985) observed that **IO** behavior is **SO** in Naor's model when the regime is LCFS-PR. This is so because under LCFS-PR the last job in the queue will remain the last one until leaving the system and therefore imposes no externalities.[11] This observation provides an alternative way to compute Naor's **SO** threshold; Simply compute the individual reneging behavior under the LCFS-PR model. This method is referred to as "the dual approach" in [672]. It is used in [1] §2.3 to compute the **SO** threshold in Naor's model and in [218] for doing the same in a model with compartmental waiting space.

Xu and Shanthikumar [672] (1993) generalize Naor's model and compute the **SO** admission-control policy for FCFS M/M/*s ordered-entry queueing systems*. In this system, the servers are heterogeneous, new customers are assigned to the lowest-indexed available server, if one exists, or join an FCFS queue otherwise.[12] The goal is to compute the **SO** admission threshold. When $s = 1$ we obtain Naor's model.

The authors compare the system with the "dual system" where the regime is a variation of LCFS.[13] Using this comparison they prove the **SO** policy is of a threshold type and present exact and approximate formulas for its computation.

---

[11]See §2.3 for other models where non-FCFS disciplines are recommended.

[12]This policy is assumed, but it may be suboptimal even when the servers are indexed in a decreasing order of service rates. The optimal strategy may hold a customer in the queue and wait for a faster server to become free.

[13]It is interesting to observe that in the dual system service can be preempted and customers reassigned. This is not possible in the original system where a fast server may be idle while a slower one is serving. For models that allow reassignments see [24, 315].

Kim, Ahn, and Righter [401] (2011) consider an M/M/s system with service rates $\mu_1 \geqslant \cdots \geqslant \mu_s$. If server i becomes idle with no customer in the queue, the server takes a vacation in the form of serving a *secondary customer*.[14] No profit is associated with serving a secondary customer. The duration of this vacation is $\exp(\mu_i)$, as would be in regular service. The goal in this model is to minimize expected waiting time of primary customers by deciding whether to start serving a primary customer when a server becomes available.

The authors apply the dual approach. They compute the equilibrium behavior when more recent arrivals have preemptive priority in choosing servers. They call this multiserver extension of LCFS *last-come first-priority* (LCFP). They show that the **IO** strategy is now of the threshold type: For $j = 1, \ldots, s$, there exist computable thresholds T_j such that if server j is available the T_jth job in the queue will use this server. It is now argued that this is the **SO** policy: (1) If a job wants to use a given server under **IO**, then it is also socially desirable to let it do so because the social alternative is not using the server at all for the exact same times as it would take to serve the customer (i.e., that the server takes a vacation). (2) Suppose that even the lowest priority job does not want to enter service at the idle server. This job causes no externalities and therefore this decision is **SO**.

Akgun, Down, and Righter [24] (2014) introduce costs in the M/M/s model of [401]. Now, customers incur identical waiting cost rates and server j incurs a processing cost rate β_j whenever processing a job.

The authors apply the dual approach assuming the same system under LCFP with preemption.[15] There are two versions of the model:

- **With reassignments:** At any *decision instance* (end of service or a new arrival) *all customers* in the system (there is no distinction between customers in service and in the queue) select their server in reverse order of arrival. In this case, at any decision instance the **IO** policy is again **SO** and of the threshold type: Index the servers by order of customer preference. For server j there is a threshold T_j such that if servers $1, \ldots, j-1$ were chosen, server j is chosen by the T_jth customer in the system, if there are that many customers in the system, where $T_j \leqslant T_{j+1}$. The other customers choose to wait.

- **Without reassignment:** As above, but a customer choosing to use a server must stay with this server until service completion. In this case, the LCFP-Pr equilibrium behavior is **SO** when $s = 2$.

Wang [637] (2015) applies the dual approach to compute the **SO** threshold in the GI/M/c extension of Naor's model. For this purpose, the author

---

[14]With secondary jobs there is never more than one available server, and this property is essential for the optimality proof when $s > 2$.

[15]Unlike [401], vacations are not required here.

constructs a priority mechanism that can be used to regulate the queue. Its description is simplified in [325] as follows: There are waiting slots numbered $1, 2, \ldots$ and the server always serves the lowest-indexed nonempty slot. An arriving customer either joins the lowest-indexed vacant slot or balks. As in [302], customers' balking behavior is **SO**.

Haviv and Oz [325] (2016) suggest another non-price mechanism for regulating the observable M/M/1 queue. Customers joining when the queue length is below the **SO** threshold are given preemptive priority over those joining longer queues. The advantage of this rule is that, in equilibrium, customers only join when it is **SO** to do so and therefore preemption does not take place.

2.2.1 Altruism and partial control

Gilboa-Freedman and Hassin [253] (2014) deal with a mixed population where a fraction α consists of *altruistic, controllable, cooperative c-customers*, and the rest are *selfish, uncontrollable, noncooperative n-customers*. The former obey the instructions dictated by a central planner who wishes to maximize social welfare while the latter act selfishly. The model bridges Naor's equilibrium **IO** strategy (when $\alpha = 0$) and **SO** strategy (when $\alpha = 1$). The goal of the paper is to investigate the central planner's ability to coordinate the system under intermediate values of α.

The authors consider two levels of control. With *admission control*, the instructions only refer to whether c-customers should join the queue or balk. With *dynamic control*, the social planner can also instruct a c-customer to renege, and always instructs c-customers to give way to n-customers, including allowing for service preemption (where service can be resumed later from the point where it had been stopped).[16]

The main results are:

- Optimal static control dictates a threshold smaller or equal to Naor's **SO** threshold n^*. Thus, c-customers compensate for the over-congestion caused by the n-customers by balking at smaller queue lengths. The threshold increases in α.

- The optimal dynamic control strategy is independent of α. In particular, the threshold is exactly n^*. More generally, this result holds whenever the **IO** threshold of each n-customer is at least n^*, including when customers are homogeneous except for their level of cooperation. The proof follows [302] by using the dual approach, arguing that a customer with the lowest possible priority causes no externalities and behaves in the same way regardless of the value of α.

---

[16]The roles of n-customers and c-customers in this model are somewhat similar to those of primary and secondary customers in other models.

- The gain of dynamic control, which is the ratio of social welfare under the optimal strategies of the two types, is typically small and therefore dynamic control is usually not justified.[17]

See [77, 662] for other models where the queue manager can partially control the customer population.

2.3   Allocation of heterogeneous items

See [698] where the focus is on herd behavior in a model similar to kidney-allocation models discussed here.

Su and Zenios [596] (2004) consider a model motivated by allocation of kidneys to patients awaiting transplant. Patient demand and organ supply are independent Poisson processes. Patients depart the queue after they either accept an organ offer or die after waiting an exponentially distributed amount of time. The cost (due to inferior quality of life) of waiting is linear and dying is quantified as a (negative) fixed reward. Rewards and costs are discounted. The organ quality is a continuous random variable with a known probability distribution. When an organ arrives, its quality is realized and it is offered to patients in the queue according to queue discipline. Should a patient decline the offer, it is offered to the next person in the queue. The organ is discarded if it is declined by all existing patients.[18]

The authors derive the following results formally, while we instead provide qualitative arguments. All patients in the queue are identical from the social point of view at any time. Therefore, the only outcome that affects social welfare is whether a given organ is accepted by any of the patients. In contrast, patients only consider their own expected welfare and ignore the possible negative effects on future patients when an organ is discarded. Clearly this means that when an organ is accepted this is the socially preferred action. Therefore, the planner cannot improve social welfare by early filtering of organs. However, it may turn out that an organ is rejected by all patients in the queue while it is socially preferred that one of them (no matter which one) would accept it.

It follows that the effective decision rests with the last customer offered the organ. Similar to Naor's model, the FCFS discipline is inefficient here because a rejection by the last patient generates negative externalities that this patient ignores and, as in [302], this inefficiency can be resolved by changing the discipline to LCFS. Under the LCFS regime the patient making the last decision is the one who arrived first among the current patients. That patient

---

[17]The value of dynamic control is also discussed in [248, 523].

[18]A similar regime is used in [551] for allocating servers to customers.

will remain the last until he accepts an offer or dies, and hence rejection entails no negative externalities. The authors observe that FCFS is optimal only if no future arrivals are expected. In this case, however, every discipline is optimal.

The authors suggest an interesting variation to alleviate drawbacks associated with the LCFS discipline: Conduct an FCFS regime but randomly prioritize a fraction p of the arrivals and assign them to the head of the queue. This regime bridges FCFS ($p = 0$) and LCFS ($p = 1$). The authors numerically demonstrate that when $\lambda \gg \mu$, a small p suffices to approach maximum social welfare. It is plausible that, as argued in [302], it suffices to assign the chosen customers to any position except for the last one in the queue.

Su and Zenios [597] (2006) consider a flow of heterogeneous candidates for organ (kidney) transplants where *candidates declare their type upon joining the system* rather than receiving offers and deciding whether to accept as in [596]. Thus, when there are n customer types there are also n (Markovian) queues with each arrival choosing one of them. Kidneys are characterized by a single quality parameter $x \in [0, 1]$ that determines the life expectancy, $m + c_i g(x)$, of an i-candidate who receives it. Here, m is a baseline reward (life expectancy), $g(x)$ represents kidney quality, and c_i represents a patient's *risk level* (low-risk candidates are those with a high c_i value). Candidates are also subject to a constant waiting cost rate while waiting in the queue. The system's objective is to allocate kidneys to the n queues and the strategy defines probabilities $p_i(x)$ of allocating a type x organ to the ith queue.

The authors formulate the system's problem using the achievable-region approach, both when customer types can be distinguished and when they are private information. They determine conditions under which the optimal (max-sum or max-min expected utility) strategy is an *assortative partition policy*: partition $[0, 1]$ into n subintervals, $A_1, \ldots, A_n$, and allocate a kidney of type x to A_i iff $x \in A_i$.

Leshno [434] (2014) assumes an infinite queue of *agents*, every agent is of type α with probability p or of type β otherwise. An agent's type is private information. In every period a single item arrives. With (the same) probability p it is an A-item, and if not it is a B-item. Items are offered to agents according to their position in the queue. An agent can accept the offer and be removed from the queue, or decline and maintain position. A declined item is then offered to the next agent and so on until taken. Agents pay c for every period they spend waiting and obtain a value of 1 from taking their matching item (A for an α-customers and B for β-customers), or a value of $v < 1$ from a mismatched item. Agents know their position in the queue as well as the kind of item offered when they decide whether to take it or decline.

It is assumed that the queue is sufficiently long and that all agents are equal with respect to social welfare. Therefore, social welfare is maximized when all agents decline mismatched items (items that are bad for them but good for other agents).

Agent types are revealed when they decline an offer, and at any moment all agents declining an offer are of the same type. The number of agents is bounded since above a threshold they would accept a mismatched item. Under the FCFS discipline an agent's incentive to decline mismatched items is low and the threshold will be small. The problem with FCFS is that incentives to decline are not distributed equally, and this is inefficient; the first agent has a strong incentive to decline, but the last one has very little. Therefore, the author suggests a *load independent expected wait* randomized-priority regime such that the expected wait for an agent who declines an item is independent of the number of other such agents in the queue. This policy reduces the misallocation rate to nearly half of the misallocation rate under the FCFS policy. The SIRO policy is shown to be a good alternative, especially when the exact system parameters are not known and under bounded rationality of the agents.

2.4   Probabilistic joining

See §11.2, §11.3 and §11.3.2 for probabilistic joining based on attraction models. The joining probabilities in these models depend on the queue length through the estimated or quoted benefit associated with joining.

Other probabilistic joining models are included in [68, 377] where joining depends on information provided by the server rather than on actual queue length, [356] where a production system operates under a base-stock base-backlog policy, [440] with dynamic service-rate control, and [168, 75] where the queue is shared by competing firms.

Whitt [660] (1999) compares two $M/M/s/r$ models. Model 1 is unobservable, and customers have delay tolerances of 0 with a given probability, or are exponentially distributed otherwise. A customer reneges once queueing time exceeds delay tolerance. Model 2 is with an observable workload and without reneging. Model 2 has the same parameters and delay tolerances as in Model 1 and customers balks if the observed workload exceeds their delay tolerance. Thus, the joining probability of a new customer in Model 2 is computed as the probability that a server becomes free before the customer would abandon according to delay tolerance.[19,20] It is numerically demonstrated that, for large s, these two models do not differ much, for example,

---

[19]The comparison between the two models makes sense especially in a single-step reward function model like that of Hassin and Haviv (1995) [1] §5.2.1, where service is of full value if started earlier than the tolerance and no value otherwise.

[20]**Jouini, Dallery, and Akşin [378] (2009)** investigate a related model with two priority classes and an exogenous constant β such that customers balk if the state-dependent β-percentile of the waiting time distribution exceeds their waiting tolerances.

with respect to the probability that an arrival is eventually served. The difference is small especially when the number of servers is large and the system load is not heavy.

Marianov, Ríos, and Barros [471] (2005) assume servers are to be allocated to a given set of sites. Customers from different locations are then routed to these sites with the goal of maximizing the total served demand. The potential demand from site i is Λ_i, and it must be routed to a single site (i.e., the demand is *unsplittable*). If i-customers are routed to a facility at site j, only a fraction β_{ij} go there. A customer arriving to a queue of length n joins with probability γ_n. This leads to the following supply-demand equilibrium equation:

$$\lambda_j = \sum_i \sum_n \beta_{ij} \gamma_n \Lambda_i x_{ij} \mathrm{P}_{nj}(\lambda_j, y_j)$$

where y_j is the number of servers assigned to site j, x_{ij} is binary and equals 1 if i-customers are routed to a facility at location j, and $\mathrm{P}_{nj}(\lambda_j, y_j)$ is the steady-state probability of n customers at site j. The authors design a heuristic procedure for computing good location-allocation solutions.

Lozano and Moreno [457] (2008) study the **SO** value of buffer size in a discrete-time single-server queue with Bernoulli arrivals and geometric service distribution. A customer arriving when the buffer is full is lost. Buffer size is constrained to be at least 2. The following costs are considered: fixed waiting-cost rates of joining customers, fixed maintenance cost rate when the server is idle, a fixed cost per balking customer, and a fixed cost per rejected customer (because of a full buffer).

In the observable case an arriving customer finding k customers in the system joins with probability r^k. In particular, the customer always joins an empty system.

In the (almost) unobservable case a customer always joins if the server is idle. When the server is busy but the buffer is not full the customer joins with a fixed exogenous probability, *independent of the buffer size*. This *discouraged rate policy* reflects a form of bounded rationality.

Hwang, Gao, and Jang [353] (2010) describe a pricing and staffing problem of profit maximization in a restaurant modeled as an observable $M/M/s/K$ queue. The arrival rate is given by a state-dependent function $\lambda(x; p, \mu) = \Lambda(p)h(\mathrm{E}[W(x)])$, where $W(x)$ denotes the *queueing* time of a customer who joins when x customers are already in the system, h is a decreasing function, and price p and rate μ are decision variables. The *price sensitivity of the demand* $\Lambda(p)$ is one of the functions $\Lambda_0 - \beta p$ (linear), $\Lambda_0 e^{-\beta p}$ (exponential), or $\Lambda_0 p^{-\beta}$ (inverse).

Singer and Khmelnitsky [579] (2010) consider an MTS firm where production rate is controlled continuously, but demand is realized at discrete points of time. Customers arriving at the same time are ordered and sequentially join in one of the following ways:

- **FCFS:** If inventory level is positive, the customer buys a unit and leaves. Otherwise, the probability the customer joins the queue linearly decreases in queue length, reaching 0 at the maximal queue length.

- **Quote before balking:** Customers are pessimistic, join according to the same probabilistic function assuming all preceding customers joined the queue.

Price, unit production cost, and linear backlog costs are exogenous. The firm dynamically monitors production rate to achieve maximum profit.

The authors obtain structural properties of the optimal strategy and develop a numerical procedure for the finite-horizon case based on these properties.

Li and Jiang [440] (2013) propose a combination of product and service provision with probabilistic joining. The firm produces at rate μ_p and applies a base-stock policy with level S. Customers require a unit of the product and non-storable service that starts only if a unit of product is available. The firm produces at rate μ_1 if the number of customers waiting for service is below a threshold, or otherwise produces at rate $\mu_2 > \mu_1$. Customers can observe the queue length but not the inventory level and the balking probability of a new arrival is proportional to the number of customers in the system. Customers also renege at a constant rate. The authors present a computational study and extend the model by defining performance measures and optimizing a social utility function.

Kesavan, Deshpande, and Lee [398] (2014) experimentally find an inverted-U relationship between the arrival rate to fitting rooms in apparel retailers and sales. This finding triggered a search for queueing Markovian models having this property. The authors investigate four models: (i) unobservable M/M/1 with a fixed arrival probability, (ii) probabilistic joining with $\lambda_k = \frac{\lambda}{k+1}$ when system occupancy is k, (iii) probabilistic joining, dynamic service rate $\lambda_k = \lambda \alpha^k$ and $\mu_k = \mu \alpha^k - 1$ where $0 < \alpha < 1$, and (iv) same as model (i), but the probability that service to a customer joining when occupancy is k is followed by actual selling is qc^k where $q, c \in (0, 1)$. The authors show the dependence of sales on λ is linear in (i), concave in (ii), and inverted-U in (iii) and (iv).

2.5 Server selection and capacity allocation

See [41] for a multiserver game with two agents, one assigning arrivals and the other allocating capacity to servers.

Rubinovitch [551] (1985) complements the results of Kumar and Walrand (see [1] §2.9) by considering a two-server Markovian system with service rates $\mu_1 > \mu_2$, and a single queue. When a server becomes free, that server is offered to waiting customers according to their order in the queue. Both **IO** and **SO** policies dictate accepting an offer from the fast server. An offer from the slow server will be rejected by the customers in the head positions of the queue until a certain threshold.

The author computes the **SO** threshold, observes that the **IO** threshold is greater, and that an appropriate toll on using the fast server can be used to coordinate the system.[21]

Tandra, Hemachandra, and Manjunath [616] (2004) consider a firm operating two observable M/M/1 servers with admission prices p_1 and p_2. When both servers are busy they operate with rates μ_1 and μ_2. However, when only one server is busy the capacities are pooled and the busy server obtains the total capacity $\mu = \mu_1 + \mu_2$. There are two customer classes distinguished by delay sensitivities $1 > a_1 > a_2 > 0$. Balking is not allowed and arriving customers join the queue with the smaller full price, which the authors define for m-customers as the sum of the expected waiting cost and $(1 - a_m)$ times the admission price. Thus the price paid by the customer is smaller than the amount received by the firm. The authors conduct a numerical study and compare the results to the same model without pooling.

Deo and Gurvich [195] (2011) consider a routing problem motivated by the diversion of ambulances to neighboring hospitals. There are two servers (hospitals). Server i faces a dedicated demand with rate λ_i^w of *walk-ins* and an independent non-dedicated (flexible) demand with rate λ_i^a of customers who can be diverted to the other server at no cost. Each server is modeled as an observable M/M/s system. The servers apply a *diversion threshold strategy*: When the number of customers in queue i exceeds a threshold K_i this server declares *diversion status* which means that, unless the other server is also on diversion status, flexible customers are diverted to the other server.

The authors observe that if the objective of the servers is to minimize expected waiting times at their location, an equilibrium with $K_i = 0$ $i = 1, 2$ results. Consequently they discuss centralized routing decisions.

---

[21]The conclusion that self-interested individuals over-congest the fast server is similar in nature to that obtained for unobservable queues in [85].

Hassin, Shaki, and Yovel [315] (2015) consider profit-maximizing capacity allocation in a system of M/M/1/1 queues. Three models are considered, and in all of them the price is optimally set to the maximal value such that customers are still ready to join the slowest server when all others are busy. The three models are:

- *All servers must be allocated the same capacity and the decision variable is just the number of servers.* The optimal number of servers is characterized. A heavy-traffic model is given and it is shown that the number of servers increases as $n^{2/3}$, where n is the scale parameter. This differs from the Halfin-Whitt regime.

- *Different capacities can be allocated to servers and an arriving customer joins the fastest free server, if one exists.* The authors give support to a conjecture that the optimal solution always has identical capacities.

- *When service terminates at a given server each customer currently served by a slower server is reassigned to the next faster server.* In this case, the authors prove that *the expected waiting time of a customer who joins the slowest server is independent of the allocation and only depends on the number of servers.* They conclude that the optimal number of servers equals Naor's **SO** threshold.

2.5.1 Polling

Altman and Shimkin [51] (1993) consider a single server which serves several observable FCFS queues. They consider the following nonpreemptive server strategies:

- **Random SLQ:** The server always serves the longest queue and if there is more than one longest queue then one is chosen at random.[22]

- **Persistent SLQ:** As above, but the queue that has just been served will be selected as long as it remains the longest queue.

The authors consider the following customer strategies:

- **JSQ:** An arriving customer joins the shortest queue.

- **JLQ:** An arriving customer joins the longest queue.

In both cases, if the choice is not unique, one of the options is chosen at random.

The main results are:

---

[22]SLQ induces sort of an FTC model. There is no threshold strategy here but customers are encouraged to follow their predecessors and join the same queue since this is the queue to be selected by the server. The authors do not attempt however to characterize the complete set of solutions nor claim that a unique equilibrium exists.

- JSQ is an equilibrium symmetric strategy under random SLQ.

- JLQ is not always an equilibrium strategy under random SLQ.

- JSQ is not always an equilibrium strategy under persistent SLQ.

- JLQ is an equilibrium strategy under persistent SLQ. However, it is not subgame perfect.

Atar and Saha [74] (2015) consider n identical servers serving a finite number of customer classes, each class with a separate FCFS queue. The service discipline is either fixed absolute priority (**FP**) or *serve the longest queue* (**SLQ**), both without preemption. It is assumed that arriving customers observe their own queue but not the other queues, and either join or balk based on this information.

The authors consider a sequence of systems at the diffusion scale heavy-traffic regime. They prove this regime has a unique feature, namely that an ϵ-Nash equilibrium can be attained with probability converging to one, with respect to costs representing the *actual* delay experienced by the customers as opposed to the more standard setting of using *expected* delay as cost. The Nash equilibrium is provided explicitly.

2.6 Dynamic control

Dynamic (state-dependent) control gives the firm more flexibility in achieving its goals relative to static (state-independent) control. However, dynamic control might prove difficult to implement. For example, implementation may involve costly *switching* or *setup costs*. The papers described here do not explicitly assume such costs, but many recognize the inconvenience associated with dynamic control. Moreover, customers may object to dynamic control since it makes the service experience less predictable. Several papers focus on replacing optimal dynamic control with second-best control mechanisms to alleviate loss of optimality. These papers often investigate the loss associated with this change. Such examples can be found in [72, 117, 119, 169, 229, 253, 298, 311, 372, 490, 523, 626].

Dynamic control provides rational customers with useful signals about the system's state or about parameters that they cannot observe directly.[23],[24]

Most papers on dynamic control assume customers are not strategic in selecting time of arrival. In other words, these models deal with policies that

---

[23]Models where customers ignore such signals are considered in §11.

[24]Dynamic capacity control is another common type of dynamic control, but it is not considered in this section because it doesn't necessarily mean the queue is observable and changes in service rate are not easy to observe.

react to the state of queue but do not affect the timing of arrivals.[25] The only exception to this seems to be [144], where strategic arrival time manipulations in a dynamic pricing setting are considered. In particular, customers may wait until the price is lowered before placing demand.

Research on dynamic pricing is described in [1] §2.8. See [299] where competing firms dynamically change capacity, [192] for a model with strategic vacations and dynamic PDTs, §9.6 for queueing games involving dynamic control of arrival and service rates, and [533] for dynamic service-rate control in a supply chain with an unobservable queue.

2.6.1   Price-sensitive customers

This section considers customers who are price sensitive but not delay sensitive. See 6.8 for dynamic pricing, admission, and searching for customers, and see [632] for a model with dynamic pricing and bounded rationality.

The earliest model of dynamic pricing control of a queue seems to be that of **Low [456] (1974)**. That Markovian model assumes price-sensitive customers and state-dependent holding costs incurred by the server. State-dependent prices can be selected from a finite set $p_1 < \cdots < p_K$ with corresponding demand rates $\lambda_1 > \cdots > \lambda_K$. The author derives an algorithm for computing the profit-maximizing prices and proves that they are nondecreasing in the number of customers in the system. **Aktaran-Kalayci and Ayhan [26] (2009)** investigate the sensitivity of the optimal prices in [456] to system parameters.

Johansen [372] (1996) considers an M/D/1 system with price-sensitive demand having uniformly distributed service valuations. and an exogenous upper bound on the delivery time. The author compares the optimal workload-dependent dynamic pricing with the optimal static pricing, and reports a numerical study in which the gain of dynamic control is relatively small and does not exceed 3%.

Paschalidis and Tsitsiklis [523] (2000) consider an M/M/s/s loss system with customer classes. Batches of size r_i of i-customers arrive according to a Poisson process whose rate $\lambda(u_i)$ is strictly decreasing in the price u_i and their service is $\exp(\mu_i)$. A batch can be wholly admitted or rejected. The utility U_i obtained by a request of class i is a random variable and the request is accepted if $U_i \geqslant u_i$, and capacity suffices to serve the request. A batch arriving when the system has insufficient capacity will be lost. The prices u_i are dynamically determined as a function of the state of the system.

- The optimal prices associated with profit and welfare maximization in the single-class case increase monotonically with the number of busy

---

[25]Strategic arrival time decisions are the subject of §4.1.

servers. However, this seemingly intuitive result does not extend to the multiclass case, as shown by a counterexample. The reason for this being the underlying *packing problem* that arises from the different batch sizes.

- Dynamic prices are difficult to manage and may also be less desired by customers who prefer a fixed (predictable) price. Therefore, a central question is whether the gain of dynamic control, measured as the added profit and welfare that can be achieved, justifies use of dynamic pricing. The authors compare optimal dynamic and static pricing, and conclude that for both revenue and welfare maximization static prices are asymptotically optimal in several limiting regimes including light traffic, heavy traffic, and a regime of many small users.

- **Paschalidis and Liu [522] (2002)** extend [523] and prove similar results for a regime with many small customers under a generalized model. For example, allowing for a (loss) network with demand substitution effects.

Maglaras [462] (2006) considers a multiclass M/M/1 model with price-sensitive customers. The rates μ_i of serving class i customers are decision variables which the firm selects once and then remain fixed. However, sequencing and prices are dynamically controlled. By choosing its price vector $\boldsymbol{p} = (p_i)$ at any point in time the server can induce a vector $\boldsymbol{\lambda}(\boldsymbol{p}) = (\lambda_i(\boldsymbol{p}))$ of instantaneous arrival rates. The firm incurs convex operating costs and a holding cost of c_i per unit time of stay of an i-customer in the system. Its objective is to set service rates and select sequencing and pricing policies so as to maximize the profitability of the system.

The author proposes a solution where the capacity vector is determined by a long-run fluid approximation in which the long-run average profit criterion focuses on optimally matching supply and demand. Subsequently, demand rates and capacity allocation decisions are dynamically selected to maximize the total profit in the fluid model along its transient trajectory from any initial condition until the buffers empty. Sequencing decisions are made according to the $c\mu$-rule. After a finite period of time, the entire workload in the system is held in the lowest priority class. Focusing on the transient problem from that point on gives a one-dimensional drift control problem for the workload process. The solution for the case of a linear demand function is solved in closed form.

The following heuristic is numerically tested for the original problem: Sequence jobs according to the $c\mu$-rule, and from any queue-length configuration make a dynamic pricing decision that optimizes the fluid model transient response for the one-dimensional workload problem. As the state evolves stochastically over time the decisions are adjusted accordingly.

Chen, Feng, and Ou [149] (2006) provide an algorithm for solving the following model: An M/M/1 firm can switch at no cost between prices $p_1 > p_2$

which induce demand rates $\lambda_1 < \lambda_2$. The firm incurs production, holding, and backorder costs. The optimal policy is determined by two thresholds, a base stock level and a price switch threshold such that the low price is chosen when inventory is above that level. The authors extend the model by considering production in constant-size batches [150], and **Chen, Chen, and Peng [147] (2011)** extend the model by considering Erlang processing time.

Plambeck and Ward [531] (2006) consider an *assemble-to-order* system where product k requires a_{kj} units of components of type j and assembly is instantaneous. The system manager sets components production capacities $(\mu_1, \ldots, \mu_J)$ and product prices $\boldsymbol{p} = (p_1, \ldots, p_K)$ which are paid when an order is filled. Orders for product k arrive at rate $\lambda_k(\boldsymbol{p})$, and components of type j arrive at rate μ_j. Each component is associated with a production cost c_j that is paid upon delivery of the component. The firm incurs a unit holding cost rate h_j, and the decision variables are the prices, the capacities, and the dynamic sequencing rule.

The authors prove that a myopic discrete-review sequencing policy is asymptotically optimal, that the optimal prices and capacities nearly balance the supply and demand for components, and that the resulting heavy-traffic solution can be approached by a diffusion approximation which adjusts the product prices and production capacities obtained in a first-order fluid approximation of the model. See [532] for a further description.

Maoui, Ayhan, and Foley [474] (2007) generalize [456] in several ways. They consider a single queue multiclass model with prices that are state and class dependent. Holding costs incurred by the server are given by a general (not necessarily linear) function of queue length, and the service rate changes according to an exogenous nondecreasing function of queue length. The demand by each class exhibits increasing price elasticity (equivalently, increasing generalized failure (hazard) rate).[26]

The authors characterize the optimal pricing policies. In particular, they show that profit-maximizing prices are nondecreasing in queue length, and if the upper bound on queue length increases, then the optimal prices decrease for every state.

Feng, Ou, and Pang [235] (2008) consider a Markovian model of an assembly system with two production facilities. Facility i produces i-components at rate μ_i. Assembly is instantaneous, demand is price sensitive, and there are two exogenous prices which induce corresponding demand rates. The firm incurs component holding costs and a per unit-time cost to hold a backorder.

The authors show how to compute discounted profit-maximizing state-dependent prices and base-stock levels. **Keblis and Feng [395] (2012)** gen-

---

[26]Let $F(p)$ denote the proportion of customers willing to pay price p. The cdf $F(p)$ has an *increasing generalized failure rate* if $\frac{yf(y)}{1-F(y)}$ is strictly increasing over the support of the pdf $f(p)$. Equivalently, the demand function has increasing price elasticity.

eralize this model so the stockout costs include both fixed and variable cost elements.

Mutlu, Alanyali, and Starobinski [490] (2009) compare dynamic and static pricing in a profit-maximizing $M/M/C/C$ loss system with primary (PU) and secondary (SU) users. Each server has a dedicated fixed arrival rate of PUs and a cost K is incurred when a PU is blocked. SUs are price sensitive and arrive at rate $\lambda(u)$ when the admission cost is u. The authors consider three pricing methods:

- **Dynamic pricing:** SUs pay price u_n when there are $n < C$ busy servers. The authors characterize the solution and in particular prove that $\{u_n\}$ is nondecreasing.

- **Static pricing:** SUs pay a fixed price, independent of the system state. The authors derive a necessary and sufficient condition for the optimal static pricing to be prohibitively large (the resulting arrival rate of SUs is 0, which generates no profit).

- **Threshold pricing:** SUs are admitted with a fixed price when the number of busy servers is below a threshold, and are rejected otherwise. Clearly, threshold pricing provides more flexibility. For example, static pricing may block all SUs, while under the same parameters threshold pricing generates a positive profit. The authors provide numerical evidence that static pricing often performs badly, but threshold pricing performs close to dynamic pricing for a variety of demand functions. Furthermore, the authors prove that if some arbitrary pricing policy yields positive profit, then a threshold pricing policy that yields positive profit will also exist.

Çil, Karaesmen, and Örmeci [169] (2011) consider an $M/M/1$ system with two price-sensitive customer classes. Given price p_j, the arrival rate of j-customers is $\lambda_j(1 - F_j(p_j))$, reflecting random service valuations with cdf F_j and density function f_j. These distributions are assumed to have an *increasing generalized failure rate*, i.e., $\frac{p f_j(p)}{1 - F_j(p)}$ is strictly increasing. Service distributions of the two classes are identical and the server incurs unit holding costs $h_j x_j$ ($h_1 > h_2$) per unit time, where x_j denotes the number of j-customers in the system. Profits are maximized by deciding at any state (x_1, x_2) which class to serve and which admission prices $p_j(x_1, x_2)$ to charge.

The authors prove that the optimal policy gives preemptive priority to 1-customers. They also prove monotonicity properties of the optimal dynamic pricing policy. Numerical results indicate, however, that static pricing policies may not perform well for certain range of parameters.

Mutlu, Alanyali, Starobinski, and Turhan [491] (2012) consider an $M(n)/G/C/C$ loss system with primary users (PUs) arriving at a fixed rate

λ_p and price-sensitive secondary users (SUs) arriving at rate $\lambda_s(u)$ when the service price is u.[27] The system is interested in maximizing revenues from admitting SUs, minus penalties associated with blocked PUs.

The authors focus on *threshold pricing policies* where SUs are admitted and charged a fixed price if the number of busy servers is below a threshold, or they are rejected otherwise. Assuming the demand function is unknown to the server, the authors analyze convergence properties of a learning algorithm that repeatedly adjusts the price and tests the resulting arrival rate.

Atar, Cidon, and Shifrin [73] (2014) consider an M/M/1/B queue with dynamic price control and price-sensitive customers. Price is dynamically selected from $C_1 < \cdots < C_K$ inducing the arrival rate $\lambda_1 > \cdots > \lambda_K$. There are no waiting or holding costs in this model, and the goal is to maximize expected discounted revenue. The authors prove that for some constants $0 = b_0 \leqslant b_1 \leqslant \cdots \leqslant b_K = B + 1$, the optimal policy is to announce C_i iff queue length is in $[b_{i-1}, b_i - 1]$.

Chen, Hao, and Wang [151] (2014) assume in their base model that an MTS M/M/1 firm charges one of two prices $p_1 > p_2$. Customers are price sensitive and arrive at rates $\lambda_1 < \lambda_2$, respectively. The firm can produce in-house at rate μ and unit production cost b, or outsource to an external facility producing at rate a and unit cost $c > b$. It is assumed that $\lambda_2 p_2 - \lambda_1 p_1 > (\lambda_1 - \lambda_1)b$. The firm also incurs linear holding and backlog costs.

The authors prove there exists an optimal policy defined by three thresholds, (R, D, S) where $S > \max(R, D)$: When inventory exceeds S there is no production; when inventory is between S and D, the firm produces in-house; when inventory is below D the firm produces in-house and outsources. The firm sells at the high price p_1 iff inventory exceeds R.

Chen, Tai, and Yang [154] (2014) consider a single-server Markovian system with two classes of price-sensitive customers. Class i's demand, $i \in \{A, B\}$, arrives at rate λ_j^i when charged price p_j^i, and the firm incurs unit holding costs rates h_i. A concavity property of the revenue function is assumed, namely, $(\lambda_{i+1}^j p_{i+1}^j - \lambda_i^j p_i^j)/(\lambda_{i+1}^j - \lambda_i^j)$ is decreasing in i.

The authors characterize the optimal price for both products in each state of the system, and specify the production policy for each state, i.e., whether to produce an A-item, a B-item, or leave the server idle.

Chen, Feng, Hao, and Keblis [148] (2015) consider an M/M/1 firm that serves subscribed demand arriving at a rate λ for an exogenous price per customer. The firm uses excess capacity to serve *free market demand* arriving at rate λ_i when the price set by the firm is p_i, $i = 1, \ldots, K$. The firm incurs holding cost rates of h_1 per subscribed customer and h_2 per free-market customer. The firm's decisions consist of the type of customer to serve

---

[27]Unlike the closely related model of [632], PUs do not preempt service of SUs.

and the price to charge given the system state (x, y), where x is the backlog of subscribed customers and y is the backlog of free-market customers. The solution is characterized by switching functions $G(y)$ and $H_i(x)$ such that for a fixed x $H_1(x) \geqslant \cdots \geqslant H_{K-1}(x)$. Subscribed customers are given preemptive priority when $x > G(y)$, free-market customers are given preemptive priority when $x < G(y)$, and the price is p_i when $H_{i-1}(x) \geqslant y > H_i(x)$.

2.6.2   Price- and delay-sensitive customers

See §11.3.2 for models with dynamic price and delay quotes under bounded rationality, and [228] for a model with dynamic due dates in a decentralized environment.

Harubi, Shechter, and Subotnik [300] (1979) offer a discrete-time model of dynamic pricing designed to maximize system efficiency. They do not assume a particular customer decision model, or specific arrival and service processes, but assume a particular form of the pricing function. We describe here an interesting generic model inspired by their ideas.

The heterogeneous waiting costs of customers are private information and admission prices increase with system load. Customers are aware of the firm's pricing policy and state upon arrival the maximum load level at which they would be willing to join the queue. If the current load in the system queue is not higher than this maximum, the customer joins the system queue. Otherwise, the customer enters a virtual queue (where the waiting cost is lower) and stays there until the system load and price decrease to the stated maximum load level. The result is an auction-type priority regime, which remains in effect until the customer joins the system queue.

Johansen [371] (1994) considers an M/G/1 system with an observable workload.[28] In particular, the job's processing time is observed upon arrival. The benefit for joining customers is stochastically decreasing as a function of the current workload u and their service time w. The server operating costs and the price also depend on these two variables.

The *opportunity cost* is the system's expected loss of future earnings caused when the customer joins the system. The author proves that this cost is increasing in u, and increasing and convex in w. The **SO** admission control consists of state-dependent threshold values which determine the maximum service time of the admitted requests. The author derives conditions for this sequence to be monotone in u, but notes that the profit-maximizing prices may decrease in u.

Plambeck [528] (2004) considers a Markovian queue with two customer types differing in price and delay sensitivity. Demand of k-customers at time

---

[28] A discrete-time version is analyzed in [370].

t is Poisson with rate $\lambda_k(t) = \left(\bar{\lambda}_k - \alpha_k P_k - \beta_k d_k(t)\right)^+$, where P_k is price charged and $d_k(t)$ is the queueing delay quote at time t for class $k = 1, 2$. The crucial assumption that $\beta_1 \gg \beta_2$ implies that 2-customers are patient and tolerate long waits. Thus, these customers can be used as a buffer against stochastic fluctuations in demand and service. The server can therefore maintain almost perfect balance between demand and service rate. Thus, this heavy-traffic condition is a *consequence* of the assumption. The decision variables in this model are the prices, dynamic delay quotes, service capacity, and service sequencing.

The heavy-traffic model assumes a sequence of systems with waiting costs decreasing to 0. The author applies an *asymptotic compliance* performance criterion which states that it is very unlikely for a customer's queueing delay to exceed his quoted lead time by a significant amount on the time scale on which he measures delay. In the solution of the model, the dynamic control reduces to giving priority to 1-customers. The lead-time quotes are $d_1(t) = 0$, while $d_2(t)$ is the queueing time resulting from the priority regime given the number of customers in the system at time t.

The author also treats the asymmetric information version where customers choose their class by adding **IC** constraints. An interesting feature of this solution is that when the queue is short, the lead-time quotation to 2-customers is higher than the actual one in order to prevent 1-customers from choosing the class 2 offer.[29]

Two aspects of the proposed solution are compared by simulation in non-asymptotic parameter regimes: the long-run average profit and the average amount by which a customer's actual lead time exceeds the quoted lead time.

Ata and Shneorson [72] (2006) consider a Markovian model where both demand and service rate are dynamically controlled. The cost rate associated with serving at rate μ is a convex function, and the value rate associated with demand λ is a concave function. Customers are homogeneous with respect to waiting-cost rate and join if their service value exceeds full price. For every queue length n the authors compute the **SO** arrival and service rates (λ_n, μ_n), from which prices p_n that induce λ_n can be easily derived. The optimal arrival rate decreases and service rate increases in queue length. In contrast, the authors observe that optimal prices need not be monotone. A numerical analysis suggests the advantage of dynamic control over static control is most significant when delay costs are large.

Plambeck and Ward [532] (2008) extend [531] by considering customers that are both price and delay sensitive. The authors consider an assemble-to-order system consisting of a *manufacturer* producing a set $\mathcal{J}$ of

---

[29]This outcome resembles the strategic delay policy introduced in [9, 12], but differs from strategic delay because it involves no trade-off: In the asymptotic regime the 2-customers become infinitely patient so that inflating their lead-time quotes does not reduce revenue from these customers.

components, and assembling from them a set $\mathcal{K}$ of *products*. Assembly is instantaneous, but the required components must be available. The manufacturer sets production rate μ_j for each component $j \in \mathcal{J}$, price p_k and lead-time guarantee l_k for each $k \in \mathcal{K}$, and the sequencing rule.[30] The production cost associated with a unit of component j is c_j and holding it in inventory costs h_j per unit time. The demand for product k is a function $\lambda_k(p,l)$ where p is the vector of prices and l is the vector of lead-time guarantees. An important example of the demand function is generated by assuming that customers differ in service values v_k for having product k, and delay-cost functions $f(l)$, and each chooses a product that maximizes $v_k - f(l_k) - p_k$, if this value is positive.

An order for product k must be filled within l_k units of time. To this end, the manufacturer can immediately obtain at any time an extra unit of component j by paying an *expediting cost* $x_j > c_j$. The manufacturer dynamically sequences outstanding orders for assembly and expedites components when necessary. The authors also expand their model to allow price, maximum lead time and production capacity to change in response to shifts in demand and supply conditions.

The model includes subtle points. The objective is to maximize expected discounted profit. Early delivery is allowed, and yet the demand depends on the quoted upper bounds and not on the lead-time distribution. Unlike the case of average profit maximization (e.g., [131]), delivery cannot be delayed at zero cost to the quoted deadline when discounting is considered. Even if a unit of demand can be satisfied earlier than its deadline because all components have been manufactured, the firm may prefer to wait with delivery and keep some components for satisfying the demand of a more urgent product. A related point is that customers agree to pay the price also in the case of early delivery. With discounting, this means paying a higher price. Therefore, the authors also consider a variation of *exact lead-time quotation* where the firm cannot deliver early, but conclude the system manager would strongly prefer the flexibility associated with quoting an upper bound on delivery rather than an exact deadline. Moreover, expediting may be worthwhile not only when it is required for timely delivery, but also for obtaining earlier payment.

The authors prove that optimal prices, lead times, and capacities *result* in heavy traffic. The high-volume asymptotic analysis considers a sequence of systems indexed by $n = 0, 1, \ldots$ with increasing demand $n\lambda_k(p,l)$ and production rates $n\mu$ leading to an asymptotically optimal discrete-review policy. In some cases, the lead-time quote should be exaggerated to encourage impatient customers to buy expensive products. However, their demand will be consistently filled earlier than the quoted lead time in order to realize the revenue early. Such policies are similar in spirit to the strategic delay of [9, 12], but here they rely instead on customers' bounded rationality.

---

[30]Two products can be physically identical but with different (p_k, l_k) values.

Çelik and Maglaras [131] (2008) solve a dynamic-control model in which a single server offers multiple service options (products) indexed $i = 1, \ldots, I$. Service options are differentiated by processing rate μ_i, price p_i, and target lead time d_i. To meet its lead-time commitments, the firm can expedite service and instantaneously produce a unit of product i at cost c_i. The firm dynamically varies prices $\boldsymbol{p}(t) = (p_i(t))$ and production sequencing to maximize long-run average expected revenue minus expediting costs.

The realized demand for each product i is (nonhomogeneous) Poisson with an arrival rate vector which depends on the prices posted at that instance and the vector of delay guarantees. The lead-time guarantees are reliable upper bounds because they are kept by expediting when necessary.[31]

The authors analyze an approximating large-scale diffusion control problem motivated from settings where both arrival and service rates are large. In this limiting model, several aspects of the problem simplify: Sequencing decisions are made according to the *least slack* policy, the system only expedites the cheapest class as measured by the $c_i \mu_i$ index, prices are increasing functions of the total workload and modulate the system so as to induce full resource utilization. The authors numerically verify the effectiveness of the resulting heuristic.

Besbes and Maglaras [90] (2009) consider profit maximization in a single-server queue processing orders from price- and delay-sensitive customers with heterogeneous service values. The main characteristic of the model is that the market size evolves stochastically over time. The authors study the queue dynamics and resulting price-optimization problem. Examples they suggest for the stochastic evolution of the market size include transitioning between predetermined levels according to some transition matrix, or affine diffusion models that are often used in modeling interest-rate behavior. The market-size process changes at a slower time scale than the transient dynamics of the queue and the authors study the behavior of this queueing system under static and dynamic pricing policies. They focus on the slower time scale where the market-size process fluctuates and show that queue dynamics can be captured via a stochastic fluid model driven by the stochastic variability of the market-size dynamics. The analysis leads to simple and implementable pricing heuristics, and numerical experiments examine the effect of arrival process variability on the solution as well as the server's policy when it does not know the instantaneous arrival rate when making pricing decisions.

Ata and Olsen [70] (2009) consider profit maximization in a model with homogeneous customers, a nonlinear delay cost function $c(\tau)$, and a finite service value R. Each arriving customer is either rejected right away or obtains

---

[31]Customers' demand is a function of these upper bounds and not of the actual state-dependent lead-time distribution. A similar assumption exists in papers on strategic delays. It may reflect bounded rationality of customers, and indeed the lead-time distribution is difficult to estimate.

a *reliable* state-dependent lead-time quote τ such that $c(\tau) < R$, and pays an admission price $R - c(\tau)$. To guarantee the quoted lead time will always be met the authors assume deterministic service duration. Service duration is a decision variable the firm sets at the beginning of the planning horizon. The central theme is the trade-off between committing future capacity now or reserving it for later higher revenue customers.

When c is convex, customers should be served in order of appearance. When c is concave it is natural to consider a solution where most customers are served soon after arrival to avoid the initial steep increase in cost, while letting others have a very long wait and exploiting the flatter section of the cost function.

The authors consider the problem in a discrete-time periodic review model with large capacity and high volume of arrivals. They propose asymptotically optimal strategies by dividing the (homogeneous) demand into high- and low-priority classes, quoting high-priority customers the shortest possible lead time, and postponing service to low-priority customers. The authors also solve the asymptotic model for a generalization where c is "S-shaped" (or, *convex-concave*).

Yildirim and Hasenbein [684] (2010) consider multiclass Poisson arrivals where the service distribution of class k customers is Erlang(k, μ).[32] All costs and rewards are discounted and the admission fee $p(i, k)$ is both state and type dependent, where i is the system workload in terms of the number of (exponential) service stages. The admission fee is paid upon admission, the reward R is obtained when service is completed, and the waiting cost of c_k per time unit is incurred continuously. For any given state i and type k, the maximum price $p(i, k)$ that induces the customer to join is computed first. A by-product of this computation is the **IO** strategy when admission is free. This behavior is characterized by thresholds k_i^* such that a k-customer joins at state i iff $k \leqslant k_i^*$. A profit-maximizing server chooses between rejecting a new customer or accepting the customer and asking for the maximal price, $p(i, k)$. The main qualitative question is then whether the profit-maximizing policy can be characterized by a threshold on $i + k$.

The authors present two numerical examples. In one example such a threshold exists, but in the other case there is a (small) deviation from a strict threshold rule.

Giloni, Koçağa, and Troy [256] (2013) consider dynamic pricing in an FCFS M/M/s/I system with customer classes distinguished by service valuations and (nonlinear) queueing costs.

The authors show that under the **SO** admission policy the set of classes admitted when the queue length is $i + 1$ is a subset of those admitted when

---

[32]The paper refers to batch arrivals with $\exp(\mu)$ service. Batches act as one entity, splitting of a batch is not allowed, and the reward is obtained only after all customers in a batch are served.

it is i. However, this property does not necessarily hold under revenue maximization, when customers are strategic and the waiting cost functions differ among classes. In fact, in this case the total arrival rate does not necessarily decrease in the queue length. This assertion is demonstrated by a simple two-class example where the class with the lower potential arrival rate is more time sensitive, but has a higher service value. When the queue is short, the optimal price is such that only this group joins. When the queue is longer, the situation is reversed.

The authors also generalize their model to include *partial differentiation*: classes are clustered into super-groups and the server can price-discriminate super-groups but not individual classes.

Akan, Ata, and Olsen [23] (2012) consider social-welfare maximization in a multiclass extension of [70]. The service values R_i are heterogeneous, as are the delay-cost functions c_i, but the two classes have the same deterministic service requirement. The function c_i is convex on $[0, d_i]$ and concave on $[d_i, D_i]$, where $D_i = c_i^{-1}(R_i)$ is the maximum acceptable delay quote for i-customers. It is assumed that $c_i'(t) > c_j'(t)$ for $i < j$ and $t \in [0, D_i]$, $d_1 \leqslant \cdots \leqslant d_N$, and $D_1 \leqslant \cdots \leqslant D_N$. The goal is to design a dynamic menu of price and lead time so that (i) all customers join, (ii) it is **IC**, and (iii) the long-run average rate of welfare is maximized.

The authors present and solve a fluid approximation and use the insights and policy derived from it to propose a policy for the stochastic system. When the system is not congested the queue discipline is FCFS. When the system is congested the fluid approximation is applicable and more impatient classes receive shorter delay quotes to achieve the same marginal delay cost.

Ata and Olsen [71] (2013) consider a two-class extension of [70] with two customer classes and asymmetric information. The firm sets queue discipline and a state-dependent **IC** menu of price and lead-time pairs. Service value R is common to both classes and the delay cost functions c_i, $i = 1, 2$, are convex-concave, where c_i is convex on $[0, d_i]$ and concave on $[d_i, \infty]$. Class 1 is the *impatient* class and class 2 is the *patient* class. The shape of the two delay costs is similar but the point of switch d_i is earlier for the impatient customers, i.e., $d_1 < d_2$. The model requires that all customers agree to one of the contracts and thus negative prices are allowed.[33]

The authors propose a sequence of heavy-traffic systems. The main result is a corresponding sequence of policies that are asymptotically optimal. The intuitive idea is that when the workload is small, both classes are kept in the convex cost region and a modified $Gc\mu$-rule is in effect (see [1] §4.6). When the workload is large, the impatient class is kept in its convex region while (a subclass of) the patient class operates in the concave region.

---

[33]The authors mention that in practice this is unlikely to occur.

Kim and Randhawa [400] (2015) consider dynamic pricing when customers are price and delay sensitive and the service-valuation distribution has a non-decreasing hazard rate. Arriving customers observe queue length and join if their valuation exceeds full cost.

The authors consider a large system with arrival rate $n\lambda$ and service rate n. The loss of revenue due to variability under the optimal pricing scheme is $O(n^{1/3})$, in contrast to $O(n^{1/2})$ under static pricing. The authors propose a simple two-price threshold policy within a logarithmic term of the optimal scale.

2.6.3 Delay compensation

Delay compensation is a form of dynamic pricing which depends on the system state through expected or realized waiting time.

See §6.2.4 for delay compensation in unobservable queueing models, [15] for discounts based on the realized lead time in the presence of risk aversion, [161] for delay compensation as part of the subscription contract, and [340] for delay compensation in a supply chain.

Banker and Hansen [79] (2002) consider a multi-period model where at the beginning of each period the server sets the capacity of the facility so it can serve all backorders and, in addition, up to N new orders in that period. Customers have heterogeneous service valuations and only those who value service above the posted price p join the queue. Let their number be s. If $s > N$ then N customers are selected to be served, and the rest are offered compensation d if they agree to be served in the next period. The cost of waiting is $a \geqslant d$ and therefore only customers whose service value is at least $p - d + a$ accept the backorder offer. The server's decision variables are the extra capacity N, the price p, and the discount d.[34,35]

From the firm's point of view this is a single-period problem and, as the authors note, price discounts behave in a similar fashion to soft capacity, i.e., expensive capacity purchased at a premium to augment existing capacity (like overtime).[36] Both price discounts and soft capacity allow the manager to serve additional customers once initial capacity has been exhausted. The authors offer three heuristics and test them numerically.

---

[34]Note that customers are rational in basing joining behavior only on p because there is no waiting involved until they learn whether they will be instantaneously served. There is no gain in speculating because $a \geqslant d$ so that the offered discount does not create additional incentive to join.

[35]A similar model is considered by **Bayram, Ismail, Abdallah, Qaraqe, and Serpedin [82] (2014)**, but the upper bound on the served demand in each period is exogenous and the solution is simple. The firm serves in each period as many customers as the bound allows and offers the smallest discount necessary to persuade the extra demand to wait service in the next period.

[36]See also papers on expediting delivery.

Keon and Anandalingam [396] (2005) compute optimal dynamic price discounts which are used to shift demand from congested to uncongested periods.[37] They consider an $M/G/c/c$ loss system with an exogenous price and an upper bound on the blocking probability. The arrival process occurs in continuous time and is non-stationary. However, time is divided into discrete periods with an approximately constant rate of potential demand λ_k in period k. These periods are long enough so that the steady-state distribution of the number of customers in the system can be used. The server offers nonnegative discounts d_k for those customers who arrive at period k and agree to postpone their demand to the next period. A customers who is delayed cannot be delayed again. Each customer is characterized by the valuations v of immediate service and v_d of deferred service, $v > v_d$. The fraction of customers agreeing to postpone service is an increasing function of the offered discount. The firm minimizes the expected value of the discounts subject to the upper bound on the blocking probability.

The authors show that the problem can be solved sequentially, using the solution for period $k - 1$ to solve for the optimal discount in period k.

Sen, Raghu, and Vinze [565] (2009) consider a multiclass multi-task $M/M/s$ priority system with service rate depending on service type. The price for type j service at priority class k is $S_{jk} - P_{jk}W_{jk}$, where W_{jk} is the expected waiting time and P_{jk} is compensation paid to the customers per unit of waiting. The authors use the term *price-penalty scheme* for this dynamic pricing mechanism. Type i customers that demand type j service select the priority class k which minimizes their full price $S_{jk} - P_{jk}W_{jk} + \delta_{ij}W_{jk}$, where δ_{ij} is the user delay-cost rate. The provider's objective is to maximize profits, consisting of the revenue obtained from customer payments minus a linear operating cost per each unit time of server activity. The decision variables are the parameters S_{jk} and P_{jk} as well as the number of servers.

The authors numerically investigate through simulation the effects of the price-penalty scheme. They conclude that although the customers' surplus or the provider's profit are often smaller, the proposed scheme is likely to improve overall system welfare as compared to a fixed-price FCFS approach.

2.6.4 Competition and networks

See [1] §7 and especially §7.3 for early research on competition in observable queues.

Hsiao and Lazar [347] (1991) study dynamic control of multiclass demand in a network. There are N_k users of class k and M service stations, each with its own queue. Demand generated by k-users is Poisson with rate $\lambda_{j_k}^k$,

---

[37]Already in 1964, **Leeman [433]** suggested that firms, such as supermarkets, charge *check-out fees* that are higher at certain times of the day or week and lower at others in order to regulate the timing of arrivals of customers who have preferences in shopping times.

$0 \leqslant j_k \leqslant N_k - 1$, $k = 1, \ldots, K$, depending on the number j_k of outstanding k-users in the network. When a k-user's demand is generated, it is routed to station $i \in \{1, \ldots, M\}$ with probability $r_{k,i}$. Upon service completion at server i, the k-user is routed to station j with probability $r_{k,i,j}$, $j \in \{1, \ldots, M\}$, or the user leaves the service facility with the complementary probability. Service at station i is exponential with a class-independent rate μ_i.

The system's goal is to establish state-dependent arrival rates $\lambda_{j_k}^k$ and maximize aggregate throughput subject to an upper bound on the average delay. The optimal controls are characterized as follows: For some integers L_k and $m_k < L_k$, if the number of outstanding k-users is less than L_k but different from m_k, then the demand rate of k-users is equal to the maximum possible value c_k. Otherwise it is 0. In state m_k the demand rate can be an intermediate value between 0 and c_k.

In the class decision version, each class maximizes throughput subject to a bound on the average delay. The best response of a given class to the controls set by the other classes and the controls in the resulting equilibrium are shown to have a similar structure as the optimal strategy.

The existence of an equilibrium in the model of [347] is proved in the follow-up paper by **Korilis and Lazar [409] (1995).**] The authors even use a generalized version of the system where the service stations have *quasi-reversible* queues. This contains, in addition to the single-server queue with exponential service, single-server LCFS queues and infinite-server systems with general service distributions.

Campos-Náñez, Fabra, and Garcia [126] (2007) consider a Markovian model of n servers with heterogeneous service rates and finite buffers. The n queues are observable, and customers are price sensitive but not delay sensitive. Suppose that an arrival occurs when the system's state is $x = (x_1, \ldots, x_n)$, where x_i is the number of customers at queue i. The servers whose buffers are not full participate in an auction, and server i offers a bid $b_i(x)$ to the arriving customer. The customer joins the server with the lowest price, but only pays the second-lowest price value. The servers' goal is to maximize discounted revenues, and the model is solved using dynamic programming.

In general, the equilibrium bidding will be such that the firm is indifferent between winning the bid, i.e., selling one unit of buffer capacity, or withholding that unit for future revenue. The authors demonstrate that the resulting equilibrium is not efficient in the sense that it doesn't maximize system throughput. They also propose an extension of the model that incorporates delay-sensitive customers.

Sundar and Ravikumar [613] (2013) consider two service providers dynamically setting prices in a market with two customer types: 1-customers that randomly select a server, and 2-customers that observe both queues.

Customers use price and delay information to decide whether to join a queue or balk. Price equilibrium is achieved through an adaptive learning mechanism where the firms update strategies according to different time scales. The slower firm waits to see the faster firm's equilibrium strategy before updating its own, whereas the faster firm considers the strategy of the slower firm as being static. The authors present experimental analysis that indicates that this process converges.

Xia [665] (2014) considers a closed Jackson queueing network with M strategic servers, N (nonstrategic) identical customers, and fixed routing probabilities. All servers incur the same fixed unit holding cost rate C_h and unit capacity rate C_o. Each server minimizes costs by dynamically controlling service rate within given queue-length dependent bounds. The author proves that the average cost of a server is monotonic with respect to its service rates, and the best response is always to set the service rate at either the maximal or the minimal value allowed. It follows that this property holds in any equilibrium. As for social optimization, since this is a closed system holding costs are the constant NC_h. Therefore, the social goal is simply to minimize capacity costs, and this is achieved by always setting the minimum possible service rate. The worst case for social optimality is when servers select the maximum possible service rate for every queue length. This outcome can be reached in equilibrium when $C_o \ll C_h$.

Chapter 3

Information

This chapter describes research focusing on the information available to decision makers in rational-queueing models. Some models deal with uninformed servers, but most focus on the information available to customers.

Obviously, when the system operates in the **SO** way, social welfare is higher when more information is available. It is also easy to show that in Naor's model with selfish customers social welfare is higher in the observable case. To see this, recall from [221] that if potential demand is high then in equilibrium social welfare is zero in the unobservable case and nonnegative when the queue is observable. Otherwise, if the potential arrival rate is not high then, as shown by Naor, too many join the observable queue but even more (i.e., all) join the unobservable queue. The question of whether a profit maximizer wishes to reveal queue length information to customers and whether it is socially desirable to force such revelation was investigated by Hassin (1986), see [1] §3.2. See [141] for more on this subject.

An important stream of research is concerned with customer information heterogeneity with regard to queue length, service quality, or even their own service duration. These types of heterogeneity are described in §3.1, §3.2, and §3.3, respectively.

3.1   Queue-length-information heterogeneity

Some papers bridge the observable and unobservable models by assuming only a fraction of the customer population can observe queue length.

The *double-ended queue* model of **Large and Norman [426] (2012)** can be described in terms of the taxi market as follows: Taxi customers are indexed by $i = 1, 2, \ldots$ in order of arrival. There are two customer types, informed and uninformed. With probability $\pi > 0$ the customer is *informed* and sees the length of the queue, otherwise the customer is *uninformed*.[1] Each customer makes an irrevocable decision of either balking ($a_i = 0$) or joining the queue ($a_i = 1$). Thus the rate of joining per unit time is $\mathrm{E}(a_i)$. Similarly, d_i is the number of taxi arrivals during period i and $\bar{d}$ its expectation. Customers incur a diminishing cost per unit time of queueing. Payoffs from getting in a taxi are drawn independently and uniformly on (0,1), but customers also pay a fixed toll p^*. They join the queue if their payoff exceeds some threshold depending on their information.

Under natural conditions it is proved that an equilibrium exists and that in all equilibria the queue length is ergodic. Consequently $E(a_i) = \bar{d}$, so there is zero drift in the process. It then follows that the average threshold used by customers is $(1 - \bar{d})$.

It can further be deduced that they all employ the average threshold of $(1 - \bar{d})$. From this fact one can infer that uninformed welfare is exactly $\bar{d}^2/2$, which does not depend on toll or queue discipline.

Hu, Li, and Wang [349] (2014) assume a fraction γ of customers in a Naor-type M/M/1 system are *informed* about the queue length prior to deciding whether to join. These customers behave exactly as in Naor's model. *Uninformed customers* join in equilibrium with probability q^*, which is uniquely determined similarly to the E&H model after taking into account the existence of informed customers.

The authors characterize q^* and investigate the sensitivity of the following performance measures to changes in γ: throughput, accessibility (the probability that an informed customer joins), welfare of each customer class separately, and social welfare. They find that these effects are uniquely determined by the type of equilibrium joining behavior of the uninformed customers, namely whether $q^* = 0$, $q^* = 1$, or $0 < q^* < 1$. For example, as a function of γ, if $q^* > 0$ then social welfare is strictly increasing, but if $q^* = 0$ then social welfare is strictly increasing for $C \leqslant R\mu < 2C$ and strictly decreasing for $R\mu \geqslant 2C$.[2]

---

[1]Informed customers serve to maintain ergodicity by joining less when they see a long queue and joining more when the queue is short. This is why it is necessary to assume that the number of informed customers is at least a positive fraction $\pi > 0$.

[2]Here, as in Naor's model, R is the service value and C is the waiting-cost rate.

3.2 Quality-information heterogeneity and signaling

As §3.6, this section relates to environmental uncertainty but here it arises when firms cannot credibly communicate quality of service to customers. The following papers provide insights on both customer and firm behavior. When and why do customers join longer queues even when waiting is costly? Why do successful firms choose to generate congestion rather than to raise prices or increase capacity?[3]

This section concerns asymmetric information. However, in contrast to the more common asymmetric information models, here customers try to infer information on the firm's relevant parameters, not the other way around.

Veeraraghavan and Debo [633, 634] (2009, 2011) consider games in which customers choose between two service facilities with observable Markovian queues and unknown service values V_1 and V_2, based on the queue lengths and on prior beliefs and information regarding the quality (value) of service at each facility. Customers cannot jockey or renege. Each customer receives a signal $s \in S = \{1, 2\}$ such that $P(s = i | V_i > V_j) = g \geqslant 0.5$. g is referred to as *signal strength*. When $g = 0.5$, the signal is uninformative. When $g = 1$, the signal reveals which service is better. A customer knows his own signal strength and the *signal strength*'s distribution but not the realizations of the signals of the other customers.

Veeraraghavan and Debo [633] (2009) assume no waiting costs, and that a customer's objective is to join the queue of the higher quality server. Customers choose which queue to join based on a prior assumption that $\Pr(V_1 > V_2) = \Pr(V_1 < V_2) = 0.5$, a private signal, and queue length.

The authors consider a class of *general threshold strategies*: Customers join the longer queue when its length is greater than $T_k + k$, where k is the length of the shorter queue and T_k is a nonnegative integer. They show that the only equilibrium strategy in this class has $T_0 = 0$. It follows that the equilibrium strategies are identical at all nonzero probability states, and the only subgame-perfect equilibrium strategy (SPE) is: choose the queue corresponding with your signal if the queue lengths are equal, and join the longer queue otherwise. Under this strategy, at any given time, one server is

---

[3]**Giebelhausen, Robinson, and Cronin [251] (2011)** examine waits as a signal of quality in laboratory experiments and conclude that when quality is important and unknown, "the increased quality signaled by a wait is enough to overcome the negative impact of that wait." **Koo and Fishbach [407] (2010)** conclude from laboratory and field studies that customers are more affected by the number of customers behind them than by the number of customers ahead of them in the queue. They provide the following explanation: "When people are part of a queue, the presence of others behind them is a proxy for accomplished actions ... signals that the queuing is more valuable. In contrast, the presence of others ahead is a proxy for unaccomplished actions ... signals required effort" (cf. [710]).

idle, and the authors show that the long-run market share of the better server is equal to the *signal strength g*. The authors assume no waiting costs in order to focus on the information value contained in the length of the queue, and their results are sensitive to the presence of waiting costs. They discuss the complexity of analyzing a model with waiting costs.

In the companion paper, **Veeraraghavan and Debo [634] (2011)** consider a model that integrates negative waiting cost externalities with positive informational externalities associated with herding behavior. There are two $M/G/1$ observable service facilities whose service values are random variables with a known symmetric joint distribution. The waiting-cost rates for the two servers are identical. Customers are homogeneous except for their private information regarding which server is of higher quality.

There are three possible strategies: **(F)** follow your signal; **(S)** join the shorter queue; **(L)** join the longer queue. The latter strategy means *herding* as customers ignore their signal and join the longer queue. The focus of this research is to demonstrate that herding is a possible rational strategy and to characterize the conditions which encourage this behavior.

In the simplest version used to demonstrate the behaviors in this model, there are two customer classes. A fraction α of the customers has perfect information (they know which server is the better one) while the other customers have no information (their signal strength is 0.5). Moreover, the buffer space of each server accommodates one customer in service and one in queue, and the service value is high enough such that balking is never optimal if there is an available space at one of the servers. Therefore, the only decision is which server to join when one is idle and the other is busy but has an empty queue.

For one and two waiting positions in queue buffer sizes, the authors fully solve this model. In particular, all three strategies can be sustained in equilibrium and for some parameter values both **L** and **S** define equilibria. A main insight is that rational customers may join longer queues despite waiting costs, especially when traffic intensity is low. Numerical examples are used to demonstrate and characterize herding when queue buffers are larger.

Since optimal rational decisions are rather complex, the authors also consider a bounded rationality version where customers behave under the assumption that all others follow their signal. In this case, less informed customers herd if the queue-length difference is greater than a threshold.

Callander and Hörner [124] (2009) investigate a model of social learning and find that the conditions that lead agents to abandon their own information and follow the minority rather than the majority are that "information is sufficiently heterogeneous and the well informed are not overly abundant." The authors illustrate their findings with the following simple queueing model. There are two observable $M/M/2/2$ queues. Service values are 1 for one queue and -1 for the other. A fraction q of the customers are informed and obtain a signal that proves to be correct with probability $p > 0.5$ on which queue is the good one. Balking comes with no cost or benefit.

The authors define a function $f(\mu, p)$ and prove that iff $q \leqslant f(\mu, p)$, the equilibrium strategy of uninformed customers is to join the shorter queue (while informed customers follow their signal). Another interesting result is that the probability a customer chooses correctly is increasing in q but non-monotonic in μ. The authors also solve the corresponding M/M/3/3 model with $p = 1$.

Debo and Veeraraghavan [187] (2009) survey results on herding behavior in service systems.

Zhang [698] (2010), motivated by applications concerning kidney allocation to transplant candidates, investigates strategic behavior in a queue where customers wait for a unit of a product. Product units are heterogeneous and when a unit arrives it is first offered to the customer at the head of the queue. Customers use their knowledge of the system parameters along with private signals on the specific unit and decide whether to accept the unit and depart the system. If a customer rejects an offer the unit is offered to the next in line. In general, when customers are offered a unit they use their private signals and also consider that the unit has already been rejected by predecessors. Customers in the queue do not share information and this may lead to herd behavior where customers ignore their own signals and reject a unit because it has already been rejected by a small number of preceding customers.

The authors add to this simplified model properties related to the kidney market. In particular, customers are heterogeneous and maximize discounted payoff. A specific utility function is assumed and its parameters are estimated based on empirical data. The empirical findings are consistent with the theoretical prediction of *observational learning*. Even identical units (kidneys from the same donor) are often received very differently as one may be accepted at an early stage while its counterpart could be rejected and travel down the queue before being accepted or not.

Debo, Parlour, and Rajan [184] (2012) consider an observable M/M/1 queue where the server has high quality (h) with probability p, and of low quality (l) otherwise. A given proportion of customers are *informed* about the server's quality, while the other customers are *uninformed* and have a prior belief p that the server has high quality. The value of a θ-service is v_θ, $\theta \in \{l, h\}$. Waiting cost is linear and common to all customers, and customers decide between joining and balking. The server chooses between a slow or a faster service, but cannot communicate the service rate to its customers. Operating at the faster service rate is more costly.

The strategy of informed customers is the standard threshold strategy with threshold n_l or n_h, $n_h \geqslant n_l$, depending on the server's type. In contrast, the strategy of uninformed customers is characterized by a *hole*, $\hat{n}$, between n_l and n_h, in which uninformed customers balk. At every other queue length between 0 and $n_h - 1$, uninformed customers join. The explanation is as follows: Clearly, if queue length n is shorter than n_l, an uninformed customer obtains positive expected utility from joining, even when the quality is low. Suppose that when $n = \hat{n} > n_l$, uninformed customers balk. This means that $n > \hat{n}$ indicates an informed customer joined it at $\hat{n}$, and hence the server's type must be h.

An interesting outcome of the customers' strategy is that a high-quality server may communicate its type by maintaining a queue of length greater than $\hat{n}$. To achieve this goal, the high-quality firm may prefer the slow rate thus ensuring queue length above the hole with a high probability. This may happen even when there is no extra cost involved in choosing the high service rate!

Debo, Rajan, and Veeraraghavan [185] (2012) study how quality can be inferred from pricing and congestion in a market with firms that cannot credibly convey the quality of their service to customers. An observable M/M/1 system is operated by a firm with either high or low-quality service. The probability a firm is of high quality is public information.

A given proportion q of the customers are *informed* and know the firm's quality. The others are *uninformed*. Customers have heterogeneous service valuations such that for $t \sim U[0,1]$ the value from high-quality service is tv_h while the value from low-quality service the value is tv_l, with $v_l < v_h$.

In the base model, customers incur positive but very small waiting cost rates. This assumption enables analytical analysis, but the qualitative results also hold under relatively high waiting cost rates, as the authors show numerically.

On arrival, each customer observes the price and queue length. An equilibrium satisfies the following conditions: (i) all customers (informed and uninformed) obtain nonnegative utility; (ii) beliefs uninformed customers have on the service quality are consistent over all possible queue lengths, and consistent with the strategies of the firms; (iii) each type of firm maximizes profit, given the customer joining strategies.

The firms may choose to differentiate themselves by *separating prices*. In this case all customers become informed of the service quality and high congestion adds no information about quality. In other cases the equilibrium is with *pooling prices*, the congestion level is informative, and more congestion is to be expected at a high-quality firm.

The authors characterize the existence of separating and pooling equilibria according to q and v_l/v_h. When both types exist, a high-quality firm would prefer the pooling equilibrium because to credibly signal its high quality it needs to charge a very high price and face reduced profits from lost customers. Therefore, in this case prefers to use congestion rather than price as a signal of quality. The low-quality firm's profits are equal in both equilibria and hence pooling is Pareto dominant.

Kremer and Debo [418] (2015) describe a laboratory experiment verifying the central qualitative conclusions obtained in [184]. Their adaptation of the theoretical model incorporates a finite number of players and deterministic service times. Since real-life customers cannot be expected to follow the predictions of the theoretical model, the authors relax the rationality assumption and compute instead a quantal response equilibrium which involves a parameter β corresponding to the degree of rationality. This theoretical model predicts, and the experimental study verifies, that under certain conditions the buying probability of an uninformed customer may locally increase in waiting time.

The study further demonstrates (both theoretically and empirically) the *empty restaurant syndrome* – short waiting times mean low waiting costs, but customers may balk because they infer low quality. This tendency to avoid empty systems increases the more informed customers are in the population. Another conclusion states, with obvious managerial implications, that the high-quality firm benefits from informed customers only when they are in sufficient numbers. If there are only a few informed customers in the population then, because most arrivals are uninformed customers who avoid empty systems, the system rarely generates the long queues that benefit high-quality firms from the presence of even a small fraction of informed customers.

Guo, Haviv, and Wang [277] (2014) solve an unobservable version of [184] assuming that service rate is fixed and not a decision variable. Service quality is a random variable and can be either high or low, and customers are informed or uninformed. A strategy defines three joining probabilities: of uniformed customers, of informed customers when service quality is high, and of informed customers when service quality is low. As the potential arrival rate increases, the solution changes from a unique equilibrium to a continuum of equilibria, and the uninformed customers first join, then do not join, and finally join with some probability.

The authors also solve a variation of their model (see §3.2) where all customers are uninformed but they can buy quality information. The equilibrium is characterized by two parameters, probability of buying information and probability of direct joining.

3.3 Processing-time information

See [583] where the firm quotes PDTs based on known service times, and [405] where new customers know their service times when bidding for priority.

Enders, Gandhi, Gupta, Debo, Harchol-Balter, and Scheller-Wolf [226] (2010) suggest a mechanism that motivates *informed customers* who know their processing times to reveal this private information, and also fairly treats *uninformed customers* who only know the service time distribution.

Each job is divisible into an integer number of unit-sized pieces that must be processed in a given order. The system maintains high- and low-priority queues, and preemption is allowed. Upon arrival, each customer receives D *tokens*; each token can be used to route one unit of the job to the high-priority queue. An informed customer places tokens on D units of his job (or on all units if the number does not exceed D) wishing to minimize his expected delay. The main result of the paper is that placing tokens on the *last D* units of the job is a strongly dominant strategy for informed users. Therefore, the resulting order of processing approximates the desired *shortest-remaining-processing-time* discipline.

A much harder task is characterizing the equilibrium behavior of uninformed customers. To do this the authors focus on the following example. The job size is one unit with probability p, or two units with the complementary probability. The arrival process is Poisson with proportion α uninformed, $(1-\alpha)p$ informed with a one-unit job, and $(1-\alpha)(1-p)$ informed with a two-units job. Each customer is given a single token. The behavior of informed customers is determined by the above rule, so what remains is to identify the strategy for uninformed customers. It turns out the situation is FTC and there are three possible types of equilibria: (i) a unique equilibrium where uninformed customers place their tokens on their first unit; (ii) a unique equilibrium where they place it on the second unit (if their job turns out to be of two units); (iii) three equilibria with the two previous pure equilibria and a mixed equilibrium. Interestingly, a numerical study reveals that **SO** solutions are always of the second type.

Haviv [322] (2014) considers an M/G/1 unobservable model where arriving customers know their service duration and base their join-or-balk decision on this information.[4] In general, if the demand has a continuous cdf, there is a threshold value x_e such that those who join in equilibrium are the customers whose demand is below x_e, and there is a similar **SO** threshold $x_s \leqslant x_e$. The system can be coordinated in many ways, for example, by charging no fee for service shorter than x_s or charging a very high fee otherwise. It might be desirable to apply pricing mechanisms that continuously change with demand x, and make an x_s-customer indifferent between joining or balking. For example, consider holding fee α, service fee β, and flat fee γ such that $(\alpha + C)W_q(x_s) + (\beta + C)x_s + \gamma = R$, where $W_q(x_s)$ is expected queueing time under threshold x_s, C is waiting-cost rate, and R is the service value. The author focuses on three special cases where just one of these values is positive: $\alpha > 0$, or $\beta > 0$, or $\gamma > 0$; and observes that a possible disadvantage of the flat fee is that less consumer surplus remains. When the regime is FCFS customers pay more under holding fees than under service fees, whereas under EPS all joining customers are indifferent towards these mechanisms.

---

[4]See [1] §3.5 for other models that make this assumption.

3.4 Information acquisition

In the *supermarket game* of **Xu and Hajek [670] (2013)**, customers are homogeneous and upon arrival each randomly inspects a given number k out of N M/M/1 queues and joins the shortest one. Customers incur a fixed inspection cost per queue and linear waiting costs. Customers choose the number of queues to inspect with the intention of minimizing expected costs.

The authors apply *mean field theory* to derive an approximate expression for the expected cost for a customer who inspects a given number of queues when all others apply a common mixed strategy. They use it to prove that a symmetric equilibrium is characterized by a mixed strategy where customers randomly choose between two consecutive integers. An interesting feature of the model is that queue inspection generates positive externalities[5] and therefore the **SO** solution inspects more queues than the equilibrium **IO** solution. The main results are:

- There exists a unique equilibrium when $\rho < 1/\sqrt{2}$, where $\rho = \lambda/N\mu$.

- Multiple equilibria are possible when ρ is very close to 1.[6]

- The equilibrium obtained by the mean field model is a good approximation if N is sufficiently large.

- In the mean field model with $\rho < \frac{1}{\sqrt{2}}$, if all other customers inspect more queues, then the variability of queue lengths is reduced, there is less incentive for a given customer to inspect queues, and we obtain ATC behavior.

- A customer inspecting only one queue would benefit if others inspect more queues. However, one who inspects more than one queue prefers large variability in queue lengths and hence could benefit if others do not inspect many queues. Thus, the act of inspection is associated with positive externalities on some customers and negative externalities on others.

Hassin and Roet-Green [313] (2014) offer another model which bridges Naor's observable and E&H's unobservable models. The queue is initially unobservable and the customer faces the three options to balk, join, or inspect the queue. If inspecting, a second join-or-balk decision is then made

---

[5]This is similar to [313].

[6]This outcome is surprising because the more queues others inspect the smaller the queue-length variability and the expected gain from inspecting a queue, so one expects ATC behavior and a unique equilibrium. The multiplicity of equilibria in heavy traffic could perhaps be attributed to the mean field approximation.

given queue length. Inspecting the queue is associated with a fixed cost. A strategy is defined by the probabilities of balking, joining, and inspecting at the first decision stage. The second decision is made according to Naor's dominant strategy.

The authors observe that fixing any of the three probabilities results in an ATC model. Thus, this is a *pairwise ATC situation*. In particular, when the probability of balking without inspection is fixed, the ATC situation reflects the *positive externalities* associated with inspection. The higher the probability that others inspect the queue, the lower any customer's need to do so. However, this observation is not sufficient to prove uniqueness of the equilibrium and a direct proof is provided for this result. The proof involves a characterization of the feasible space of utilities obtained from joining without inspection and from inspecting the queue.

Interestingly, the authors demonstrate that the equilibrium average joining rate when customers incur inspection costs can be higher than the rates of both the observable and unobservable cases.

Hassin and Roet-Green [314] (2014) consider a multi-queue system with customers sequentially inspecting multiple queues to observe their lengths, and after each inspection decide whether to join one of the inspected queues or to continue the search. Customers incur a fixed inspection cost and linear waiting costs, and seek to minimize expected total cost.

Numerical evidence with two queues indicates that the equilibrium strategy is unique, and in general, it is not of a threshold type and may involve *cascades* even in the case of two identical queues. That is to say, a certain action (continue or stop the search) can prevail in the equilibrium when i or $i + 2$ customers are observed in the first inspected queue, but the opposite action is taken if $i + 1$ customers are observed. In general, as the inspection cost increases, customers tend to inspect fewer other queues. However, for a given observed first queue length customers may inspect more queues as the cost increases.

Guo, Haviv, and Wang [277] (2014) solve a variation of their model (see §3.2) where all customers are uninformed but can buy quality information. The equilibrium is characterized by two parameters, the probability of buying information and the probability of direct joining.

Hassin and Snitkovsky [316] (2016) study a queueing system with Poisson arrivals, two $\exp(\mu)$ servers S_L and S_Q, and a single queue. Upon arrival, a costumer chooses between paying a *sensing price* and trying to attain service by S_L, or joining the queue and waiting to be served by S_Q. A costumer who senses S_L and finds it idle is immediately accepted to service, but if S_L is busy the customer is sent to the queue and waits to be served by S_Q.

The authors prove that there exists a unique equilibrium sensing strategy in this system. The equilibrium sensing probability is below the **SO** value in most cases, but the opposite inequality is also possible. As a consequence, **PoA** is not always a unimodal function of the system utilization.

3.5 Information control

The level of information available to customers and servers is an important part of every strategic queueing model. We describe here models that focus on this issue, but of course there are many other models scattered elsewhere in this survey also dealing with this subject.

Dobson and Pinker [203] (2006) consider an M/M/1 queue and a sequence of strategies $S_0, \ldots, S_\infty$. In S_k, if the system's load is less than or equal to k, the firm announces the τ fractile of the lead time, where τ is exogenous. Otherwise, the firm announces the expected lead time conditioned on the state being k or higher.[7] The strategy S_0 corresponds to the unobservable model; the other extreme S_∞ corresponds to the observable model. The joining probability is an exogenous decreasing function of the quoted delay.

The authors assume the firm chooses the S_k strategy which maximizes the throughput of the system, and investigate the effect sharing increasing amounts of information has on the system's throughput and average waiting time.

Guo and Zipkin [284, 285, 286, 287] (2007-9) compare different levels of information and investigate impact on system performance. They consider throughput and average customer utility as the system's performance measures. There are no prices and the server's decision is the level of information: unobservable, queue observable, or workload observable. The service value is R and a customer's waiting cost is $\theta E[c(W)]$, where W is waiting time, $c(W)$ is a basic cost function common to all customers, and θ is a customer-specific weight that measures sensitivity to delay. The θ-customer's expected utility is then $u(W, \theta) = R - \theta E[c(W)]$.

A simple example demonstrates that the system's throughput may increase or decrease when moving from observable queue to observable workload. Suppose service time is 1 or 2 with equal probability, and the waiting-cost rate is 1. Denote service value by R. (i) $R \in (1, 1.5)$. No customer joins when the queue is observable even when the system is empty, but when the workload, in particular the arriving customer's service time, is observable a customer joins if his service duration is of unit time and the queue is empty. Therefore, in this case the throughput is larger when the workload is observable.

---

[7]In a rational model this is equivalent to informing customers of the exact queue length if it is below k or otherwise only informing them the length is at least k.

(ii) $R \in (1.5, 2)$. No customer joins when the system is not empty. When the system is empty all join in the observable queue case, whereas only half of the customers join when the workload is observable. Therefore, in this case the throughput is larger when the queue is observable.

In [284], the authors show the primary factor determining whether information is good or bad for the service provider and the customers is the shape of the cdf H of the customers' delay sensitivities and not the waiting cost function. The authors find sufficient conditions to ensure that more information benefits the provider and the customers. For example, when $H(\theta) = \theta^\alpha$ for $\theta \in [0, 1]$ and some $\alpha > 0$, the average customer utility is proportional to the throughput, the two system performance measures are aligned, and more information benefits both. The authors also give examples where more information reduces the customers' expected utility due to externalities.

In [285] it is shown that the results of [284] with respect to comparing the unobservable and observable workload models still hold with phase-type service times. The observable queue model is not discussed here because its analysis is much more involved, as has been demonstrated in [49, 397].

In [287] the authors investigate the effect of customers' delay and risk sensitivities on the system's performance measures. Some of their main results are:

- If customers are less heterogeneous with respect to delay sensitivities, they obtain a smaller average utility, and the system throughput is larger under light traffic and smaller in heavy traffic.

- In the unobservable case, when customers have a smaller cost function their average utility is larger. However, this property does not necessarily hold in the two observable versions.

- A system with more risk-averse customers need not have a smaller throughput.

- The value of information is not necessarily larger for customers who are less patient or for more risk averse customers.

In [286] the authors incorporate into their model two other information models. In the case of *partition information*, the nonnegative integers are partitioned into intervals and a customer is informed about the interval that contains the system occupancy at the time of arrival.[8] In the case of *phase information* the service time of each customer consists of a random number K of exponentially distributed phases. Assuming that K has a geometric distribution, the unconditional service time is again exponentially distributed. The available information is the number of remaining phases of all present customers. This is more information than just queue length but less than

---

[8]This model is similar to the compartmental model of [218] and contains as special cases the unobservable and the observable queue length models.

the workload. It is shown that more information can increase or decrease throughput depending on the shape of the distribution of customer delay sensitivities.

Economou and Kanta [218] (2008) consider two models with partial information that bridge the extremes of observable and unobservable queues. In their setting the queue is composed of *compartments* of fixed size a. Joining customers obtain partial information on their positions in the queue. Suppose that the number of customers in the system is n. Model **N** assumes a new arrival knows the number $\lfloor n/a \rfloor + 1$ of the compartment to which he will enter. Model **P** assumes that he knows only his position $(n \bmod a) + 1$ in the compartment. In the **N** case, the information with $a = 1$ is the exact queue position, as in Naor's observable queue, for which there exists, in general, a unique equilibrium threshold strategy. When $a \to \infty$ we obtain E&H's unobservable system for which there exists a unique equilibrium strategy, which in general is a mixed strategy. The **P** case yields exactly the opposite conclusions. Hence, both models bridge the two extreme cases but in opposite ways. However, the dependence of the value of the information on the compartment size a is not straightforward. In particular, whether the compartment number is more significant when a is large or when it is small.

The authors analyze both cases, but their analysis of the equilibrium in the **N** case is less complete as it only deals with pure strategies. The authors apply the *dual approach* (see §2.2): Assuming customers know continuously the compartment in which they reside but not the position within it, and assuming an LCFS-PR service regime, a customer has incentive to renege only when moved to another compartment because of a new arrival. When this happens, all customers who were behind him have already left the system and therefore this customer imposes no externalities and behaves in the **SO** way. The authors use this insight to compute the **SO** solution.

The analysis of the **P** case includes a detailed solution of the equilibrium threshold strategies and shows the existence of an equilibrium threshold strategy such that the threshold is a decreasing function of the arrival rate, while the social and profit-maximizing thresholds exhibit unimodal behavior. The authors also show that the objectives of profit maximization and welfare maximization coincide only when $a = 1$, i.e., when the queue is unobservable.

An interesting question is the dependency of the social and profit gains on the compartment size a, especially when this parameter can be controlled. The authors find that in the **N** case these gains can reach maximum values at intermediate levels. In the **P** case, the authors show an example in which the optimal values increase with a, indicating that more information increases both profits and social welfare. They also obtained examples where profit decreases with compartment size while social welfare increases.[9,10]

---

[9]S. Kanta, private communication.

[10]Recall that in the **P** model the information is minimal when $a = 1$ and maximal when $a \to \infty$.

Armony, Shimkin, and Whitt [68] (2009) consider non-price control of a queue by supplying new arrivals with information that affects their join/balk/renege decisions. The goal of the delay announcements is to induce balking when the system is heavily loaded and retrying at a later time. Specifically, the server announces the *delay of the last customer to enter service* (DLS). Given an announcement w, arriving customers respond by two exogenous functions: Balk with probability $B(w)$, or otherwise renege before t time units with probability $F(t|w)$, for $t > 0$. In equilibrium, announcements conform with customer responses which affect the system's performance that leads to these announcements. The results are compared to the *fixed delay model* (FD) where the customers only know the long-run average delay (as E&H). However, it is assumed that the *customers respond by the same functions B and F*. Simulation shows that DLS announcements are more effective in causing customers to renege and reducing the delay of served customers, and that the difference can be significant. Heavy-traffic approximations are also used to show that a unique equilibrium fluid delay exists under very general conditions, but multiple equilibria are possible when these conditions are not satisfied.

Shone, Knight, and Williams [574] (2013) prove a necessary and sufficient condition on the system parameters under which the equilibrium average joining rate is the same when the queue is observable and when it is unobservable. They also prove a similar condition with respect to the **SO** solutions. However, it is not possible that both of these conditions hold simultaneously.

Hassin and Koshman [311] (2014) consider limited dynamic pricing in a profit maximization model that combines features from both the observable and unobservable versions of Naor's model. For an exogenous threshold N, customers are informed of whether the queue length is less than N (state L) or at least N (state H).[11,12] Admission prices are p_L and p_H.[13]

The authors numerically solve the model and find that:

- $N = 1$ (meaning customers join iff the server is idle) always guarantees at least half of the maximum value that can be generated by the system (see §6.1).

Yu, Allon, and Bassamboo [688] (2015) provide empirical evidence that a longer estimated delay not only affects customers waiting cost estimates but may also give them more freedom in selecting activities while waiting and hence reduce their waiting-cost rates.

To explore this phenomenon the authors construct a model where customers receive delay announcements. To be credible, announcements must

---

[11]See §6.1 for a discussion when N is a decision variable.
[12]See [440] for another paper that assumes a similar service rate strategy.
[13]The special cases $N = 0$ and $N = \infty$ are equivalent to the E&H unobservable model.

take into consideration the effect on customer behavior. The authors use an iterative procedure to compute the equilibrium such that the waiting estimate offered by the announcement matches the one experienced by the customers.

The balance of the contradicting impacts of a longer expected waiting time with a smaller waiting cost rate may *reduce* customer motivation to renege when a longer wait is predicted. The authors find that, according to their call center data, informing customers about anticipated delays can improve their surplus, and that for this purpose it might suffice to announce partial information messages, for example low, medium, and high expected waiting times.

Boudali and Economou [102] (2015) investigate a bulk-service system where a compartmental model naturally arises.[14] The system is Markovian, customers have identical service values and waiting cost rates and are served in batches of size K. Let $q = Km + j$ where $0 \leqslant j < K$ be the queue length. Thus, m denotes the number of complete batches in the queue and j is the number of customers in the incomplete batch.

An interesting feature of the model is that it combines ATC and FTC behavior. The authors determine the equilibrium joining strategies in four cases, depending on whether customers observe each of m and j:

- **The observable case:** There exists a unique equilibrium strategy, which is of threshold type with respect to m and of *reverse-threshold* type with respect to j, i.e., for $m_0 \leqslant \cdots \leqslant m_{K-1}$ join iff $m \leqslant m_j$.

- **The almost observable case:** In this case m is observable but j is not. The authors characterize the set of *equilibrium m^* pure threshold strategies* where customers join the system iff $m \leqslant m^*$. Multiple equilibria are possible and their m^* values constitute an interval of integers.

- **The almost unobservable case:** In this case only j is observable. The equilibrium strategies are of the *mixed reverse-threshold* type, i.e., joining if j is above a threshold and mixing at the threshold. The authors conduct a detailed case analysis and compute for each case the (generally unique) equilibrium.

- **The unobservable case:** In this case all balk is always an equilibrium and there may be one or two equilibria with a positive joining probability. This case reminds of a similar situation that has been noted regarding the customer behavior in the unobservable $M/M/1$ queue with the N-policy (see [272]).

The authors numerically compare the four information settings and reach many interesting conclusions. For example, when the system load is low social welfare increases when j is concealed from customers, whereas when the load is high it is important to reveal m.

---

[14]See [218] for a similar situation for a single-service system.

Simhon, Hayel, Starobinski, and Zhu [577] (2016) consider a variation of Naor's model where the firm reveals the queue length if it does not exceed a threshold D. Customers arriving when the queue is longer than D know this fact. The authors compute the equilibrium joining probability when the queue is longer than D. The authors show that equilibrium throughput is a monotone function of D and therefore if the firm's goal is to maximize throughput then the optimal policy is one of the extremes, either the observable queue or the unobservable queue.

3.6   Environmental uncertainty

This section considers systems that operate in an uncertain environment. The users or operators in the system may be uninformed on the state of the environment at the time they make a decision and therefore some of the system *parameters* may become random variables with unknown realization.

3.6.1   Customer uncertainty

See [542] for an uncertain number of customers that can be included in a single batch of service; [438, 168, 75] for models of competition where servers set service rates but do not inform their customers about the selected values; [216] for environments that differ in both arrival and service rates; see [217, 352, 519, 699] for other models where the service rate is unknown.

Hassin [305] (2007) considers an M/M/1 queue where the service value (quality) R, the waiting-cost rate C, or the service rate μ is a random variable obtaining one of two given values with known probabilities. For each case, the author analyzes three subcases: (i) Customers are uninformed about the realization of the random variable and the server sets a single price independent of the realization; (ii) customers are informed, and the server sets a price depending on the realization; (iii) customers are informed, but the server is restricted to setting the price before the random variable is realized.

Some of the main qualitative results are:

- **Uncertain μ:** Informing customers is desired even if the price is set exogenously and not optimally by the server. The difference between profits obtained with informed and uninformed customers increases in the amount of uncertainty.

- **Uncertain C:** For prices that induce positive demand in the case with no information, the same level of social welfare is attained whether customers are informed or not. However, informing customers may be ben-

eficial when one of the possible values of the waiting-cost rates is so high
that without information no customers will show up.

- **Uncertain quality of service:** When restricted to a single price, for
  small C the server is motivated to conceal the realized service value from
  the customers. For large C, the opposite holds.

- **Increased uncertainty** may increase or reduce profits and welfare.

- **Sun and Li [604] (2014)** extend [305], allowing for the random parameter to obtain $n > 2$ values. They also introduce further variations
  where the server can set distinct prices when one of the $k < n$ lowest (or
  highest) possible values is realized, and set a common price otherwise.
  The results are qualitatively similar to those obtained for $n = 2$ in [305].

Guo, Sun, and Wang [282] (2011) consider an unobservable queue
where the server provides partial information on the service time distribution
to its customers. It is assumed that customers interpret this information according to the *maximum entropy principle*. This means that customers behave
as if the service time distribution is the one leaving the largest remaining uncertainty (i.e., the maximum entropy) consistent with the partial information
revealed by the server. Examples of partial information considered by the authors are the range, mean, mean and range, mean and second-order moment
of the logarithm, and mean and variance of the service time distribution. In
each case the equilibrium joining/balking strategy is computed both when the
waiting costs are linear and when they are quadratic increasing. The authors
conclude that supplying only partial information may increase welfare, but for
profit maximization it is beneficial to reveal more information on the service
time distribution.

Sun and Li [602] (2012) consider an almost-observable Markovian queue
with multiple vacations (see §10). Customers are informed that the service
rate has one of two possible values (or that it belongs to a given interval), but
the exact value is concealed from them. The authors consider several ways
of determining the equilibrium threshold queue length for the join-or-balk
decision; pessimistic (max-min), optimistic (max-max), a combination of the
two, min-max regret, and highest average payoff. All of these criteria ignore
the signal the queue length provides on the service rate.

Sun, Li, and Tian [608] (2013) obtain analogous results for the unobservable version of [602]

Debo and Veeraraghavan [188] (2014) consider an observable M/M/1
queue where service value $V(t)$ is a linear function $V(t) = V_0 + rt$ of the expected service time t, and where $r > 0$, implying that longer service is associated with a higher value on average. Customers cannot observe the realization

of these variables. Customers incur linear waiting costs and a customer's strategy defines for each queue length n a joining probability $\alpha(n)$. An important feature of the model is that the positive correlation between service duration and quality means a long queue indicates high-quality service but also a long waiting time.

In an interesting special case, there are only two possible service values. The authors characterize the possible types of equilibria in this case. The equilibrium can have a (pure or mixed) threshold strategy, or a *sputtering strategy*. A sputtering strategy is defined by two thresholds $n_0 < n_1$. When the queue is shorter than n_0 or strictly between n_0 and n_1, the strategy dictates joining, there is a randomization at n_0, and balking occurs if the queue length is n_1. Thus, joining probabilities are not decreasing in queue length.

In the general case where the priors about the expected service times and value are more uniformly distributed over some interval, the authors show that randomization may occur in multiple queue lengths.

Cui and Veeraraghavan [178] (2015) consider an observable M/M/1 system where customers hold heterogeneous beliefs about the service rate. These beliefs can differ arbitrarily from the true rate. Price is exogenous, and therefore the server's goal is to maximize throughput. The server's decision is whether to supply customers with service-rate information.

Customer beliefs induce heterogeneous individual thresholds. Customers do not update their beliefs given the queue length they observe. The insights obtained include:

- As expected, greater individual thresholds increase the firm's revenue, but higher revenues also result when thresholds are less spread-out.

- When beliefs are pessimistic, the average threshold is lower than the real one and the firm can increase profits by revealing the service rate. This revelation typically, but not always, reduces welfare.

Zheng [709] (2015) considers two customer classes. *Optimistic* customers constitute a fraction α of the customer population and believe the service rate is μ, while *pessimistic* customers believe the service rate is $\mu' < \mu$. Customers also differ by waiting cost rates $c \sim U[0,1]$ but have the same service value and pay the same price.

The author solves the resulting equilibrium joining probabilities and shows that while the equilibrium joining probability of optimistic customers is non-increasing in price, this is not always so with respect to the pessimistic customers.

3.6.2   Server uncertainty

See [358] for an adaptive learning algorithm when the demand function is piecewise linear with a single breakpoint whose location is unknown to the firm; [90] for stochastic evolution of the market size with a server not knowing its instantaneous value; [231] for a case where the capacity cost of a server is unknown to its competitor; [390] for a supply chain game where each of two players applies min-max optimization to cope with the uncertainty about the parameter set by the other player; [331, 491] for cases where the server does not know the arrival rates; [380] where an MTS server has imperfect information on the system length.

Gayon, Talay-Değirmenci, Karaesmen, and Örmeci [248] (2009) study an M/M/1 system operating in an exogenously changing demand environment that evolves according to a continuous-time Markov chain with small transition rates as compared to production and demand rates. Customers are price sensitive, and those arriving when there is no stock on hand are lost. The firm maximizes the long run average rate of revenues net of its (nonlinear) holding costs.

The authors present MDP equations for three models: static pricing, environment-dependent pricing, and dynamic pricing (varying both with the environment and queue length). In each case the optimal replenishment policy is an environment-dependent base-stock policy. The authors prove monotonicity properties of the dynamic prices and base-stock levels and conduct a numerical study which reveals that the gain of dynamic pricing is quite modest in a wide range of parameter values.

Afèche and Ata [14] (2013) study a learning-and-earning problem for an observable queue serving a population with an unknown proportion q of *patient* customers with waiting-cost rate of c_L, while the remaining customers, with waiting-cost rate c_H, $c_H > c_L$, are *impatient*. There are two scenarios: *optimistic*, where $q = q_o$, and *pessimistic* where $q = q_p < q_o$. Service value is R for all customers. The server uses dynamic pricing to maximize discounted expected revenue. The server cannot distinguish patient from impatient customers and updates its belief of q by observing the behavior of arriving customers, *including those who balk*. Specifically, learning is possible only under a high price that deters H-customers from joining, so there is a tradeoff between immediate earning and learning. Waiting costs apply while queueing but not during service. Therefore, when there are n customers in the system it is optimal to charge either $P_L = R - n(c_H/\mu)$ or $P_H = R - n(c_L/\mu)$, or reject the new arrival.

The main results are:

- For the restricted problem of an M/M/1/2 queue, where the only decision is the price at state 1, when q is known to the server there exist thresholds $R_L, R_H, \underline{q}$ such that:
  If $R \leqslant R_L$ then reject customers for all values of q;
  If $R_L < R < R_H$ then set price P_H for all values of q;
  If $R > R_H$ then set price P_L if $q \leqslant \underline{q}$ and P_H otherwise.

- Suppose now that the prior probability for the optimistic scenario is α. If the high price P_H is charged the server can observe whether arrivals join or balk and update α in a Bayesian way. It turns out that in the restricted problem there is a threshold α^* such that the optimal price is P_H iff $\alpha > \alpha^*$. Clearly, once the outcome is to charge P_L the learning process stops. As a result, *learning is potentially incomplete*. The authors characterize the probability the server ends up charging P_L indefinitely while the true scenario is optimistic.

- For the unrestricted case and known q, there exists an optimal *nested threshold* strategy where the server sets price P_L if the queue is short, P_H when the queue is of intermediate length, and rejects all arrivals when the queue is long.

- For the unrestricted case and unknown q, the optimal pricing policy partitions the queue state into five zones. In the first (lower) part, the price is low and all customers are admitted. In the second, the price is high or low depending on the updated belief. In the third zone, the price is high and only L-customers are admitted. In the fourth, the price is high or all are rejected depending on the updated belief, and in the fifth, all customers are rejected.

Haviv and Randhawa [326] (2014) consider an M/M/1 system with heterogeneous service valuations. Let $R(\lambda, p)$ be the equilibrium revenue when price is p and the potential arrival rate is λ. Let $R^*(\lambda) = \max_p\{\Pi(\lambda, p)\}$. The authors impose regularity conditions on the customer valuation distribution to ensure that optimal demand-dependent pricing is well defined and increasing in λ.

Suppose the queue manager does not know λ and applies *demand-independent pricing* which maximizes $\inf_\lambda \frac{R(\lambda, p)}{R^*(\lambda)}$. The problem is first simplified by proving that worst-case performance occurs either when $\lambda \to 0$ or when $\lambda \to \infty$. After introducing general bounds on worst-case performance, the authors show that when the distribution of customer valuations is uniform over $[0, v]$, demand-independent pricing performs very well. Let h denote the waiting-cost rate.

The main results include:

- The lowest value of the bound $\max_p \inf_\lambda \frac{R(\lambda, p)}{R^*(\lambda)}$ equals 75% and is achieved when $h = 0$.

- The performance of demand-independent pricing improves as customers become more delay sensitive (equivalently, they have higher service values). When $h/(\mu v)$ grows without bound the performance approaches 100%.

- Demand-independent pricing works well for other distributions of service valuations, for example the exponential.

- Demand-independent pricing works better for revenue optimization than for social optimization.

3.7   Delayed information and cheap talk

Each of the two seemingly unrelated topics of this section is introduced separately, and then combined in [28].

The model treated by **Wolisz and Tschammer [662] (1993)** is non-strategic but raises interesting questions concerning strategic variations. The following part is of particular interest. Consider a system of identical parallel servers. Workload information at the servers is announced every Δ units of time. There are two extreme selection strategies. In one, each arriving customer randomly and uniformly selects a server. In the other, customers use the queue-length announcement and all join the least loaded queue. Which alternative is socially preferred depends on the information delay parameter Δ. When Δ is small, the expected waiting time is smaller when customers use the information. When Δ is large, following the information leads to batches of customers jointly joining the server with the smallest workload, thus causing high congestion in the queue.[15]

Suppose now that Δ is such that social optimization prefers random selection. In this case it might be that individuals could profit from using the delayed information (random choice by all is not an equilibrium). Suppose a proportion p of the population consists of *clever boys* who use state information, whereas the other customers are *obedient*. The authors show that if p is small, the clever boys may considerably profit from using state information, and even the obedient customers will slightly benefit from the behavior of the clever boys' behavior. However, when p is large both types suffer from the lack of obedience.[16] Let p^* be the fraction which minimizes expected waiting time. The queue manager may then control the system by letting an arriving customer use the information with probability p^*, and instructing him to randomly select the server otherwise.

Allon, Bassamboo, and Gurvich [31] (2011) consider an unobservable M/M/1 system where customers value service at R and incur waiting-cost rates of c. The server obtains a value v per served customer and incurs convex holding costs $h(w)$ per customer who waits w in the system. The server provides announcements regarding queue length information (referred to as *cheap talk*) which customers cannot verify.

---

[15]Hence, customers are irrational.

[16]Another model where only a portion of the population is obedient is discussed in [253].

With symmetric information customers would behave, as in Naor's model, according to the self-optimizing threshold strategy $n_s = \lfloor \frac{R\mu}{c} \rfloor$. If the server had *full control* over the customers, they would follow a different *revenue-maximizing* threshold n_m.[17]

The main results are:

- There exists an optimal strategy for the firm with only two announcement types. The firm may randomize its announcement type in at most a single state of the queue. The rest of the discussion refers to pure firm strategies.

- There are two possible types of two-signal equilibria: The firm announces **High** when the queue length exceeds a threshold and announces **Low** otherwise.

    - *Join or Randomize:* Customers who receive **Low** join the system while those who receive **High** join with probability $\theta \in (0,1)$.

    - *Randomize or Balk:* Customers who receive **Low** join the system with probability $\theta \in (0,1)$ while those who receive **High** balk.

    The authors provide a full characterization of both types of equilibria.

- There always exists a *babbling equilibrium* where the firm provides meaningless signals and customers ignore them. (Such an equilibrium does not always exist if customers are restricted to pure strategies.)

- Clearly, if $n_s = n_m$ the server will provide correct information and customers will trust this information. The following results hold when customers are restricted to pure strategies:

    - When $n_s > n_m$ there is no pure-strategy equilibrium. Intuitively, if customers adopt n_m as their threshold, an announcement of $Q \geqslant n_m$ actually means $Q = n_m$. In such a case an arriving customer prefers joining, contradicting the equilibrium requirement.

    - Suppose $n_s < n_m$: If the expected cost of joining the system is positive when it is known that $Q < n_m$, then customers will co-operate with an announcement threshold n_m and the system is in equilibrium when customers join if the server announces $Q < n_m$ or balk otherwise. In the opposite case, no pure-strategy equilibrium exists.

Allon and Bassamboo [28] (2011) investigate the advantages of *delaying the delay announcements* in the M/M/s model of [31] by adding a first M/M/∞ stage thus creating a system of tandem queues. *Upon completion of a first-stage service* the firm makes its (non-verifiable "cheap talk") delay

---

[17]The authors apply different notation for the thresholds.

announcement to the customer who then decides whether to enter the main service phase or to balk. The first stage could be associated with some direct benefits, but it also allows the firm to improve its main stage admission policy by using information from the first queue. The desired admission policy of the firm in the second stage is a threshold policy where the threshold depends on the number of customers currently in the first stage (i.e., a switching curve). Customers' joining behavior to the first stage is assumed to be non-strategic, with a fixed rate independent of the firm's delay announcement policy.

An influential cheap talk equilibrium in which the firm can affect customer behavior exists for a range of normalized customer service values. The authors show that adding the first stage may increase or decrease this range, and they investigate its effect on the firm's profits and customer welfare.

3.8 Ticket queues

Xu, Gao, and Ou [671] (2007) consider an M/M/1 system with partial information on queue length given by means of ticket technology. Before entering the system, newly arriving customers obtain, consecutively, numbered tickets. These customers then observe a display showing the number of the currently served customer and decide whether to stay or balk. Customers cannot see the actual queue length. A customer joins if the difference between his number and the displayed number, is less than a threshold value. To compute the threshold the customer needs to compute the expected queue length given his position, which is not an easy task. The authors describe a set of states and the associated birth-and-death equations, but the number of states grows exponentially with the threshold. Therefore, they also suggest an approximation algorithm.

The authors solve a bounded rationality version of the model where customers are naive and assume their position reflects the exact queue length. To eliminate the resulting customer decision errors the authors suggest displaying, in addition to the current number in service, the conditional expected queue length as well. Of course a simpler alternative would be to give a ticket only to customers deciding to join the queue *after* observing the displayed number. However, even in such cases, customers often obtain a ticket and then change their minds and abandon the queue. Therefore, the model suggested here can be viewed as an interesting prototype of a queue with partial information obtained with ticketing technology.

The authors do not discuss the impact of the bounded rationality assumption on social welfare. As in Naor's model, the **IO** threshold tends to be greater than the **SO** one and thus customers' overestimation of queue length can improve social welfare. Therefore, concealing queue-length information may be socially desirable.

Chapter 4

Customer decisions

The vast majority of the strategic queueing literature focuses on server-customer interaction; customer equilibrium behavior is just one component of the game. This chapter surveys models focusing on customer decisions.

4.1 Temporal decisions

Most of the literature considered in this survey assumes steady-state conditions of a queueing system. In contrast, this section mostly deals with non-stationary models related to decisions of customers who are aware of the timing of some special events, like opening and closing times of a facility. Similarly, in queues with retrials customers deciding to retry take into account information they have on the state of the system from their previous trials and therefore such a model is intrinsically non-stationary.

See [396, 450, 452] for models where customers select the time to submit demand requests facing waiting costs and dynamic pricing.

4.1.1   Arrival-time decisions

Early research on strategic timing of arrivals to queues is described in [1] §6.1-3. Holt and Sherman (1982) analyzed the equilibrium arrival process when prizes are allocated on an FCFS basis at time 0 and with linear waiting costs. In the observable version, customers balk if the number of earlier arrivals already matches the number of available prizes, whereas in the unobservable version customers only discover this at time 0 after having already incurred waiting costs.[1] Glazer and Hassin (1983) considered the ?/M/1 model where customers independently choose arrival times to an FCFS single-server system with exponential service that opens at time 0 and closes at time T. The model assumes a monopolistic server queue with no option for customers to balk. Therefore, the only concern of customers is the minimization of expected waiting time. It is assumed that the distribution of the total number of arrivals is Poisson and early arrivals, before time 0, are possible. Glazer and Hassin (1987) considered a similar model in a queueing system with bulk service at predetermined instants, much like that of a bus schedule.

Lariviere and Van Mieghem [427] (2004) consider M strategic customers each choosing a time period $t \in \{1, \ldots, T\}$ to receive service. The base model assumes sufficient service capacity so that all M customers can be served in one time period. Customer m seeks to minimize his delay cost $W_m(\alpha)$, which is assumed to be monotone increasing in the number of customers choosing the same period. Customers spread out as much as possible and therefore a pure asymmetric equilibrium is such that the number of customers in various periods differs by no more than one. Similarly, there exists a unique symmetric mixed equilibrium in which customers independently, uniformly and randomly choose their arrival times. As the number of customers and time periods grows, an independent Poisson arrival process to each time period is generated.

The authors consider several extensions. An interesting one, similar in spirit to the model of Glazer and Hassin (1987) (see [1] §6.3), is when the system can only serve a limited number of customers in each period, with customers unserved in period t carrying over to period $t + 1$. The authors develop conditions under which the above results, and in particular the Poisson limit, still hold.

Wang and Zhu [650] (2005) assume a service period divided into evenly spaced *shifts*, and the waiting time in each shift is determined by the number of customers who choose it. Customers are heterogeneous, and in particular have different delay costs per shift and per wait during the shift. The price is the same in all shifts, so in equilibrium later shifts experience shorter expected waits such that the marginal customer in a given shift is indifferent

---

[1]**Bell [84] (1985)** independently solved a similar model.

between that shift and the next. Also, the marginal customer in the last shift is indifferent between choosing it or not showing up at all.

The authors prove, under additional assumptions, that there exists a unique equilibrium and that later shifts will be selected by customers with a lower ratio of delay cost per shift to waiting cost within the shift.

Wang and Zhu [651] (2007) consider profit-maximizing decisions made by the firm in a demand model similar to that of [650], but under asymmetric information.

McCain's *queueing game* **[481] (2010)** §11 is another observable version of Holt and Sherman's model with n customers and n prizes of different values. Prizes are simultaneously awarded at a specified time. Customers can join the queue for the next best prizes, or stay out and obtain a random prize from those remaining. Joining decisions are sequentially made in some arbitrary order and queue length is observable during the process so customers know exactly the prize they will receive should they join the queue. A fixed waiting-cost rate is incurred while in the queue, but there is no such cost for those who stay out. The **SO** policy is obviously for all to stay out and prizes to be arbitrarily distributed. In general, in equilibrium some customers join while others stay out and, compared with the optimum, some of those who join are better off while the others are worse off.

Hassin and Kleiner [310] (2011) follow the ?/M/1 model assuming service durations are exponentially distributed and the total number of arrivals is a Poisson random variable. All customers arriving before or at T are served. Customers wish to minimize waiting time and have no preference about when they are served. The main results are:

- Unless the system is very heavily loaded, *the elimination of queueing before opening time does not significantly reduce (in equilibrium) expected waiting time.*[2] This surprising result is explained as follows: Customers who would otherwise show up before time 0 arrive now at 0, which reduces expected waiting time. In addition, approximately the same number of customers who would otherwise show up in some interval $(0, t')$ now arrive at 0, which increases expected waiting time by approximately the same amount.[3]

---

[2] Random order at time 0 is clearly optimal when the system closes at $T = 0$; see Bell [84].

[3] **Yoshida [686] (2008)** observes a similar result in a non-stochastic model where the decision variables of a mass transit authority are the number, capacity, and schedule of trains. Commuters decide on arrival time wishing to minimize costs of early and late arrival, and of waiting. It is assumed the cost of arriving early is the smallest of the three and it is shown that the implications of random access to trains depend on whether the cost of arriving late is smaller or larger than the cost of waiting for the train. In the former case, replacing FCFS by random access has no effect. In the latter case, it reduces customers' aggregate costs. See [321] §3 for a similar phenomenon in the fluid approximation of the model.

- The authors present a simple accurate approximation to the **SO** solution. According to this approximation, customers arrive with positive probabilities at times 0 and T, and uniformly on $(0, T)$.

- Numerical comparison of the equilibrium and **SO** solutions shows that the suboptimality of the equilibrium is greater when λ is large and μT is small. For example, with $\lambda \leqslant \mu T \leqslant 20$, **PoA**< 1.62.

- Social welfare can often be increased in equilibrium if arrivals are restricted to $\{0, t_1, T\}$, where t_1 is an appropriately selected internal point.

Jain, Juneja, and Shimkin [359] (2011), consider the *concert queueing game*. Customers choose arrival time, which can be before or after opening time, preferring both early service completion and short waiting time. A fluid approximation greatly simplifies the analysis and enables obtaining closed-form solutions. The authors show how to obtain the fluid model as the limit of a sequence of stochastic queueing systems. They note that this analysis does not show convergence of the equilibrium solutions of the underlying systems to the fluid model's equilibrium.

In the fluid model, a server starts operating at time 0 with a constant rate μ. The cost function for a class i customer who waits w units of time and obtains service at time τ is $\alpha_i w + \beta_i \tau$. The class i arrival profile is a cdf F_i with total mass Λ_i. The aggregate arrival profile is $F(t) = \sum F_i(t)$, and it uniquely defines the queue-size process $Q(t)$. Therefore, an arrival at t is associated with waiting time $W(t) = Q(t)/\mu + \max\{0, -t\}$. With an equilibrium profile F, the cost $C_F^i = \alpha_i W_F(t) + \beta_i(t + W_F(t))$ is minimal over the support of F_i. The main results with a single class are:

- There exists a unique equilibrium profile given by a uniform distribution on $\left[-\frac{\Lambda}{\mu}\frac{\beta}{\alpha}, \frac{\Lambda}{\mu}\right]$.

- The equilibrium queue size increases linearly for $t < 0$ and decreases linearly for $t > 0$.

- Assuming the **SO** solution sets customer arrivals to the instant their service is due to start, then **PoA**=2.

For the multiclass extension with I classes, let $m_i = \alpha_i/(\alpha_i + \beta_i)$ and assume $m_1 < m_2 < \cdots < m_I$. Let $T_I = \Lambda/\mu$ and $T_{i-1} = T_i - \Lambda_i/(\mu m_i)$. The equilibrium profile is unique with F_i being a uniform distribution on $[T_{i-1}, T_i]$ with density μm_i, and **PoA**$\leqslant 1 + \sqrt{\beta_{\max}/\beta_{\min}}$. The authors also discuss possible **PoA** reductions by priority assignment, time-dependent tariffs, and class-dependent restrictions on time of service.

Juneja and Shimkin [379] (2013) consider the stochastic (non-asymptotic) single-class version of the concert queueing game with a general (not necessarily Poisson) number of customers. A central finding is the non-existence of asymmetric equilibria: A unique equilibrium exists and it is symmetric.[4] The authors also study the convergence of the equilibrium distribution to that of the fluid model as the number of users increases. They show that the **PoA** is larger than 2 and converges to this value for a large population size.

Haviv [321] (2013) unifies the ?/M/1 and the concert queueing game models by considering the arrival patterns of customers in both stochastic and fluid models, with a finite or infinite closing time T, with and without early arrivals, and with and without tardiness costs. The total number of arrivals is Poisson(Λ). The author derives the equations necessary to solve the stochastic versions, solves the fluid approximations, and derives the **PoA** for the latter case. The main result is that in the fluid model with $T < \frac{\Lambda}{\mu}$, with or without early arrivals, when the waiting-cost rate increases the **PoA** approaches $\Lambda/(\Lambda - \mu T)$ and therefore is unbounded. However, when the other parameters are fixed, **PoA**=2 in both extreme cases $T \to 0$ and $T \to \Lambda/\mu$.[5]

Ravner [544] (2014) considers a server that starts operating at time 0. Customers are indexed by arrival times and incur three types of costs: a *queueing cost* α per time unit in the system; a *tardiness cost* βt if service starts at t; and an *index cost* γN for the $N + 1$st arrival. In fact, the index cost is in many cases natural, with earlier models often using tardiness costs as a proxy.

The author provides closed-form solutions for the case where two customers choose their arrival times, and compares several variations depending on whether arrivals before time 0 are allowed, and whether there is a given closing time. The solution is then characterized for more than two customers. The equilibrium arrival process is qualitatively similar to models that do not assume index costs. For example, these models also include a uniform arrival distribution before opening and a decreasing arrival density after opening.

A notable case where the model with index costs is qualitatively different from previous models is when $\beta = 0$. In this case, the equilibrium distribution has infinite support, the density for $t > 0$ decreases exponentially fast, and numerical analysis suggests that this rate depends on the cost parameters and service rate but not on population size. As a result, the individual equilibrium cost incurred by customers is determined straightforwardly by the order cost parameter.

Breinbjerg, Sebald, and Østerdal [108] (2014) consider a non-stochastic discrete-time queueing variation of the concert queueing game with three homogeneous players independently choosing when to arrive. Only one customer can be served during each period and early arrivals are excluded.

---

[4]This result raises similar questions regarding other models of equilibrium arrival patterns where only symmetric equilibria were analyzed.

[5]There is a typo in Remark 6 of the paper where the bounds are given, Δ there should be replaced by Λ.

The authors compute the (pure) asymmetric equilibrium solutions under FCFS, LCFS, and SIRO for all possible values of the ratio of waiting to lateness costs. They conclude that LCFS provides the highest social welfare among these disciplines, whereas FCFS provides the lowest. The authors also present results of lab experiments where players simultaneously and independently choose arrival times but the game is not repeated and therefore does not converge to the asymmetric equilibrium.

Platz and Østerdal [534] (2015) study a non-stochastic (fluid) concert queueing game model where service starts at time $t = 0$, early arrivals are excluded, service value decreases over time, and waiting costs are linear. The system has a finite per unit time service capacity such that T time units are needed to serve the given demand.

Suppose the service discipline is FCFS. As in [310], either all customers arrive at $t = 0$ or a fraction arrive at $t = 0$ followed by a period without any arrivals and then a smooth stream of arrivals until time T. The authors prove that:

- Equilibrium under FCFS is unique and *minimizes* aggregate welfare among all work conserving disciplines.

Under LCFS, if agents arrive at a rate greater than capacity, service time for a customer will be a lottery with two possible outcomes: either the customer is served immediately (with a probability corresponding to the ratio of capacity to arrival rate) or the customer is served later when all customers arriving after him have been served. In equilibrium, customers will arrive at a faster rate than capacity, hence everyone except for the last arrival faces a lottery over service times, and the arrival rate will be such that all agents have the same expected utility (with no jumps). The authors prove that:

- Equilibrium under LCFS is unique and maximizes the aggregate welfare among all queue disciplines.[6,7]

Honnappa and Jain [346] (2015) extend the concert queueing game model, assuming that service is rendered by several parallel servers with different opening times $T_{s,i}$ and service rates μ_i. They consider a fluid approximation of the model and show that all active servers finish serving customers at the same time, and that equilibrium arrivals to server i are uniformly distributed over an interval $[-T_{0,i}, T]$. These results are also extended to more general networks and heterogeneous customers, and give a **PoA**$\leqslant 2$ as in the single-server model of [359].

---

[6]Note the similarity of these results and those related to Naor's model. The optimality of LCFS was proved in [302] (see [1] §2.2), and it is intuitively clear, though no formal proof has been given, that FCFS is the worst discipline. See this observation on page 27 of [338].

[7]Jouini [376] proves the following interesting related result in a model with non-strategic customers. Consider a GI/GI/s queue with a constant rate of reneging, For the class of work-conserving non-preemptive scheduling policies, the expected waiting time of served customers is maximal (minimal) under FCFS (LCFS). The opposite relations hold for reneging customers.

4.1.2 Arrivals to a loss system

Mazalov and Chuiko [478] (2006) present another problem concerning strategic choice of arrival times to a queue. They consider a single exponential server in a loss system such that an arriving customer is rejected if the server is busy. Customers have identical preferences with regard to start of service. This is expressed by a function $C(t)$ giving profit resulting from service starting at t. Player strategy is a density function of arrival times on a finite interval $[t_0, T]$ and players try to maximize expected profit. The authors obtain a closed-form solution for two players and $C(t) = t(1 - t)$.

Haviv, Kella, and Kerner [323] (2010) consider an $M/M/s/s$ loss model in which a customer pays for trying to get service but is rewarded only if the server is available. The model is non-stationary, starts from an empty queue at time 0, and that fact is public knowledge.

The queue is unobservable but customers can infer the probability of the server being available. Specifically, the authors investigate the equilibrium arrival pattern under two information scenarios:

- Suppose customers know their time of arrival. When $s = 1$, there exists a time t_e such that until t_e all customers try with probability 1, and after t_e try with probability p_e, which is exactly the equilibrium joining probability in the E&H model. When $s = 2$, there exists a time t_e such that all try with probability 1 until t_e, and for $t > t_e$ they will try with probability $p_e(t)$, which is monotonically decreasing to p_e.

- Suppose customers do not know their arrival time but they know their serial number. Let $s = 1$. Then, the equilibrium strategy is *periodical* such that for some integer m_e the ith customer tries iff $i = 1 \pmod{m_e}$.[8]

Haviv and Ravner [327] (2015) consider an $?/M/m/m + c$ version of the $?/M/1$ model where the number of arrivals N is deterministic, early arrivals are forbidden, arrivals are rejected if there are no free servers or waiting positions, and customers maximize their admission probability.

The equilibrium solution resembles that of [310] with an atom at 0, an interval $(0, t_e)$ with no arrivals, and a continuous arrival density on $[t_e, T]$. The authors characterize the equilibrium and **SO** solutions. An interesting finding is that the PoA is not monotone with respect to N and is higher for intermediate values of N.

4.1.3 Retrials

As explained in the introduction to this chapter, the solution of a rational model of retrials is difficult, as it involves non-stationary analysis. For this

---

[8]This policy is reminiscent of the "cascades "in [314].

reason, the literature on strategic retrial models is quite scarce and the models that can be solved are subject to restrictive assumptions.[9]

See §6.8 for other models with orbit queues where instead of customers retrying, when the server becomes idle it is the server that starts a search for a customer in the orbit queue. See [694, 703, 642, 643] for models combining retrials, breakdowns, and vacations.

Kulkarni [419] (1983) was the first to solve a strategic model of retrials. Two players (or customer classes) with different waiting costs submit service requests to an $M/G/1/1$ system. A blocked request returns after an exponential time, and a customer's decision variable is the retrial rate. The author computes the best response functions and finds there can be at most three equilibrium solutions. The equilibrium in a similar system, but with nonatomic customers, is solved and shown to be unique by Hassin and Haviv (1996), see [1] §6.4.

Brooms [110] (2000) considers a Markovian model where customers arrive to a multiserver loss system. A customer seeing an available server joins to obtain immediate service. If all servers are busy, the new arrival either joins a buffer and retries after an exponentially distributed time with an exogenous parameter, or obtains the service elsewhere with a given fixed expected waiting time.[10] The system is observable to new arrivals who can see both the number of busy servers and the number of customers "in orbit." However, this information is not available to the customers in orbit who reside there and retry until successfully connecting to a server. The customer's objective is to minimize the expected sojourn time, including time spent in orbit.

The author proves that a customer's best response is of the threshold type and that a unique (mixed) symmetric equilibrium exists. Intuitively, this conclusion follows because if more customers are orbiting then a new arrival is less inclined to join the buffer, and this is an ATC situation.

Wang and Zhang [647] (2013) investigate the single-server case of [110]. They characterize the equilibrium and **SO** threshold strategies and numerically compare them.

The authors also solve the unique equilibrium probability q_e of joining when the server is busy in the almost-unobservable case where arriving customers know the state of the server but not the number in orbit. They also compute the **SO** probability q^* and show that $q^* \leqslant q_e$. The authors also conclude that social welfare increases when the orbit pool is observable.

---

[9]From about 300 papers in an extensive bibliography on retrial queues, only [110, 445, 660] deal with strategic behavior. See J.R. Artalejo, "Accessible bibliography on retrial queues: progress in 2000-2009," *Mathematical and Computer Modelling* **51** (2010) 1071-1081.

[10]The author refers to the servers as *public servers* and to the alternative option as getting service from one of an infinite number of slower *private servers*. The situation resembles the *shuttle model* in [1] §1.5.

Cui, Su, and Veeraraghavan [177] (2014) assume customers who retry will do so after a much longer time than the service cycle and therefore the decision is not when to retry but only whether to do so, and the model becomes stationary. The queue is observable and customers choose among joining the queue, joining an orbit queue, or balking. Customers incur linear waiting costs and a constant retrial cost.

- When the retrial cost is low, customers join if queue length is below a threshold or retry otherwise. When the retrial cost is assumed to be more significant, customers join short queues and balk when queue length exceeds a threshold. Equilibrium strategies may include mixing at the threshold queue length.

- The option of retrying may increase or decrease customer welfare. Retrials spread the workload and therefore induce positive externalities. When the retrial cost is low, customers do not retry enough as compared to the **SO** solution. Customers retry too much when the retrial cost is high.

Wang and Zhang [644] (2015) consider a Markovian model that alternates between ON and OFF states. Customers do not observe the state of the system, with the exception that if the server is available upon arrival they know this and start service. Customers finding the server busy or OFF can balk or join an orbit queue from which they retry. The unique feature of the model is that if the state changes to OFF during the service of a customer, the customer waits at the server until the state is ON again and then resumes service. Thus the server can be at one of four states: idle, serving, or OFF with or without a preempted customer. The authors compute the unique equilibrium joining probability and the (smaller) **SO** joining probability.

4.1.4 Restarting

Restarting is the combined action of reneging and re-arriving. Clearly, this is not a rational action in a single FCFS queue. On the other hand, it may be rational when, for example, the regime is different from FCFS. This possibility is discussed by Hassin [302] where the optimality of the LCFS regime in Naor's model is conditioned on the system's ability to prevent such actions.

Note the difference between a restart and the more common retrial. Restarting is an elective action while retrials are often forced. For example, customers may restart to obtain a more favorable position in the queue or a faster server. Retrials are typically associated with customers arriving to a queue already at maximum length and having to wait outside the queue to retry later.

Restarting is also closely related to jockeying with the main difference being that a jockeying customer usually has full information on the queue he joins whereas a restarting customer does not have this information.

Libman and Orda [445] (2002) consider a general model of accessing network resources and describe a queueing example. Consider an $M/M/1/m$ loss system. A customer sends a request for service but does not know at the time whether it is accepted. Only after processing of a request is complete will the customer obtain an acknowledgment. When no acknowledgment is received within a given time frame the customer estimates the probability of rejection to be sufficiently high as to justify resending the request.[11] A customer cannot hold more than a single request and sending a new request voids the prior request, even if it had been accepted.[12] In such a case the initial request is still processed to completion but no acknowledgment is sent. The customer incurs linear waiting costs and a fixed cost per request. Balking is not permissible and therefore the customer keeps retrying until receiving an acknowledgment. The customer's decision variable is the (identical) waiting time between consecutive requests, which the authors denote as *timeouts*.

Short timeouts increase the server's load and, moreover, shorter timeouts adopted by other customers induce a shorter best response. This FTC behavior naturally leads to the possibility of multiple equilibria, and the authors solve an example where three equilibrium solutions are possible: (i) A stable solution with long timeouts and high probability of admission; (ii) a stable solution where customers retry with no timeout at all, servers are overloaded, and requests are lost with probability 1; (iii) an intermediate unstable solution.

Zhang, Wu, and Huberman [699] (2008) consider a system of m FCFS $M/M/1$ parallel queues. The service rate of each server is drawn from a known probability distribution. Customers do not see queue length or know the service rate of the server, but do know when they are being served. Each customer aims to minimize expected waiting time. Waiting customers have the option of restarting, i.e., costlessly canceling and re-submitting a service request.[13] A new request, as well as a restarted one, is randomly allocated to one of the servers. The customer's decision is how long to wait before restarting.

Proving the existence of an equilibrium solution and computing it is difficult because of the need to keep track of the information accumulated by a customer and using it in the restart decision (this is similar, and seemingly even harder, than in the context of retrials). The authors treat an interesting special case in which each customer chooses either to be *patient* and never restart or to be *impatient* and restart continuously. An equilibrium is characterized by the probability p that a customer chooses to be patient. Intuitively, one expects that when the system load is low, it is better to be impatient and discover an idle server when one exists, whereas it is better to be patient

---

[11] A similar model was considered by Mandelbaum and Shimkin (2000) (see [1] §5.2.3), but in that model customers renege and do not resubmit service requests.

[12] See §4.5 for models that allow duplicate orders.

[13] This is different from [445] where an accepted old request is not canceled but rather remains in the queue to be processed though the customer would not benefit from it.

when the system is highly loaded. Moreover, this is a single parameter ATC model and therefore it should have a unique equilibrium. Indeed the authors compute bounds $\rho_1 < \rho_2$ such that the equilibrium probability is $p = 0$ if $\rho \leqslant \rho_1$ and $p = 1$ if $\rho \geqslant \rho_2$.[14] When $\rho_1 < \rho < \rho_2$ there exists a unique mixed equilibrium.

It is noted that in the **SO** solution all customers are impatient and the system acts like a single $M/M/m$ queue. Thus the equilibrium, in general, is not **SO**.

4.1.5 Laboratory experiments

This section reviews experimental studies on queues with endogenous arrival times. The papers [543, 564] concern single-server FCFS queues where neither balking nor reneging are allowed. The basic theoretical ideas underlying these two studies, having been modified to satisfy the constraints of the experimental design, can be found in the $?/M/1$ model of Glazer and Hassin. The models in [591, 542] are variations of the model of Holt and Sherman. (See [1] §6.1-3 for early literature mentioned in this section.)

The experiments were conducted with groups whose members interact anonymously with one another. Participants were paid in cash contingent upon their decisions and they attempt to maximize individual payoffs. These experiments provide information about the way people decide whether and at what time to join a queue and uncover systematic patterns of behavior specifying under which conditions aggregate behavior may or may not converge to equilibrium play.

The experiments reveal a contrast between individual and aggregate behavior. Individual behavior is heterogeneous with participants exhibiting different patterns of adaptive learning over time. For example, some participants switch arrival times on almost every round, whereas others hardly switch at all. At the same time, the experiments show steady patterns of aggregate behavior that approach equilibrium play. An attempt to explain the dynamics of play can be found in **Bearden, Rapoport, and Seale [83] (2005)**.

Unlike in the theoretical studies, in the experimental models service is assumed to be deterministic and equal for all players, the size of the population is finite and commonly known, time is discrete, and customers in the queue at the time of closing do not obtain service.

In the first experiment of **Rapoport, Stein, Parco, and Seale [543] (2004)**, early arrivals are not allowed. Four groups of 20 players repeatedly played the stage game 75 times. In each round, each individual player had to choose an arrival time for service in a day of 600 minutes, with service lasting 30 minutes. Thus, with perfect coordination, all customers could be served without any waiting. After each repetition players received a fixed reward for

---

[14] $\rho = \lambda/(m\mu)$.

successfully receiving service and were charged a fixed amount per unit time of staying in the system.

The aggregate equilibrium arrival pattern was characterized by many entries at opening time followed by very low arrivals in the next 90 minutes, and then a stabilizing to a nearly uniform pattern. This aggregate behavior is accounted for remarkably well by the theoretical equilibrium distribution.

The subsequent experiment by **Seale, Parco, Stein, and Rapoport [564] (2005)** allowed for early arrivals. This proved to be a good fit to the theoretical equilibrium. In particular, the arrival rate before opening was constant. The experiment was repeated in a "public information" setting where after each round players were accurately informed of the arrival times chosen by all participants. This information setting resulted, as expected, in faster learning and convergence. In the *private information* setting players were only informed of their own payoff. With an extended service time of 45 minutes, the average payoff remained negative throughout the experiment in spite of the balking option. Another interesting result was that information about the decisions and payoffs of the other members of the population mattered only when congestion was unavoidable.

Stein, Rapoport, Seale, Zhang, and Zwick [591] (2007) conducted an experiment where, rather than a sequence of service provisions, the experiment provided only a single batch of service and thus is closely related to the *waiting-time auction* model of Holt and Sherman. Customers not included in the batch leave the system without obtaining service.[15] The authors state their model in terms of a ferry departing at a fixed time T, serving a population of size n, and having capacity $s < n$. In the observable case balking is possible if there are already s customers at the queue. Customers receive a reward for obtaining service, and incur a fixed cost for arriving at the location of the ferry and linear waiting costs. In the experiment $s = 14$ and $n = 20$.

Four settings were investigated, with and without balking, and with and without public information. The stage game was repeated 60 times. The main outcome was that in all four settings aggregate behavior shifted towards equilibrium. The other results were consistent with [543, 564].

The authors compared social welfare under the symmetric mixed-strategy equilibrium solutions to two alternative strategies. In one strategy, s customers arrive at T and the others stay out. This strategy requires central coordination. An alternative strategy that requires less coordination is when all n customers arrive at T so that s will be served and $n - s$ incur the arrival cost. The **PoA** for the data used in the experiment was greater than 3.[16]

---

[15]Rather than waiting for the next batch as in Glazer and Hassin (1987), see [1] §6.3.

[16]We note a third option that should give a better result than the second one, in which players compute a probability p of arrival at T that maximizes the expected overall utility (in the second option, $p = 1$ which, in general, is not optimal).

Rapoport, Stein, Mak, Zwick, and Seale [542] (2010) repeated the experiment of [591] using *real time* so subjects could experience the time pressure before joining the queue and waiting time after arrival. They also considered a condition where the batch size assumes one of two values with equal probabilities. When s was fixed the authors obtained a monotone decreasing density of the arrival time, but with a batch of random size the density has another peak caused by the possibility that a large batch size is used (hence giving a higher chance of service for late arrivals). At the end of each round the players were provided with full (public) information on the choices made by the others. The main outcome of the experiment was that when s is deterministic, the aggregate behavior approaches equilibrium. However, with variable batch size participants behaved in a way that prevented convergence.

4.2 Joining, reneging, and jockeying

4.2.1 Reneging

Zohar, Mandelbaum, and Shimkin [719] (2002) present numerical support for the hypothesis that reneging is determined through a *patience function* of expected system delay. An equilibrium is obtained when the delay assumed by customers when deciding on their reneging time and that which results from their behavior coincide. Specifically, let x denote the expected delay (unconditional, or conditioned on finding all servers busy) in an M/M/s system. The queue is unobservable, but a customer knows when service starts. The probability a customer abandons within t units of time from arrival if service does not start by then is given by the parametric patience distribution $G(x,t)$. Let $v(x)$ be the expected time a customer has to wait until admitted to service when customers assume the expected waiting time to be x. Equilibrium is obtained by equating x to $v(x)$.

The authors note that rational behavior should lead to decreasing patience in x, because expected return per unit wait becomes smaller as time progresses and, in this case, a unique equilibrium exists. However, in practice one often observes the opposite tendency of customers to comply with the expected waiting time in the system. The authors derive sufficient conditions for the existence of a unique equilibrium in both cases. When these conditions are violated, multiple equilibria are possible. The authors also give conditions for the abandonment rate to be approximately constant, which conforms with data.

Zhou and Soman [710] (2003) investigate psychological aspects of reneging. They describe two empirical studies which test whether the number of people *behind* a customer in a queue influences that customer's decision to renege. The experiments showed that as that number increases, the likelihood of reneging decreases. The authors do not explore strategic implications of their findings, but their findings do give rise to some natural and interesting strategic models.[17]

4.2.2 Joining decisions when the service rate, value or costs are queue-length dependent

Dimitrakopoulos and Burnetas [196] (2011) assume an unobservable queue where the service rate dynamically changes according to queue length. Specifically, there is a threshold T such that whenever queue length exceeds it, the service rate is changed from a slow value μ_l to a faster one μ_h. These three parameters, namely T, μ_l and μ_h, are exogenous to the model. The only variable is the arrival rate λ. The authors compute the equilibrium and **SO** values of this variable and investigate their sensitivity to the parameters.

The model involves contradictory elements of ATC and FTC behavior, which lead to interesting outcomes. The authors prove that there are at most three equilibrium arrival rates. When multiple equilibria exist the **SO** arrival rate is between the two extreme equilibrium rates, whereas in all other cases it is less than the equilibrium rate.

Więcek, Altman, and Ghosh [661] (2015) consider an M/M/∞ system where customers are homogeneous, have a fixed service value, and their time cost rate decreases with the number of users in the system. The authors consider (mixed) threshold joining strategies such that customers join the system when the number of users is *above* a given threshold.[18,19] Clearly, this is an FTC situation. The authors obtain closed-form formulas for the equilibria in a fluid approximation model and show that, unless the equilibrium prescribes never or always to join, there are infinitely many equilibria.

Lachapelle, Larsy, Lehalle, Lions [425] (2015) propose a mean field game approach to several queueing models motivated by financial market trading. One of these models considers an observable Markovian FCFS queue of traders. Traders with a unit for sale join the queue if the expected value received will be positive. An arriving buyer buys from the first in queue at a price depending on queue length. The price vector is exogenous. Traders also incur a constant waiting cost rate.

The authors study the dynamics of the proposed models, derive numerically solvable equations, and compare the results with empirical data.

---

[17]See [229] for a theoretical model where the length of the queue behind a customer affects his welfare by providing protection against being overtaken.

[18]If the threshold is at least 1, then all states except for the empty system state will be transient. However, the authors allow for an initial state or, alternatively, there is an additional sufficiently small uncontrolled inflow.

[19]The model assumes that the joining decision precludes a customer from reneging when the number of users drops below the threshold.

4.2.3 Jockeying

Ganesh, Lilienthal, Manjunath, Proutiere, and Simatos [244] (2012) define the following strategic *random local search* queueing model. There are m processor-sharing servers with capacities $\mu_1, \ldots, \mu_m$. Let N_i be the number of customers at server i so the service rate a client gets is μ_i/N_i. Every customer in intervals of $\exp(\beta)$ time randomly picks a server j, apart from his current server i, observes N_j, and jockeys if $\frac{\mu_j}{N_j+1} > \frac{\mu_i}{N_i}$. Customers are myopic in the sense that they ignore past information they might possess on congestion in some of the queues.[20]

The authors first consider a *closed system* with n customers, no new arrivals and no departures and show that the expected time to balance the system, i.e., achieve $|N_i - N_j| \leqslant 1$ for all i, j, is $O\left(\log m\left(\frac{m^2}{n} + \log m\right)\right)$. They then consider an *open system* with arrival rates $\lambda_1, \ldots, \lambda_m$ to the respective servers with departures after service terminations and derive stability conditions and asymptotic estimates for the steady-state distributions when $m \to \infty$.

Dehghanian, Kharoufeh, and Modarres [190] (2015) consider a model with a *single strategic customer*.[21] There are two observable M/M/1 queues with different arrival and service rates. The strategic customer chooses a queue upon arrival and can jockey between queues at any time, even while being in service. This customer incurs linear waiting costs and a cost of jockeying. The authors characterize the optimal jockeying policy.

4.2.4 Joining and reneging in a shared facility

When the congestion in an EPS facility is high, customers may suffer negative expected utility. Thus, some may wish to leave the system. In such a case those who remain enjoy positive expected utility. This situation raises interesting questions about the adequate concept of equilibrium. The decision model is simpler when it distinguishes between the new arrival and current users of the system by excluding the reneging option. Literature on strategic reneging is surveyed in [1] §5.1-2, and further results and ideas can be found in [308] §2.1 and §3.[22,23]

Buche and Kushner [113] (2000) follow Altman and Shimkin (see [1] §2.6.2) and consider joining decisions of customers in an observable EPS queue with a state-dependent service rate. Reneging is not possible, balking is associated with a fixed cost, and there is also a stream of customers that cannot

---

[20]This is a bounded-rationality simplification common to many models with jockeying, retrials, etc. In fact, the resampling in this model is a form of retrials.

[21]See [662] and [1] §2.9 for similar models.

[22]Similar questions arise when the actions of noncooperative agents must jointly satisfy a constraint. See, for example, [129].

[23]For a non-queueing treatment of similar situations, see "A theory of exit in duopoly," by D. Fudenberg and J. Tirole, *Econometrica* **54**(4) (1986) 943-960.

balk. Arriving customers do not know the strategies of the other customers but have access to sample data on past performance which they can use to estimate the cost associated with joining and compare it with the cost of balking. The authors suggest a learning process and show that it converges to an equilibrium.

Ben-Shahar, Orda, and Shimkin [87] (2000) generalize the model of [113]. They consider an observable Markovian EPS multiclass service system. The service rate is $\mu(x)$ when x customers are present, and the service rate for each user, $\mu(x)/x$, is assumed to be a strictly decreasing function of the load, x. Customers of class i find the offered service acceptable iff its expected duration, given the system's state upon arrival, is at most θ_i. A *threshold profile* is specified by a (mixed) threshold strategy for each customer class. The authors prove the existence and uniqueness of such an equilibrium profile.

Suppose customers are unable to make complex predictions regarding the decisions of other users. A naive approach for estimating expected service time would be based on the assumption that the initial load persists throughout a user's service period. However, customers will find that this approach does not accurately reflect acquired experiences. The authors therefore consider a learning process.[24] The system starts at time 0 without any prior data. A customer arriving at time $t \geqslant 0$ has access to the average service times of customers who have already left the system, conditioned on the load at their time of arrival. The authors demonstrate convergence of a simple heuristic joining rule that relies on this information to the (fully rational) equilibrium solution.

Blume, Duffy, and Temzelides [94] (2010) generate a dynamic entry game played by rational agents with the potential of exhibiting *self-organized criticality*, i.e., where "tension from small shocks, gradually built into the system without notable aggregate implications, reaches a critical level," and then "a small disturbance might have a disproportionately large effect." In essence, they envision a group of EPS facilities, or *pools*, with independent demands. When a pool becomes crowded, customers independently exercise randomized reneging and may move to another pool in the group which may in turn trigger a progression of reneging decisions.

Consider first a single pool. When x agents are in the pool, each enjoys a strictly decreasing expected utility $u(x)$, such that $u(x) \geqslant 0$ if $x \leqslant \bar{x}$, and $u(x) < 0$ otherwise. When a new agent arrives and x exceeds threshold $\bar{x}$, every agent in the pool independently and simultaneously stays with probability $p(x)$ or otherwise balks. The equilibrium probabilities are such that the expected discounted value resulting from this lottery is 0. Of course the realized outcome may be that many agents renege simultaneously and those choosing to remain enjoy positive expected utility, or it may be that nobody

---

[24]Learning is not a result of a lack of information, but of *bounded rationality*, i.e., using a simpler algorithm.

leaves and all suffer negative expected utility. The authors prove the existence of unique equilibrium reneging probabilities.

The authors simplify the multi-pool case by assuming the pools are linearly ordered and there are no explicit moves from one pool to another, but still, the payoff to agents in pool $j \geqslant 2$ is negatively affected by the number of agents in that pool and also by the number of agents leaving pool $j - 1$. The pools operate sequentially within each period and the number of agents leaving pool $j - 1$ is known to agents in pool j before they make their choice. The authors prove the existence of an equilibrium. Indications to the possibility of self-organized criticality in this model are numerically demonstrated.

Brooks [109] (2014) considers customers' entry to a network of queues with a given transition matrix. There is a special node such that completing service there is associated with a fixed reward R. Staying in the system costs the customer C per time unit. As in Naor's model, due to the existence of negative externalities the number of customers joining the system under self-optimization exceeds the **SO** number and the system can be regulated by appropriately taxing the reward R. All joining customers face identical expected utility rate, which a profit maximizing system owner can fully extracts. Therefore, the profit-maximizing toll induces the **SO** joining behavior.

4.3   Benchmark effects

See [32] and [413] for models where demand depends on the difference between the delay or service rate at a server and a market standard, respectively.

Yang, de Véricourt, and Sun [679] (2014) consider two service firms facing an exogenously fixed price p. The demand captured by each firm depends on the difference between the expected waiting time at that firm and a market waiting-time *benchmark* r. Specifically, firm i, $i = 1, 2$, is an M/M/1 system with arrival and service rates λ_i and μ_i, unit capacity cost c_i, and expected waiting time w_i. The benchmarks considered here are the minimum expected waiting time $r_m = \min\{w_1, w_2\}$ and the average expected waiting time $r_a = (w_1 + w_2)/2$. The *satisfaction* level of a customer with a realized waiting time t is a decreasing concave piecewise linear function $s(t, r)$ with a single breakpoint at r. The demand of a given firm increases linearly with the expected satisfaction of its customers and decreases linearly with the expected satisfaction at the other competing firm. Firm i wishes to set capacity to maximize profit $p\lambda_i - c_i\mu_i$, where μ_i is a decision variable and λ_i is the equilibrium arrival rate.

The main results are:

- With benchmark r_m, the equilibrium is either unique or there is a continuum of equilibrium solutions with $w_1 = w_2$. The latter *stickiness effect* case occurs when the benchmark effect is strong enough.

- With benchmark r_a there exists a pure equilibrium associated with a *reversal effect*. That is, the firm with the shortest waiting time at the equilibrium in the absence of benchmarks may end up with the longest delay when benchmark effects are present.

- In both cases, expected waiting time decreases as a result of the existence of a benchmark.

Yang, Guo, and Wang [678] (2014) assume customers have *reference-dependent utility* composed of the sum of an *intrinsic* component and a *gain-loss* component. Specifically, they add to the E&H model gains and losses incurred when customers compare themselves to a reference point represented by a random customer. If the customer waits less (more) than the random customer, a gain (loss) proportional to the difference in waiting costs is incurred. A similar effect also exists with regard to the nominal amount the customer pays. Risk aversion arises because losses are weighted more heavily than gains.

An interesting feature of this model is that utilities associated with joining and balking depend on the strategy of other customers. Let $g(\delta)$ be the difference between the utilities of joining and balking when other customers join with probability δ. The authors prove that for small service rates g is concave; otherwise it is monotone increasing. These properties imply an FTC behavior and multiple equilibria. Intuitively, the FTC situation arises because customers are risk averse with respect to losses they incur relative to others and one who deviates from the common behavior risks an even greater difference. However, there is at most one *stable* non-zero equilibrium. The authors compute the **SO** joining probability and the profit-maximizing price, assuming the server can induce the desired equilibrium when it is not unique. They also consider an extension where customers choose between two servers and prove the existence of a duopoly price equilibrium.

4.4   Priority purchasing, overtaking, and line-cutting

For early contributions on customer priority purchase decisions see [1] §4. While the result of "overtaking" or "line-cutting" is similar to that of priorities, these terms are often associated with a more negative connotation. The literature on cutting in lines usually concentrates on the sociological and psychological aspects of such actions. Little has been said about strategic implications of line-cutting, though it is a natural and common aspect of queueing.

Marbach [470] (2001) mostly deals with a non-queueing model, but also describes the following queueing application. A finite number of heterogeneous customers submit demands to an unobservable $M/M/1/m$ preemptive priority system. Preemption here is not only with respect to service but also with respect to queueing. Thus, in this case when the buffer is full and a new service request arrives the system may expel a lower priority request from the queue. Customers are loss sensitive, i.e., their utility is a function of the expected rate of service completions of their jobs. The price u_i for submitting a request in priority i is independent of whether the request is eventually served. Customers select the rates of requests in each priority wishing to maximize utility after deduction of payments. The author solves an example suggesting that equilibrium properties which are proved for the non-queueing model also hold in the queueing model. In particular:

- For some index i^* the active priority classes are i^* and the classes with lower priority.

- The success probabilities associated with these classes relate to each other in the same ratio as their prices (while for i^* this ratio is a lower bound).

Allon and Hanany [38] (2012) assume each individual from a finite population of identical customers generates a Poisson stream of service requests. A request is of type H (High) with probability α or otherwise it is type L (Low). Type is private information. A request of type $X \in \{H, L\}$ is associated with service rate μ_X and waiting-cost rate c_X, such that $c_H \cdot \mu_H > c_L \cdot \mu_L$. Balking is not allowed and customers minimize expected discounted waiting costs.

New requests can adopt a line-cutting strategy, asking those in the queue for permission to overtake them. This is done by starting at the end of the queue and, if permission is granted, proceeding towards the head of the queue.

The total system cost is minimized by the $c\mu$-rule. This means a request should choose the line-cutting behavior iff it is of type H, and that L-requests in the queue accept all cutting requests. The authors consider a punishment strategy. Once a deviation from the $c\mu$-strategy is observed, all future cutting requests are rejected, resulting in an FCFS regime. Thus customers face a trade-off between optimizing short-run and long-run expected utility.

The authors provide a necessary and sufficient condition for the $c\mu$-rule to define an equilibrium when queue length is unobservable. Qualitatively this condition is satisfied when $c_H - c_L$ is large, $\alpha(1 - \alpha)$ is large (i.e., α is neither very small nor close to 1), and the discount factor is high (customers are patient). When the queue is observable the strategies may be state-dependent, but the $c\mu$-rule still defines an equilibrium if customers are sufficiently patient. The authors also consider a variation where a deviation from the $c\mu$-strategy is observed only by a small number of customers in the queue but the strategy could depend on a commonly known public signal.[25]

Erlichman and Hassin [229] (2015) consider an observable M/M/1 queue with homogeneous customers choosing upon arrival, the number of customers in the current queue they wish to overtake while paying a fixed cost per overtaken customer. An overtaking customer saves the wait during service of customers overtaken. Moreover, customers more towards the end of the queue are more likely to be overtaken by future arrivals and therefore the more customers expected to be overtaken by future customers, the greater the incentive of an arriving customer to overtake those already in the queue. Therefore, this is an FTC type game. Some of the main results are:

- The set of equilibrium overtaking strategies is very rich and contains unexpected solutions.

- Analyzing the pure equilibrium strategies Σ_k of overtaking k customers (or all, if their number is less than k) for a constant k, the authors define two thresholds. If the normalized service value is below the lower threshold, then the unique equilibrium is Σ_∞. If it is above the higher threshold, then Σ_0 is the unique equilibrium. If it is between the two thresholds, then all Σ_i $i = 0, \ldots, \infty$ are equilibrium strategies.

- In some cases, under a Σ_k equilibrium, an arriving customer's expected waiting time decreases with queue length.

- If the server can induce customers to choose the equilibrium which maximizes profits, then the overtaking regime brings higher profits than would the respective single-price two-priority regime.

Haviv and Ravner [328] (2015) consider in their base model an unobservable M/G/1 queue with homogeneous customers incurring linear waiting costs. The system applies non-preemptive delay-dependent dynamic priorities. Each arriving customer decides on the amount to pay for priority. The priority at time t of a customer arriving at time s, $s \leqslant t$, and paying $b \geqslant 0$ is $b(t - s)$. The authors prove that though the best-response function of a customer is first increasing (FTC behavior) and then decreasing (ATC behavior) as a function of the payment offered by other customers, the equilibrium payment is unique. The authors also extend their analysis allowing for multiple classes differing in waiting costs, but in this case the uniqueness of the equilibrium remains unresolved.

Alves [53] (2015) describes a model in which customer behavior determines the number (one or two) of queues in an M/M/2 system. A customer

---

[25] A related non-strategic model is analyzed in "Analysis of queueing systems with customer interjections," by Q-M He and A. A. Chavoushi *Queueing Systems* **73** (2013) 79-104. There, the success of an interjecting request depends on the kindness of customers already in the queue as measured by an exogenous probability of the request being accepted.

arriving when the servers are busy and there is no queue can choose between forming a single common queue or separate queues. A new customer arriving to a single queue can form a second queue, but this intention is understood by his predecessors who can react before he implements it and divide the single queue into two queues. The outcome depends on customer attitudes to risk and the possibility of jockeying.

4.5   Duplicate orders

See §4.1.4 for models where customers can resubmit service requests while canceling their previous ones.

The first part of **Li [439] (1992)** considers a single server model we describe in §2.1.1. The author also extends this model by modeling lead-time competition among n identical MTS firms. Demand is satisfied with equal probabilities by one of the firms having positive inventory. If the product is not available from any of the firms, the customer places the same order at each firm, and obtains it from the firm which produces the finished product first. Remaining orders with the rest of the firms are then canceled. No cost is associated with duplicate orders, neither is there any cancellation fee. An important assumption of this model is that *each firm only observes its own inventory level*. The model's parameters are the number of firms, arrival rate, service rate, unit price, production cost, and holding cost. The main results are:

- An explicit condition for the existence of an MTO equilibrium (i.e., an equilibrium where all firms have zero base-stock levels).

- An MTS equilibrium is more likely to exist under competition than in a monopoly market, and is more likely to exist in a monopoly market than under competition where duplicate orders are forbidden.

- The likelihood firms make to stock increases in the number of the firms.

- Allowing duplicate orders increases buyers' welfare and decreases producers' welfare.

- Duplicate ordering could be socially desirable if buyers' value of time is much higher than the value placed on time by the producers (i.e., the waiting-cost rate is much higher than the interest rate). In other situations, a cancellation fee may be appropriate as a means of deterring duplicate orders.

Calbert [121] (1999) models a command center which obtains reports from many sources that may be reporting on the same event, thus causing a replication of messages. The reward from a serviced message decreases with waiting time. Sending a message is costly and each source selects the sending rate of its messages hoping that one will be serviced while aiming to maximize expected net utility. To prevent over-congestion in the resulting equilibrium it is suggested that the command center regulates the cost/reward structure.

Armony and Plambeck [67] (2005) consider a *manufacturer* producing a product at a finite rate and selling it through two independent *distributors* with base-stock levels B. A customer arriving to an out-of-stock distributor orders from this distributor with probability $1-\alpha$ and waits for the order to be ready. The customer tries the other distributor with probability α. If the other distributor is also out of stock, the customer orders from *both* distributors, waits until the first of these orders is satisfied, then cancels the other order. Customers renege after an exponentially distributed time. The distributors identify customers placing duplicate orders and give them priority to avoid losing sales to the other distributor. Assuming the manufacturer is unaware of duplicate orders, the authors prove that both the demand and reneging rates are overestimated by the manufacturer who consequently over-invests in capacity. The customers and distributors in this model are not strategic, in particular α and B are exogenous. However, the model has potentially interesting variations in which these parameters are endogenously determined.

Ata, Skaro, and Tayur [69] (2015) consider a Markovian model motivated by organ transplantation. There are K servers and K customer classes. Class k customers arrive to server k, $k = 1, \ldots, K$. A given fraction $1 - \pi_k$ are dedicated to this server, while the other fraction π_k can costlessly submit J duplicate orders to servers from a set $A(k)$. Customers of all classes abandon the system at a common rate γ. Customers choose the servers to which they submit requests so as to minimize expected waiting time.

The authors solve a fluid approximation and show that if every class has sufficiently many customers who can submit multiple orders the equilibrium choice results in equal waiting times at all servers. They also characterize the equilibria in a diffusion approximation obtaining second-order perturbations of the waiting times under fluid approximation. A simulation study demonstrates the advantages of duplicate orders, particularly in lowering the effective rate of abandonments.

Guo and Hassin [275] (2015) consider a single-server unobservable SIRO queue.[26] Customers can submit multiple (duplicate) orders and when any one of these orders enters service the customer cancels the others. The key observation is that the SIRO queue with duplicate orders is equivalent to a queue with relative priorities. The authors show that, as in the absolute

---

[26] An alternative regime with similar features is EPS.

priority model (see [307]), this is an FTC case and in general the equilibrium number of duplicate orders by a customer is not unique. However, it turns out there can be at most two pure equilibrium solutions, which are consecutive integers, and one other solution obtained by mixing these two strategies.

Note that [27, 205] deal with the closely related problem of purchasing relative priorities in a single-server queue. A main difference is that in these papers the decision variable is continuous.

Guo and Hassin [276] (2015) consider a Markovian model with two identical servers and homogeneous customers. The service fee is exogenous and server i, $i = 1, 2$, designates a part O_i of that fee as an ordering fee. Customers can submit service requests to both servers and when one of the orders enters service the customer cancels the other, but the ordering fee is non-refundable. Submitting duplicate orders thus shortens the expected system time at the expense of losing the ordering fee.

Customers choose among submitting requests only to server 1, only to server 2, to both servers, or balking. The authors derive expected waiting time for each of these possibilities and a given strategy of the other customers and use it to compute the symmetric equilibrium, which they show to be unique. It turns out the more duplicate orders the less a single-order customer waits and the more a duplicate-order customer waits. The latter property arises despite the positive effect of duplicate orders. The system also demonstrates a pairwise ATC property. In particular, for a fixed probability of balking, the more customers submit duplicate orders the less attractive it becomes. A monopoly would set ordering fees low enough so that all joining customers submit duplicate orders and we obtain the equivalent of an M/M/2 queue. However, this is not the case when servers compete by setting ordering fees and competition actually reduces customer incentive to place duplicate orders.

4.6   Choosing the arrival rate

This section surveys models focusing on arrival rates decisions of atomic customers. We note that these models and others where customers decide on service duration (§4.7) are related in the sense that both assume customers decide on the size of their demand.

Additional models where customer arrival rates are decision variables are scattered in other chapters, particularly in §8. For models where customers choose arrival rates and a central planner acts to induce social optimality by controlling the queue discipline or by pricing, see [570] and [593], respectively.

Some papers on routing demand (for example, [46, 554] also assume that the total demand from a user is a decision variable. Thus there is a combined decision of determining the arrival rate and allocation to servers.

Dutta, Goel, and Heidemann [211] (2003) assume that a finite number of homogeneous customers decide on their demand rates in an unobservable M/M/1 system. Customers are not delay sensitive and their goal is to maximize individual throughputs.

The authors discuss the effect of admission mechanisms on the customers equilibrium choice of arrival rates. In some of these mechanisms a customer equilibrium does not exist, for example if a request is rejected when the buffer of an M/M/1/s queue is full. In other cases an equilibrium exists but the aggregate arrival rate approaches 0 as the number of customers increases. Lastly, the authors introduce an admission mechanism such that an equilibrium exists and, furthermore, the arrival rate is bounded from below when the number of customers grows.

Menasché, Figueiredo, and de Souza e Silva [485] (2005) assume a finite number of users, with each one choosing an arrival rate. Users' utility depends on their demand rate and on the congestion at the server. The authors investigate the dynamics by which users adapt their rates and the convergence of this process to an equilibrium. They present an example with an M/M/1/s system and a specific utility (QoS) function.

Jin and Kesidis [368] (2005) consider a general model with N customers. Customer n sets arrival rate λ_n wishing to maximize a performance measure $\theta_n(\lambda_1, \ldots, \lambda_N)$ that depends on the actions taken by other users. The network charges $M\theta_n$ and the utility of customer n is a function $U_n(\theta_n) - M\theta_n$, where (i) θ_n is nondecreasing in λ_n, and (ii) U_n is nondecreasing and concave. The authors design an algorithm for computing the equilibrium arrival rates. They describe in detail two queueing applications: An M/G/K/K queue, and a queue where the demand of customer j is rejected if the *expected* queue length exceeds a threshold $\gamma(j)$.

4.6.1 Maximizing power

A customer in a queueing system often faces conflicting objectives of maximizing throughput λ and minimizing expected delay W. A performance measure that heuristically takes into account both of these goals is the *power*, which is the ratio λ^β/W for a constant $\beta > 0$. The parameter β indicates the relative emphasis placed on throughput versus delay. This measure is a reasonable rule-of-thumb since it increases in the customer's throughput λ and decreases in its expected delay W. See also [594] §2.5.

Douligeris and Mazumdar [207] (1992) consider a single-server Markovian model where demand is generated by two users with homogeneous service requirements. User i submits demand at rate λ_i to maximize the power $\lambda_i^{\beta_i}/W(\lambda)$, where $W(\lambda)$ is the expected delay when the total demand rate is $\lambda = \sum \lambda_i$.

- A unique equilibrium point exists where $\lambda_i = \mu \beta_i / \left(1 + \sum \beta_k\right)$.

- A Stackelberg version of the game brings about higher throughput, power and delay than the equilibrium of the simultaneous game.

- **Sahin and Simaan [555] (2008)** consider a variation of [207] proving the existence of a unique *interior equilibrium* in a similar flow control problem, but with exogenously fixed routes on a general network.

Altman, Başar, and Srikant [46] (2002) consider N users generating demand requests and distributing them among M/M/1 servers. For demand rate λ_{ij} sent by user i to server j, $\Lambda_i = \sum_j \lambda_{ij}$ is the throughput of user i, $\lambda_j = \sum_i \lambda_{ij}$ is the demand served by server j, and $W_j = 1/(\mu_j - \lambda_j)$ is the expected delay at server j. The objective of user i is to maximize the power function $(\Lambda_i)^{\beta+1} / \sum_j \lambda_{ij} W_j$, for $\beta \in (0,1)$, quantifying a trade-off between the goals of maximizing throughput and minimizing delay. The authors define a refinement of asymptotic equilibrium that applies to finite but arbitrarily large values of N and prove that such an equilibrium exists.

Inoie, Kameda, and Touati [355] (2006) consider an M/M/s system where customers choose demand rates to maximize their power values. The authors demonstrate the following "paradox." If several such M/M/s systems are aggregated into a single system serving the total demand and employing the total number of servers, then, as expected, customers' power can be improved if their choices are centrally controlled. However, the equilibrium solution may be associated with reduced power.

Gai, Liu, and Krishnamachari [240] (2011) consider the system of [207] for the case of a common parameter $\beta_i = \beta$. They compute the **SO** arrival rate both when social utility is the sum of individual utilities and also when it is the sum of their logarithms. They bound the **PoA** in terms of the number of users and β, showing that it degrades linearly (or worse) with the number of users.

As expected, the equilibrium arrival rates exceed the **SO** values. The authors suggest non-price mechanisms for controlling the arrival rates. A *dropping policy* is characterized by a function $P(\lambda)$ such that when the total arrival rate is λ, every request is accepted with probability $P(\lambda)$. The authors investigate functions with thresholds $r_1 < r_2$ such that $P(\lambda) = 1$ for $\lambda \leqslant r_1$, $P(\lambda) = 0$ for $\lambda \geqslant r_2$, and the function linearly decreases from 1 to 0 in the interval $[r_1, r_2]$. They show how to compute, for any given $\epsilon > 0$, thresholds which lead to a unique stable equilibrium such that **PoA** $\leqslant 1 + \epsilon$.

4.7   Choosing the service duration

Here we describe models where service duration is set by the customer (see [1] §8.6). This is in contrast to expert systems where it is *the server* who determines service duration (see §6.3).

Kim and Hwang [404] (2009) consider revenue maximization in an $M/G/N/N$ model where customers choose the duration of service (or "call"). Service fee is a piecewise linear function of service duration with a single breakpoint from which a discount rate applies. The server's decision variables are the location of the breakpoint and the discount rate. The discount encourages long calls, but also increases blocking probability. Customer behavior is guided by an exogenous function. When customers with demand that exceeds the breakpoint arrive, the duration of their calls beyond the breakpoint is assumed to increase by a factor of $1 + \alpha \cdot d^{\beta}$, where $d \leqslant 1$ is the discount rate and α and β are positive constants.

Tong [625] (2012) considers an $M/G/1$ system where customers choose their service duration τ. The service value to a customer of type α is $v(\tau) = \tau - \frac{\tau^n}{\alpha}$. This is a concave function which is first increasing and then decreasing. The parameter n represents the degree of concavity. If customers are charged a fixed rate per service, they choose τ such that $v'(\tau) = 0$. If they are charged a fixed cost of r per unit time, they will choose a service duration such that $v'(\tau) = r$. A customer's type is realized from a given pdf upon arrival. The author shows that when there is ample demand, the time-based scheme induces higher revenues, a higher throughput, and less congestion. Fixed pricing tends to be chosen only when congestion is less important: either market size is not large or customers are not highly sensitive to waiting time.

Chapter 5

Social optimization and cooperation

As mentioned in the introduction, computing the **SO** solution defines an optimization problem, which is not the subject of this survey. Our focus is on topics related to the gap between the **IO** and **SO** behavior and on ways for inducing an **SO** equilibrium, i.e., *coordinating the system*.

We also include in this chapter research concerning degradation of system performance due to lack of cooperation, in particular the **PoA**. Other related subjects are the Downs-Thomson and Braess paradoxes, which concern routing decisions, and are surveyed in §8.7 and §8.6.

5.1   Coordination by pricing

Nadiminti, Mukhopadhyay, and Kriebel [492] (2002) consider a general model with a finite number of customers. The benefit derived by customer i from using service at rate λ_i is an increasing and concave function $v_i(\lambda_i)$. The waiting cost of customer i is $\lambda_i C(\lambda)$, where $\lambda = \sum \lambda_j$, and $C(\lambda)$ is increasing and convex.

Let λ^* be the **SO** total arrival rate, and let λ_i^* be the corresponding demand rate of customer i. The first-order **SO** condition for $\lambda_i^* > 0$ is $v_i'(\lambda_i^*) = C(\lambda^*) + \lambda^* C'(\lambda^*)$.

The authors suggest that under symmetric information the firm charges a nonlinear price $p(\lambda_i) = C'(\lambda^*) \int_0^{\lambda_i} (\lambda^* - \lambda) d\lambda$. Customer i wishes to maximize $v_i(\lambda_i) - \lambda_i C(\lambda) - p(\lambda_i)$, and when $\lambda_i = \lambda_i^*$ the first-order conditions for individual and social optimality coincide.

Under asymmetric information the firm does not know the value functions $v_i(\lambda_i)$ and cannot compute λ^*. The authors suggest the following mechanism. The firm announces that if total demand is λ then customers will be charged prices $p(\lambda_i) = C'(\lambda) \int_0^{\lambda_i} (\lambda - q) dq$. Furthermore, customers are required to supply *functions* $\lambda_i(\lambda)$ giving their demand if the total arrival rate is λ. The authors prove that λ^* is the unique value of λ that achieves $\sum \lambda_i(\lambda) = \lambda$, and therefore the firm can use this information to induce the **SO** solution as in the case of symmetric information.

Stidham [593] (2004) discusses a general model in which users set their demand rates to a set J of servers, and each unit of demand from user r requires *complementary services* from a subset $S_r \in J$. A user setting demand at rate x_r obtains utility $U_r(x_r) - h_r x_r \sum_{j \in S_r} D_j(y_j)$, where $y_j = \sum_{r: j \in S_r} x_r$, U is concave, and D is a convex function giving the delay at server j. The goal is to induce an **SO** equilibrium by pricing the different services. The author shows that:

- If demand of each user is very small, an equilibrium point satisfying the first-order **SO** conditions is induced by tolls equal to the users' external effects, which are defined as the marginal increase in total delay caused by a marginal increase in that user's flow.[1] However, this doesn't guarantee global optimization.

- An algorithm dynamically setting a price equal to the current external effect of a customer converges to an equilibrium point.

- In a simple example with a single M/G/1 processor-sharing (or M/M/1 FCFS) queue with two users, if utility functions are linear then there exists an **SO** solution whereby only one of the users submits any demand. If the utility functions are nonlinear the **SO** solution can be internal, i.e., both demand rates are positive. The suggested pricing mechanisms may converge to an equilibrium which is not globally **SO**.

---

[1]Cf. [1] §3.3.

5.2 Positive network effects

In some cases congestion has opposing effects. On one hand customers are delay sensitive, while on the other hand they also enjoy *network effects* which increase with the number of other customers using the system. See §9.5 for Internet-related models with network effects. See [495] on profit maximization in a model with network effects.

Johari and Kumar [373] (2009) study a system with N identical users, negative congestion effects, and positive network effects. Specifically, user i's utility is

$$u_i(N) = \alpha \lambda_i + \beta \lambda_i \cdot f\left(\sum_{j \neq i} \lambda_i, N\right) - \lambda_i \ell\left(\sum_j \lambda_j, N\right) - d(\lambda_i),$$

where λ_i is the *usage rate* of customer i (i.e., the arrival rate of its demand). The four terms refer to the linear utility from usage, positive network effects (f is nonnegative, nondecreasing, and concave), congestion effects (ℓ is nonnegative, increasing, and convex, such as the M/M/1 expected waiting time), and personal cost of usage (d is nonnegative, nondecreasing, and convex).

The main results include:

- A unique symmetric equilibrium.[2] The corresponding rate $\lambda^{\text{EQ}}(N)$ is derived as a function of the number of users N, and is an increasing function.

- The authors derive the **SO** arrival rates for a given N and also derive the **SO** number of users N^* given that for any number N the users follow the equilibrium $\lambda^{\text{EQ}}(N)$. It turns out that N^* is larger than the number M^* of users that maximizes the utility of a single user when usage levels are according to the equilibrium (of course, without network effects $M^* = 1$).

- For a given N the **SO** individual usage rate may be smaller or larger than the corresponding equilibrium value.[3] The authors characterize these cases.

- The analysis is complemented by discussing special cases with M/M/1 delay when each of the congestion and network effects depends either only on N or only on total usage.

---

[2]This is not obvious because the model contains FTC elements.
[3]This is a result of the coexistence of positive and negative externalities.

Nair, Subramanian, and Wierman [494] (2014) assume that, given a total arrival rate λ, the utility of each user is proportional to $\lambda^{1+\beta}$ for some $\beta \in [0, 1]$. Two types of delay costs are considered; fixed waiting-cost rates with M/M/1 expected waiting time, and a convex increasing function of the system utilization ρ. The firm gains an exogenous amount per unit demand and incurs linear capacity costs. The firm selects capacity to maximize net profit.

The authors compute the equilibrium in the monopoly case, and in [493] also consider two types of competition. In one, network effects are industry-wide and user utility depends only on aggregate demand. In the other case these effects are firm specific and user utility depends on demand at the firm they choose. The authors also solve a variation in which users cooperate to maximize total welfare (this is not the social problem as it ignores the firm's costs). The main qualitative result is that competition does not significant improve user welfare. If network effects are firm-specific, then the firm with greater network effects will nearly become a monopoly.

5.3   Priorities

Kim and Mannino [403] (2003) generalize a result of [487] (see [1] §4.4.3.2) by considering general service distributions. Consider a multiclass M/G/1 model with asymmetric information, nonpreemptive priorities, heterogeneous service valuations, heterogeneous delay costs, and heterogeneous service time distributions. The system can be coordinated by charging a price $p_i(t) = A_i t + B t^2$ from a customer who declares to be of type i and with a realized service time t.

Van den Berg, Mandjes, and Núñez-Queija [88] (2007) consider profit maximization in an M/G/1 priority system with jobs whose size x is a random variable. The size of the job is observable and the server charges size-dependent prices for priority. We observe that because of the ability to discriminate according to size the resulting prices leave no customer surplus and therefore the profit-maximizing solution is also **SO**.

There are two priority classes. The service discipline within each class is EPS and low-priority customers are served only when there are no high-priority customers in the system. The value received when the expected delay of an x-job is D is a function $w(D/x)$. The server sets unit-prices $p_h(x)$ and $p_l(x)$ for admitting an x-job to the high- and low-priority classes, respectively.

The main result of the paper is the existence of thresholds $0 \leqslant t_h \leqslant t_l$ such that under the optimal solution (short) jobs with $x \in [0, t_h)$ obtain high priority, (long) jobs with $x \geqslant t_l$ obtain low priority, and those in the middle interval balk.

Hsu, Xu, and Jukic [348] (2009) generalize the model of [487] by assuming customer classes are subject to exogenous bounds on expected delay. Let v_i be the unit waiting cost, c_i mean service time, and B_i the upper bound on the expected system time of i-customers such that $\frac{v_i}{c_i} \geqslant \frac{v_{i+1}}{c_{i+1}}$, $i = 1, \ldots, n-1$. It is assumed that $B_i - c_i \leqslant B_{i+1} - c_{i+1}$, $i = 1, \ldots, n-1$. Thus, the class given higher priority under the $c\mu$-rule is also less tolerant of queueing delays. The **SO** scheduling rule partitions the user classes into groups with fixed priorities among these groups, and randomized priority within each group.

The authors also design optimal **IC** pricing and scheduling rules for the decentralized version of the model, i.e., where customer types are private information. An interesting feature of the solution is that the price charged to users within a group can be attributed to externalities imposed on the first class within that group. When service requirements are nonhomogeneous, as in [487], the **IC** pricing depends on the actual processing time of the customer.

Gavirneni and Kulkarni [247] (2015) examine an M/G/1 variation of the two-class non-preemptive priority model (see [1] §4.2) with random customer waiting-cost rates having the Burr cdf $G(x) = 1 - \left(1 + (x/a)^d\right)^{-k}$ for given positive parameters a, d, k. Service fee c is exogenous and priority can be purchased at a price K set by the firm. In equilibrium, customers join according to a threshold *level of participation* α such that the α fractile of the customers with the highest waiting costs buy priority. The queue manager can induce any level of customer participation by controlling the priority price.

Denote the expected system cost when customers are served in order of arrival by C_{FCFS}, by $C_{\min}$ when served non-preemptively in decreasing order of waiting costs, and by C_{SOC} when there are two service priorities and priority is allocated in an **SO** way. The efficiency of the system is $\frac{C_{\text{FCFS}} - C_{\text{SOC}}}{C_{\text{FCFS}} - C_{\min}}$.[4] The authors find that the efficiency of the system is above 73% and increases further with system utilization ρ.

5.3.1 Auctions

Literature on equilibrium priority auctions is described in [1] §4.5. These models are applicable in cases of asymmetric information. In a *highest-bid-first* (HBF) regime customers offer payments (also called *bribes*) to the queue organizer with those offering higher payments obtaining priority over those offering smaller payments. See §6.6.3 for the use of priority auctions to maximize profits.

Stahl [590] (2002) considers social-welfare optimization in a Markovian unobservable multiclass system with heterogeneous waiting-cost rates and service valuations. There exists symmetric information about customer type and it is possible to control both admission and priorities. For example, a class

---

[4]One can refer to this measure as *differential* **PoA**.

with higher service value and waiting costs may be given admission preference but lower priority in the queue. The author derives a closed-form solution
when customers have identical service valuations or identical delay costs. The
author then derives the equilibrium bidding for priority strategy assuming a
nonpreemptive priority regime and therefore, unlike in [303], the solution is
not **SO**.

Kittsteiner and Moldovanu [405] (2005) consider an auction for priority in an $M/G/1$ queue with nonpreemptive priorities where customers pri-
vately know their processing time.[5] The waiting-cost function is nonlinear and
common to all customers.

Deriving bidding equilibria turns out to be complex because the bid of
a customer depends on the entire distribution of queueing time. Indeed, an
increase in the cost incurred by a customer waiting an additional time unit
depends on his specific processing time. If the waiting cost function is convex
(concave), this dependence is increasing (decreasing) in the processing time.
This effect determines the form of the equilibrium bidding function which may
be increasing in the processing time (leading to the longest-processing-time-
first (LPT) discipline) or decreasing (leading to the shortest-processing-time-
first (SPT) discipline).

Generally, when customers are homogeneous and service value is finite,
some balk. In such an equilibrium that implements SPT, only customers with
processing times below a threshold join. The equilibrium threshold is greater
than the **SO** threshold, which can be explained by the existence of negative
externalities. However, the authors make the interesting observation that in
their model there are also *positive* externalities. Suppose two customers arrive
consecutively while a third is being processed, and assume the first of the
two has a long processing time. The externality on the customer arriving last
might be overall positive because he will be served before his predecessor. This
would not be the case had the machine been idle upon the first customer's
arrival.

The authors also discuss the following stronger form of control. Customers
bid a price per unit time of processing when arriving, and pay this price after
their service is completed according to the realized processing time. Numerical
calculations indicate the possible advantage of this scheme.

Courcoubetis and Dimakis [174] (2009) consider a single Markovian
unobservable queue shared by n types of strategic customers. Aggregate arrival rate is assumed to be sufficiently high so balking is induced to keep the
load below C. An i-request has value r_i, incurs waiting costs of c_i per time
unit, and has average size $1/\mu_i$. Customers choose weights w which determine their *congestion-control mechanism*. If at a given instance t there are
requests $j = 1, \ldots, a$ in the queue, then the shares $x_j(t)$ of the capacity C

---

[5]See §3.3 for related models.

that the requests obtain are such that the sum $\sum_j u_j(x_j(t))$ is maximized where $u_j(x) = -w_j x^{1-\alpha}/(1-\alpha)$ and $0 \leqslant \alpha < 1$. The case $\alpha = 1$ corresponds to $u_j(x) = w_j \log x$. Requests are charged at $\lambda(t)$ per unit of capacity per unit time where $\lambda(t)$ is the shadow price (or Lagrange multiplier) of the capacity constraint, which is equal to the probability that demand exceeds capacity.

The authors observe that when $n = 1$ the case $\alpha = 0$ corresponds to a HBF bidding regime where preemptive priority is given to the highest bid. This regime maximizes social welfare, as shown by Hassin (1995) (see [1] §4.5, [594] §2.6). On the other hand, the case $\alpha = 1$ reduces to a model with relative priorities analyzed by Haviv and van der Wal (see [1] §4.3.2) and leads to a unique symmetric equilibrium which is not **SO**. The authors conjecture that this inefficiency holds for every $\alpha > 0$. A similar inefficiency is demonstrated when $n = 2$ and one of the customers is only slightly sensitive to delay (c_2 close to 0). However, efficiency can be achieved in this case through a two-part tariff where, in addition to $\lambda(t)$, customers also pay a fixed price per unit demand.

5.3.2 Discriminatory processor sharing (DPS)

The survey on DPS by **Altman, Avrachenkov and Ayesta [43] (2006)** deals in its last section with DPS models of non-cooperative queueing games. Much of that material is also covered in [1]. We describe here more recent contributions.

The following intriguing question arises in unobservable queueing systems with heterogeneous customers. Suppose customers arriving to an M/M/1 facility differ in service value, delay cost and service rate. Clearly it is desirable to encourage the entry of customers with high service values as well as low time costs and short expected service lengths. This can be done by giving these customers priority over others. However, *for a given composition of the queue*, the optimal service discipline is the $c\mu$-rule that allocates customer priorities without regard to service value. **Mendelson and Whang [487] (1990)** solved this problem, showing that **SO** joining can nevertheless be obtained while implementing the $c\mu$-rule by setting appropriate class-dependent entry fees.[6]

In some cases, restrictions imposed on the firm's policy prevent it from implementing the first-best solution. **Hayel and Tuffin [334] (2005)** and **Hassin and Haviv [309] (2006)** independently considered second-best solutions in such cases in models with two customer types, identical service rates, and symmetric information. Hayel and Tuffin assume the utility of customers is a type-dependent negative exponential function of their waiting time, while Hassin and Haviv assume a linearly decreasing function. These papers consider social-welfare and profit optimization when service prices are exoge-

---

[6]Their result is even stronger as they show that social optimality is achieved even if customer types cannot be observed by the queue manager.

nously fixed and demand is determined by the equilibrium conditions; they prove that both objective functions can benefit from the use of DPS relative priorities. The authors of [309] also consider a variation where service price is a decision variable, but it must be the same for both classes. They show that giving absolute priority to one of the classes is profit maximizing, but this class is not always the one selected by the $c\mu$-rule.

Sun, Tian, Li, and Zhang [611] (2007) extend [309], allowing the customer classes to have different service rates and give sufficient conditions for the existence of a unique equilibrium when the service fees are exogenously set. They describe a degenerate case where multiplicity of equilibria is possible, leaving open the question of whether multiplicity of equilibria is possible except in such cases.

Wu, Bui, and Johari [664] (2012) consider a multiclass M/M/1 DPS model where classes differ in waiting-cost rates and customers buy relative priorities. The cost of purchasing relative priority weight β is β^α, where $\alpha > 0$. The authors prove that for any $\alpha \geqslant 1$ there exists an equilibrium such that all jobs of the same class choose the same priority level. The existence of an equilibrium when $\alpha < 1$ is supported by numerical computations but has no analytical proof. The authors also consider heavy-traffic approximations for which they bound the **PoA** by a function of arrival and waiting-cost rates.

Doncel, Ayesta, Brun, and Prabhu [205] (2014) consider an M/G/1 DPS model with atomic customers buying relative priorities.[7] The price of relative priority is proportional to weight, and a minimum allowable weight ϵ is given. The service capacity is normalized to 1, customers have heterogeneous expected service durations B_i and heterogeneous expected delay upper bounds c_i, and they buy minimum relative priorities that guarantee these requirements. A customer is *fair* if $R_i = \frac{B_i}{c_i} \leqslant 1 - \rho = 1 - \sum \lambda_j B_j$, which means that the QoS requirement would be obtained under EPS. The main results are:

- If the game is feasible then the social optimum is also an equilibrium.

- In the case of two customers, except for very a particular case, the equilibrium is unique. The authors present a closed-form solution in the case of exponential service. In particular, the equilibrium is (ϵ, ϵ) if the customers are fair.

- Assume $R_i = k < 1$ for all i. If $\rho \leqslant 1 - k$, then the unique equilibrium is the EPS solution $(\epsilon, \ldots, \epsilon)$ and if $\rho > 1 - k$, then the game is not feasible.

---

[7]See [275] for a related model where customers submit duplicate orders in a single-server queue.

- The authors provide a heavy-traffic approximation in the general case. This approximation is numerically verified to be accurate when all R_i values are similar.

Ali, Bodas, and Manjunath [27] (2014) consider equilibrium and **SO** sale of relative priorities when customers have heterogeneous waiting-cost rates. As expected, customers with higher delay costs buy higher relative priority weights in both solutions. The authors show that prices can be used to coordinate the system.

Monsef, Anjali, and Kapoor [488] (2014) approximate equilibrium routing of demand by K atomic users to parallel M/M/1 servers under a *generalized processor sharing* (GPS) regime: service capacity is divided proportionally to given weights $\phi_1, \ldots, \phi_K$ among users having requests in the queue.[8]

Oz, Haviv, and Puterman [508] (2014) show that relative priorities can reduce the equilibrium social cost in systems having flexible customers. In their models, a server serves at most two customer classes. Class i customers have waiting-cost rate C_i, server j is an M/M/1 system with service rate μ_j, and the total demand of class i is λ_i. Flexible customers select their server and servers that serve more than one customer class apply relative priorities. The authors consider two models. The *W-shaped model* has two customer classes and three servers, such that i-customers can obtain service from server i, $i = 1, 2$, or from server 3. The *M-shaped model* has three customer classes and two servers such that server j, $j = 1, 2$, can provide service to j-customers and to class 3.

The authors derive the expected waiting time of each customer class at each server for given priority parameters. The main results are:

- In the W model there exists a unique equilibrium routing of demand, social cost cannot always be minimized with absolute priorities.

- In the M model both servers apply relative priorities. For any set of relative priority parameters there exists a unique equilibrium routing of demand. At least one of the servers allocates absolute priorities in the **SO** solution.

5.4 Strategies using memory

A central planner can adopt a strategy of randomly admitting each arrival with probability 0.5, admitting every other customer, or admitting a customer

---

[8]This is in contrast to DPS where the fraction depends on the number of user requests in the system.

only if the time elapsed since the last admission exceeds a given threshold. The first strategy is more common in strategic queueing literature; the latter options may be advantageous but require a stronger means of control, i.e., using *memory*.

Lin [446] (2003) considers a Markovian unobservable single-server loss system with multiple user classes, each represented by a *gatekeeper*. The gatekeeper of class i cannot observe the server, the other arrival processes, nor the other gatekeepers. If the gatekeeper admits a customer and the server turns out to be idle, the customer enters service and the gatekeeper obtains reward α_i. If the server turns out to be busy, the customer balks and the gatekeeper pays β_i. If the gatekeeper blocks a customer, no cost or gain is incurred. In an equivalent formulation, the gatekeeper receives no reward when a customer is served, there is a unit cost when a customer is admitted while the server is busy, and there is a cost $c_i = \alpha_i/(\alpha_i + \beta_i)$ when the customer is blocked. The latter formulation has a single parameter, c_i, and its cost is a linear transformation of the former formulation. The objective of each gatekeeper is to maximize long-run average net gain (minimize long-run average cost).

The assumptions of exponential service and single server mean the only information a gatekeeper can use is the amount of time s elapsed since last admitting a customer.[9] A randomized strategy is a function $\pi : s \rightarrow [0,1]$ which gives the probability of blocking a customer arriving at state s. Under a pure threshold strategy (or, a *call-gapping policy*) with a threshold t (the *gap size*), $\pi(s) = 1$ if $s \leqslant t$ (the arrival is blocked), and $\pi(s) = 0$ otherwise.

The main results of the paper are:

- Regardless of the policies other gatekeepers use, the optimal policy of gatekeeper i is of the threshold type, which also includes the case $t = \infty$ (i.e., block all arrivals).

- A solution is found for the best response threshold of a gatekeeper given the thresholds used by the other gatekeepers.

- A counterexample with two (possibly symmetric) gatekeepers to the intuitive conjecture that the best response gap size is always monotone decreasing in the other gatekeepers' gap size.

- When gatekeepers are homogeneous there exists a unique symmetric equilibrium and, possibly, other asymmetric equilibria.

- A solution is derived for overall minimization of the system cost. As expected, the resulting threshold is greater (more customers are blocked) than the equilibrium threshold because admission of a customer is associated with negative externalities.

---

[9]The model resembles one with retrials, with gatekeepers playing a role similar to customers in orbit having information about the server being busy at a given time.

Anselmi and Gaujal [61] (2011) consider the use of memory in the routing of a single Poisson arrival stream to N servers with general heterogeneous service distributions by a *broker* who cannot observe the state of the queues. The traditional probabilistic memoryless broker attaches a routing probability to each arrival (for example, [85]), but here the broker attaches a routing probability that can change for different arrivals depending on the knowledge of previous dispatching decisions. The ratio of expected delays without and with memory is *the price of forgetting* (**PoF**).

- The authors provide a lower bound on the optimal expected delay.

- When service is exponentially distributed, **PoF**$\leqslant 2$.

- By definition, **PoA** in the system with memory is bounded above by **PoA** in the probabilistic system times **PoF**. Using the bound of N derived in [330] for the **PoA** for a probabilistic broker, it follows that with memory **PoA**$\leqslant 2N$.

5.5 Decentralized systems

This section considers coordination of different divisions of a firm inducing them to adopt overall optimal behavior. Similar models are surveyed in Chapter 9 on supply chains. The difference is often related to the motivating scenario. Chapter 9 describes models of competing agents in a supply chain and ways to coordinate the system while sharing profits. However, the distinction is often arbitrary. See [1] §4.6 for a decentralized model by Radhakrishnan and Balachandran (1995).

Wang and Barron [639, 640, 638] (1995, 1997, 2000) present queueing examples to variations of a decentralized model where operating costs depend on a random parameter θ which is the private information of the *information systems* (IS) department of an organization. The *central management* announces a strategy $(\lambda(\hat{\theta}), \mu(\hat{\theta}), T(\hat{\theta}))$ where $\hat{\theta}$ is the value reported by IS, and T is a transfer payment. IS reports $\hat{\theta}$ so as to maximize the sum of direct revenues and transfer payments minus operating costs. The objective of the central management is maximizing aggregate service value minus the sum of waiting costs and a convex combination of IS operating costs and transfer price.

Radhakrishnan and Balachandran [540] (2004) consider an FCFS system where user i generates demand at rate λ_i and has a waiting-cost rate v_i, $i = 1, \ldots, n$. Mean service time is $(s - k)$ and variance is $\sigma^2 = 1/p$. The parameter s is fixed, but the variables k and p are controlled by the queue manager at a price given by a function $C(p, k)$, that is separable in p and k. The system costs consist of waiting costs and the cost $C(p, k)$. A time interval is available for service and the expected waiting time is approximated by the Pollaczek-Khinchin formula.

Suppose an optimal pair (p, k) is chosen and user i is charged a fraction $v_i \lambda_i / (\sum v_j \lambda_j)$ of the cost $C(p, k)$. The authors show that with this cost allocation the optimal pair (p, k) would also be the expected utility-maximizing choice of each user, and that the allocation is the only one with this property.

Suppose now the arrival rates λ_i are private information and the queue manager sets the values p and k according to customer arrival rate declarations. The authors design an **IC** allocation mechanism that achieves truthful usage reports with the **SO** (p, k) values.

Nasrallah [499] (2006) suggests a model of *interaction value analysis* where members of an organization interact to maximize the organization's goals. The decision variables are the fractions p_{ij} of i-messages directed to user j in the organization. Messages possess predetermined priorities, expire at a common rate, and the benefit of a successful i-to-j message is h_{ij}. The organization sets routing probabilities to maximize the total value of successful messages. A decentralized solution is obtained when individual parties in the organization act to maximize the organization's goal by modifying their choices and react to the choices of other parties. The author numerically solves an example under different "organization climates" (denoted capitalist/disciplined/fraternal) and compares the self-organizing outcomes to the centralized optimum.

Lynn and Balachandran [460] (2007) consider an M/G/1 system with two divisions, $i = 1, 2$, which submit service requests at constant rates λ_i, obtain profits of g_i per request, and incur constant waiting costs of H_i per unit time. Requests of division 1 have priority over those of division 2. The service means s_i are fixed, but the firm wants to reduce waiting costs by controlling the service variance. Service with second moment $1/p_i(k)$ costs $C(k)$. The functions $p_i(k)$ are monotone increasing and concave and $C(k)$ is monotone increasing and convex where k is the control variable.

When the firm has full information on the divisions' parameters, the cost $C(k)$ is allocated to divisions in proportion to their expected waiting-cost rates. In this case the value of k which minimizes the firm's waiting and investment costs also does so for each division separately. The authors also investigate the effects of asymmetric information, in particular when H_i or λ_i is the private information of division i.

Erkoc, Wu, and Gurnani [228] (2008) deal with a general model of cost minimization with promised delivery time (PDT) and capacity buildup, where decisions are made by the marketing and engineering divisions of a firm. We describe here a queueing-related special case of their model where definitive results are obtained. Customers arrive to an M/M/1 system. Each arrival is quoted a state-dependent PDT. In one version of the model, capacity is built at the beginning of the planning horizon and cannot be changed. In the other version capacity can be dynamically controlled. Capacity costs are linear. If the firm quotes PDT l and service is completed at time x then the firm incurs a *tardiness penalty* $\tau(x - l)$ if $x > l$ and a *quotation cost* $\eta(l - x)$ if $l > x$, where $\eta < \tau$. PDTs are quoted and the associated quotation costs are paid by the marketing division, while capacity is decided and financed by the engineering division. Both divisions are responsible for the tardiness cost which they share in a given proportion. This setting generates a game between the two divisions, each minimizing its expected costs.

In general, the equilibrium solution of this game differs from the firm's optimal policy. When capacity is static, the authors give an explicit solution and show that:

- The optimal and equilibrium PDTs are independent of arrival rate.

- The deviation between optimal and equilibrium capacity levels declines with the arrival rate.

- The system can be coordinated in both dynamic and static cases by instituting transfer payments from the marketing to the engineering divisions.

Pekgün, Griffin and Keskinocak [525] (2008) analyze the inefficiencies created by decentralized decisions in an M/M/1 service firm. Demand is a linear function of the price p set by the *marketing* division and the lead time L quoted by the *production* division. The firm incurs a fixed unit production cost. The setting is a Stackelberg game where one of the departments serves as leader and the other follows. Two scenarios are considered, each with a different leader. In both cases, production acts to satisfy the quoted lead time, which marketing ignores. Moreover, production is motivated to maximize the firm's net profits whereas marketing only considers income and ignores unit production costs.

The firm's profit turns out to be greater when marketing leads. However, this profit is still smaller than would be in a centralized solution. Therefore, the authors suggest contracts such that from every served customer, marketing obtains $\alpha_1 p - w$ and production obtains $\alpha_2 p + w$, for some w and $\alpha_1 + \alpha_2 \leqslant 1$. They prove that contracts that guarantee the centralized optimal solution exist for each of the leader scenarios,

Xu, Dai, Sycara, and Lewis [674] (2012) consider a Markovian model where a *principal* assigns demand to an *operator*. The *quality* θ of the operator is high with given probability p, and low with complementary probability $1 - p$. A θ-operator operates at rate μ_θ, where $\mu_H > \mu_L$. The value of θ is not observable by the principal who bases its decision on the type $x(\theta) \in \{H, L\}$ claimed by the operator. The principal then assigns the operator demand at rate $\lambda(x)$ and payment $I(x)$ per unit time. The principal maximizes the expected value of $\lambda(x) - \frac{C}{\mu_\theta - \lambda(x)} - I(x)$, and the decision variables are $\lambda(x)$ and $I(x)$, $x \in \{H,L\}$.[10] The operator reacts to the offered rates and prices maximizing utility $U(x,\theta) = I(x) - \frac{e \cdot \mu_\theta}{\mu_\theta - \lambda(x)}$ for a constant e. The ideal benchmark solutions cannot be achieved in this decentralized setting. To achieve a truth-telling equilibrium, the H-type operator is given an extra rent cost to prevent him from mimicking an L-type and obtaining less assigned jobs. The authors derive an explicit solution for the optimal **IC** strategy by the principal.

5.6   Systems with public and private service facilities

Stenbacka and Tombak (1995) (see [1] §8.2) consider a market with one private server and one public server, and examine the effects of privatizing the public server. See [283] for a model where customers choose between a private firm and a public multiserver system that provides cost-free service and inactivates some servers at periods of low congestion.

Larsen [428] (2005) considers a market where customers face two service options: either be served at no cost by a public M/M/1 system where queueing is associated with a linear cost function, or be served by a profit-maximizing *subcontractor* that provides immediate service, but for a fee. Customers choose where to go in a decentralized way. The system performance measure is the expected full price for the customers. The equilibrium consists of a joining probability in the unobservable case and a threshold joining strategy in the observable case, and a price set by the subcontractor.[11]

The main results are:

- The subcontractor's profit is higher when the public server is observable and customers act independently. In this case, the subcontractor obtains lower demand but charges a higher price.

- The existence of the subcontractor increases average customer welfare in both models.

- No definitive conclusions can be made as to whether one of the settings leads to higher customer welfare over the other.

---

[10]The authors' goal was to design a system with both high throughput and low waiting time so the objective function is a combination of these two measures. The first two terms have different dimensions. Multiplying the constant C by expected queue length rather than by expected waiting time seems a reasonable alternative.

[11]The author's terminology is different from ours, referring to the equilibrium decision in the unobservable case as "centralized" and to the performance measure as "total social cost" although it includes transfer payments to the subcontractor.

Guo, Lindsey, and Qian [279] (2015) assume customers choose between a toll-free public facility and a congestion-free private facility charging P per service. A central planner controls a budget B to minimize the sum of customer costs.[12] The budget B is treated as a sunk cost.

Two subsidy schemes are compared.

Conditional subsidy: customers who wait at the public system τ units of time are sent to the private server and their service fee is borne by the planner.

Unconditional subsidy: An amount $S \leqslant P$ is given to any customer who chooses the private server. The paper provides insights about the preferred subsidy scheme when the public queue is observable and when it is not, and about the impact of providing delay information to customers. For example:

- When the public queue is unobservable the best choice of subsidy type depends on the available budget. Conditional subsidy is the preferred option when the budget is large.

- When the queue is observable, unconditional subsidy is preferred.

- Providing delay information to customers improves customer welfare.

Andritsos and Tang [56] (2013) consider a model motivated by healthcare insurance applications. This is, sort of, a supply chain consisting of a *funder* that outsources treatment of patients to a nonstrategic M/M/1 *public hospital*. The utility of a patient is $U = V - \alpha d$, where $V \sim \mathrm{U}[0,1]$ is the value of treatment, d is the expected delay, and α is the waiting-cost rate. Patients seeking treatment are those for whom $U \geqslant 0$. An exogenous average *welfare requirement level* τ must be achieved, i.e., the rate of surplus generated by the public system must be at least τ. The funder's decision variables are the price paid to the hospital per treated customer and the hospital's capacity. The authors derive explicit formulas for the values of these variables that minimize the funder's total cost rate.

Suppose now a patient can obtain treatment at a *private provider* without delay. The private provider incurs a fixed cost κ per customer and charges the customer a price p, which is not covered by the funder. The authors derive the equilibrium solution in this game between the funder and the private provider; they conclude that the entry of the private provider reduces waiting time at the hospital and that the funder's costs may increase or decrease. The possibility that the funder's costs increase when the private firm enters the market might look surprising. This results from the economies of scale inherent in the queue and the requirement to achieve a given total surplus.

Andritsos and Aflaki [55] (2015) assume a Hotelling type model with two competing M/G/1 servers located at the ends of the interval. The utility

---

[12]Customer costs consist of queueing costs at the public system and payments to the private system. This is different from social welfare which doesn't consider the transfer payment P as a cost.

of a customer located at l and served by a server located at l_h is $U = V - \alpha W - \gamma p - \tau|l - l_h|$, where V is the service value, W is expected waiting time, and p is the price. The authors compare two scenarios:

- **Non-profit servers:** Servers obtain a reimbursement amount r from a public funder per customer served. Servers incur heterogeneous operation costs per customer and unit capacity costs. The server's goal is to set capacity for achieving maximum total customer welfare subject to a self-financing constraint. The authors show there exists a unique equilibrium and derive a closed-form formula for it.

- **A non-profit and a for-profit server:** The waiting time at the profit-maximizing server is negligible, but a price for service is charged. The server also obtains from the public funder a subsidy $s \leqslant r$ per customer. The authors characterize the equilibrium and provide an explicit formula for the M/M/1 case.

The authors conduct sensitivity analyses for each scenario and compare the cases. They obtain several interesting results, including the following:

- There exists a threshold $\bar{r}$ of the reimbursement r such that the expected waiting times with two non-profit servers is lower than in the mixed case iff $r \leqslant \bar{r}$.

- For low rates of demand, non-profit competition yields a lower total customer cost. Otherwise, there exist thresholds $r_1 \leqslant r_2$ such that mixed competition yields a lower total customer cost iff $r_1 < r < r_2$.

Guo, Lindsey, and Zhang [280] (2014) consider a Markovian *two-tier service system* where customers with homogeneous linear delay costs choose one of two unobservable queues: a private queue for a profit-maximizing server and a public toll-free queue for a server with a fixed capacity μ_f. Balking is not an option, but it is assumed that $\mu_f > \lambda$ so that the public server can serve the entire population. There are two decision variables related to the private server: The service rate, μ_c (that comes at a linear cost $c\mu_c$), and the price, p.

- The authors derive an explicit **SO** solution such that the toll service is *self-financing*, i.e., covers its expenses. Three types of solutions are possible: All customers use the toll service, both systems are used, or all customers use the toll-free service.

- The maximal social welfare of a self-financing system is *decreasing* in the public capacity μ_f. This is the counterpart to the Downs-Thomson paradox in transportation systems.

5.7   Cooperation in service systems

5.7.1   Sharing of revenues and costs

Our interest in this section is in models where queue operators are willing to cooperate by pooling resources and sharing costs. Most of these papers assume customers are not strategic and waiting costs, if they exist, are incurred by the servers. Some papers on supply chains with a common supplier are relevant here, for example [35], as well as the concept of economies of scope. Also, [682] describes a situation where two users share the costs associated with lost demand. Models of competition under limited cooperation are described in §7.3. See §8.1.2 for routing decisions under partial cooperation, [558, 559] for a model where *customers* share information, [75, 168] for resource sharing under competition, and [92, 93, 527] for cooperation in routing games.

Haviv [319] (2001) considers a single-server system with n customers generating Poisson demand requests at rates λ_i, $i = 1, \ldots, n$. Customer service requirements have different first and second moments. Customer cost is defined as the mean number of requests in the system.

The author considers ways of sharing system costs. Specifically, charging customers for the congestion costs they impose according to the *Aumann-Shapley mechanism*. The author examines four types of queue disciplines, FCFS, LCFS with and without preemption, and EPS, and computes for each the resulting cost allocation. The special case with only two customers, where one has a zero arrival rate, suggests a charging scheme in the single-class M/G/1 queue.

González and Herrero [259] (2004) consider n agents running M/M/1 systems with arrival and service rates λ_i and μ_i. Agent i is committed to a guaranteed expected waiting time $(\mu_i - \lambda_i)^{-1} = t_i$. The capacity cost is linear and, without loss of generality, is $c(i) = \mu_i = \frac{1}{t_i} + \lambda_i$. It is assumed that if a subset S of agents cooperate and maintain a common server, this coalition must commit to expected waiting time $t^S = \min_{i \in S} t_i$, and their capacity cost will be $c(S) = \frac{1}{t^S} + \lambda(S) < \sum_{i \in S} \frac{1}{t_i} + \lambda(S) = \sum_{i \in S} c(i)$, where $\lambda(S) = \sum_{i \in S} \lambda_i$. Therefore cooperation reduces costs. We note that this property is not intuitively obvious; on one hand there are economies of scale which coalition S enjoys, but on the other hand it provides better but costly service to its aggregate demand. The authors observe that the core of this game is not empty and provide a core allocation based on the Shapley value.

Hayel and Tuffin [335] (2006) apply the Aumann-Shapley price mechanism to Mendelson and Whang's model [487] with homogeneous service rates. In particular, they determine which fraction of the total congestion must be borne by each of the customer classes and show that these prices are **IC**.

The prices are derived also for heterogeneous service rates and for a related network model.

García-Sanz, Fernández, Fiestras-Janeiro, García-Jurado, and Puerto [246] (2008) analyze variations of the model in [259]. In the first one, the commitment of each agent specifies delay standard ω_i and reliability level α_i. This is equivalent to $\mu_i = \lambda_i - \frac{\ln \alpha_i}{\omega_i}$, and the cost of a coalition S is $c(S) = \lambda(S) + \max_S \left\{ \frac{-\ln \alpha_i}{\omega_i} \right\}$. As in [259], the core of this game can be fully described and the authors present a core allocation based on the Shapley value. The authors interestingly observe that if the delay guarantee refers to queueing time only (excluding service time), agents may prefer not to cooperate. In another variation, preemptive priority schemes are allowed to exploit the fact that different players have differing commitments to their populations. The authors prove that the use of preemptive policies allows for fulfilling the requirements of all customer types at a lower cost. Again, the existence of a core allocation is constructively proved.

A cooperative game is *totally balanced* if the core of all its subgames is non-empty. **Anily and Haviv [57] (2010)** assume that by pooling queues a coalition S serves the demand $\sum_{i \in S} \lambda_i$ at rate $\sum_{i \in S} \mu_i$. The coalition cost is the mean number of customers in the pooled system. The game is neither monotone nor concave. To prove the game is totally balanced the authors define an auxiliary game that is both monotone and concave, and its core coincides with the non-negative parts of the core of the original game. In addition, except for degenerate cases, there exist infinitely many core allocations in which at least one server pays some of the other servers to persuade them to join the coalition. The authors introduce further insights into this model in [59].

Özen, Reiman, and Wang [510] (2011) consider two types of queueing systems. A *type-μ system* has a fixed number of identical servers and capacity can be varied by adjusting the service rate. A *type-K system* has a fixed service rate and its capacity can be controlled by changing the number of servers. The authors also consider two cooperation modes. Under the *resource sharing mode* cooperating operators combine their service capacities and optimize the amount of demand to be served. Under the *demand pooling mode* operators combine their customer bases and optimize the system capacity in which they invest.

- The authors consider an allocation scheme based on distributing coalition profits in proportion to each participant's contribution and provide sufficient conditions for it to be a core element. This is the case in both cooperation modes in type-μ systems if the underlying queueing model is Erlang-B (M/G/s/s model using blocking rates as the performance measure) or Erlang-C (M/M/s model using the probability of waiting, mean delay, or the probability that delay exceeds a given threshold as the performance measure). Similarly, this is the case in resource sharing games in type-K systems.

- The core of a demand pooling game may be empty in type-K systems and the authors present a sufficient condition for proportional allocation to be in the core when the underlying model is Erlang-B.

Anily and Haviv [58] (2013) consider a class of cooperative games called *regular games*, where each player is associated with a vector of quantitative properties and the cost of a coalition is a function of its vectors of properties but otherwise is independent of player identities, as in the model of [57]. They define a class of regular games called *regular market games* which are totally balanced and possess core allocations. Three queueing examples falling within this class are demonstrated:

- Servers have dedicated demand streams and customers arriving to a busy server are lost. The servers in a coalition can reallocate service capacity but they cannot redirect demand. They can also rent out capacity for a given fixed price per unit time. The cost of a coalition is the cost of lost servers minus revenues from capacity rental.

- Server i owns an unobservable M/M/1 queue with demand and service rates λ_i and μ_i. Waiting cost rates are homogeneous. A coalition can redirect demand among its members and also outsource some demand at a fixed unit cost. The cost of a coalition consists of waiting and outsourcing costs.

- Similar to the previous model, but instead of redirecting customers among servers the capacity is reallocated among the servers. In addition, the authors allow the option of shifting some of the capacity to other tasks in the firm. The cost of a coalition consists of the total congestion minus a linear savings per capacity unit that is not used by the coalition.

Timmer and Scheinhardt [623] (2013) consider cooperation among N service providers while preserving the autonomy of the individual queues. It is assumed that demand for server i is λ_i, and initial capacity is μ_i. Cooperating servers may redistribute capacities among each other. The cost $c(S)$ of a coalition S is the minimum sum of delays of its customers over the possible allocations of aggregate capacity $\sum \mu_i$.

The authors describe a mathematically equivalent description of the system as a Jackson network for which $c(S) = \dfrac{(\sum_{k \in S} \sqrt{\lambda_k})^2}{\sum_{k \in S}(\mu_k - \lambda_k)}$. In particular, when $\lambda_k = \lambda$ for all $k \in S$, the minimum cost is obtained by equally distributing the capacity $\sum_{i \in S} \mu_i$ among its $|S|$ members, and $c(S) = |S|\lambda/(\bar{\mu}_S - \lambda)$, where $\bar{\mu}_S = \sum_{i \in S} \mu_i/|S|$. The main result is the explicit formulation of fair cost allocation x, which also shows that the core of the game is not empty. In the special case of equal arrival rates, the following cost allocation is in the core: $x_i = \left(2 - \dfrac{\mu_i - \lambda}{\bar{\mu}_N - \lambda} \right) \dfrac{c(N)}{|N|}$.

Ernez-Gahbiche, Hadjyoussef, Dogui, and Jemaï [230] (2014) analyze a Stackelberg game with a single customer and multiple MTS M/M/1 suppliers. A *coalition structure* is a partition into *coalitions* of the supplier set, each pooling service capacity. For a given coalition structure, the customer leads a Stackelberg game allocating demand to coalitions while anticipating their resulting behavior. The customer can refuse to participate in a game that doesn't guarantee nonnegative utility or otherwise must allocate the entire demand. Given the allocation, each coalition selects a base-stock level (or refuses to participate). The profit of a coalition is distributed among its members in proportion to their capacities. The customer gains a fixed reward from service and incurs backorder costs partly reimbursed by the coalition. In addition to backorder compensation the coalition incurs unit production costs and holding cost rates while earning a fixed reward per served unit.

The authors numerically solve an example where the size of coalitions in stable structures grows with system load, and the grand coalition remains stable only for high loads.[13]

Karsten, Slikker, and van Houtum [391] (2014) extend results for the M/G/s/s type K model of [510] allowing for the cost rate $H(s)$ associated with operating s servers to be concave increasing and unbounded. They also treat a variation where blocking penalty costs are replaced by a maximal blocking probability constraint, and prove that allocation of costs in proportion to the server's own demand is in the core of the game.

Karsten, Slikker, and van Houtum [392] (2015) consider a set of players, each associated with an M/M/s system with a given rate of demand. Players form a coalition by pooling servers and demand.

- In the FIX-queueing game, each player brings a predetermined number of servers to a coalition. The coalition's costs consist of a fixed cost per server and a fixed cost rate per customer in the system. The game is shown to be strictly subadditive and therefore servers benefit from forming coalitions. It also benefits all customers as a whole, who experience less expected delays in total when a larger coalition is formed. A core allocation is guaranteed to exist for this game. Moreover, if the ratio of servers to arrival rates is the same for all players, then a cost allocation where each player pays proportionally according to personal demand is shown to be in the core and nonincreasing in the size of the coalition.

---

[13]Stability is defined assuming coalitions are *farsighted* and consider the possibility that "once they act, another coalition may react, a third coalition might in turn react, and so on, deterring then the first move."

- In the OPT-queueing game, a coalition chooses the optimal integer number of servers to satisfy demand. The cost of employing a given number of servers is nonincreasing as the coalition grows. The authors demonstrate that the core of this game may be empty but there are approximately stable allocations.

Yu, Benjaafar, and Gerchak [689] (2015) consider a cooperative game among n firms each with its own dedicated demand. Firms set their service rates, each paying the same linear capacity cost. However, firms have heterogeneous holding cost rates and are subject to heterogeneous delay standards, namely, the probability that the delay is below a given value w_0 is at least α_i for firm i. A coalition pools the demand of its members and serves them in FCFS order while being subject to the maximal delay standard of its members. A crucial observation is that the *extra capacity*, $\mu - \lambda$, required to fulfill a delay standard is independent of λ. Therefore, the extra capacity for a coalition is the same as the maximum extra capacity required independently for any of its members. Consequently, as in [259], although the shared facility guarantees a better standard relative to what these customers would obtain without pooling, this is done with smaller capacity and cost. The authors provide an explicit cost allocation that belongs to the core of the game.

The advantages of allowing the coalition to prioritize customers according to their original affiliations are discussed under a more general setting where the bound w_0 is replaced with a bound w_i for firm i, $i = 1, \ldots, n$. The resulting game is shown to be submodular and a core allocation is derived.

As in [246], these results do not hold for variations of the model, for example when the guarantee constraint is on queueing time rather than on sojourn time.

5.7.2 Pareto optimality

The common assumption is that social utility is additive and hence the objective of the central planner is to maximize the sum of individual utilities. A weaker notion is that of Pareto optimality, which states no user can be made better off without hurting some other user.

Courcoubetis and Varaiya [175] (1983) consider a Markovian model of a closed system with two players sharing an FCFS server and maximizing *resource utilization*, which is the time they spend in service. The service rate μ_i and expected time from a service completion to submission of a new service request $1/\lambda_i$ are the ith player's decision variables, but their ratios $\rho_i = \lambda_i/\mu_i$ are exogenously fixed. If player i were alone, the proportion of time in service is $u^i = \frac{(1/\mu_i)}{(1/\lambda_i)+(1/\mu_i)} = \frac{\rho_i}{1+\rho_i}$, which is independent of the decision variables. The authors show that the sum of resource utilizations is maximized when $\frac{\mu_2}{\mu_1} = \sqrt{\frac{\rho_1}{\rho_2}}$. However, for any values (μ_1, μ_2) the resulting utilizations (u^1, u^2) are Pareto maximal. That is:

- No cooperation between players can simultaneously increase both utilizations.

Shenker [570] (1995) considers a queue in which N users select demand rates. The utility of user i is a convex function $U_i(\lambda_i, c_i)$, increasing in λ_i and decreasing in c_i, where c_i is the user's average queueing demand. The queue manager wishes to induce the socially best equilibrium from the interior of the achievable set[14] by selecting a queue discipline from a class of work-conserving disciplines that include FCFS, LCFS, EPS, and polling. The first result states that no discipline in this class is guaranteed to induce a Pareto optimal solution for every convex utility function.

The author examines a *fair share service discipline* that is best explained by the following example. Consider three users choosing rates $\lambda_1 < \lambda_2 < \lambda_3$. Then, all users get highest (preemptive) priority on a portion of size λ_1 of their demand. Users 2 and 3 get second priority on a portion of $\lambda_2 - \lambda_1$ of their demand, and the third user obtains the lowest priority on the last portion of $\lambda_3 - \lambda_2$ of its demand. This mechanism has desirable features. For example, it is the only one in the class of disciplines considered here which guarantees a unique equilibrium for every admissible utility function.[15]

Liu and Simaan [454] (2005) consider a two-class model of routing demand to parallel M/M/1 servers with a finite number of users in each class. *Class objectives differ from those of their members.* While users wish to minimize expected wait, the class objective is either to maximize the class demand served by the fastest server, or for a given set of server- and class-dependent service fees to minimize the total cost of its users. The authors offer a solution concept called *Non-inferior Nash Strategy* (NNS), which means that the routing strategies adopted by the classes define an equilibrium of the class games and, in addition, the routing of a class must be Pareto optimal with respect to its individual users' objectives.

5.7.3 Bargaining solutions and routing

See [199] for a related model in a system with breakdowns.

Mazumdar, Mason, and Douligeris [480] (1991) look at random loop-free routing in a Jackson network of M/M/1 queues. There are N players, player i wishes to maximize the *negative inverse power* $-P_i^{-1}$ where the power P_i of player i is defined as the ratio between throughput and delay (see §4.6.1). The authors show that the flow scheme S^* which maximizes the product of

---

[14]I.e., absolute priority is not given to any subset of users.

[15]The fare share service discipline is similar in some aspects to the token system of [226]. In both cases different parts of a customer's demand are given different priorities, but the way in which they are used is different.

user powers corresponds to a Nash bargaining solution with respect to the initial agreement point $-aP_i^{-1}(S^*)$, $i = 1, \ldots, N$, for a being sufficiently large.

Dziong and Mason [212] (1996) examine an M/M/s/s loss system with two users, where user j sends a Poisson(λ_j) stream of connection requests (calls), each requiring d_j servers (channels) and service of duration exp(μ_j). The user receives utility $\bar{\lambda}_j d_j / \mu_j$, where $\bar{\lambda}_j$ is the acceptance rate of j-type calls. The system state is defined by the number and type of connections established, and the admission strategy is state-dependent.

From amongst the Pareto efficient strategies, a *fair* one is desired. For this purpose it is assumed that players have preference functions that depend on their own utility as well as on that of other players. The dependence on others' utilities can be positive,[16] negative, or non-dependent.[17]

Traditional solutions maximize either the average or the minimum utility among all players. The authors consider several bargaining solutions and compare them to each other and to the traditional strategies of (i) *complete sharing policy*, in which a new call is accepted if there is sufficient free capacity; (ii) *coordinate convex policies*, in which the maximum number of type j calls in the system is bounded; (iii) *trunk reservation policies*, where a jth type call is rejected if the number of free trunks is smaller than a threshold; (iv) *dynamic trunk reservation policies* where these thresholds are state-dependent.

Cao, Shen, Milito, and Wirth [127] (2002) consider a Stackelberg game between an M/M/1 server and a user. The user generates service requests with random *maximal acceptable response time s*. If service of an s-request is completed within the required time s, the user gains an amount $g(s)$. Otherwise there is no gain or loss. The server leads the game by announcing the service fee, and the user reacts by choosing a range of s values of requests it submits for service. The server maximizes revenue while the user maximizes net gains (after paying for service).

The authors present examples for a cooperative version of the model with a parameter-dependent bargaining solution concept that generalizes the Nash bargaining point and where the disagreement point is either the solution of the Stackelberg game or the origin $(0,0)$.

Also offered is a noncooperative version with the same demand model and two competing servers with heterogeneous capacities. The user applies a two-thresholds policy such that small s-valued requests balk, high values join the less congested server, and the rest join the more congested server.

La and Anantharam [424] (2002) consider a *repeated game* where each repeated *stage* is equivalent to the game analyzed in [506]. Thus, cooperation

---

[16]Positive dependence expresses a form of altruism.

[17]The authors use a generalization of several well-known arbitration schemes. The case of independent utilities corresponds to the Nash bargaining point.

can be enforced by policies that penalize users if they deviate from the desired behavior. The main results are:

- When the network consists of parallel links there exists a subgame perfect equilibrium that achieves the **SO** solution and no user is worse off relative to the unique stage-game equilibrium.

- In general directed networks with a common source-destination pair there exists a subgame-perfect equilibrium that is also **SO**. This property does not always hold when source-destination pairs are different.

Grosu, Chronopoulos, and Leung [263] (2008) compute the Nash bargaining solution when N M/M/1 servers with heterogeneous service rates $\mu_1 \geqslant \cdots \geqslant \mu_N$ allocate demand of rate Φ amongst themselves. Servers minimize their rate of demand and at the disagreement point incur infinite cost. The following algorithm computes the bargaining point: Let $n = N$. While $c_n = \frac{1}{n} \left(\sum \mu_j - \Phi \right) > \mu_n$ set $\lambda_n \leftarrow 0$ and $n \leftarrow n - 1$. Lastly, set $\lambda_i = \mu_i - c_n$ for the n remaining servers. The authors observe that the resulting overall expected sojourn time is not very far from the **SO** value derived in [85].

Subrata, Zomaya, and Landfeldt [598] (2008) compute the Nash bargaining point in a game routing given demand to M/G/1 servers with heterogeneous service distributions, communication costs, and predetermined committed guarantees on the sum of communication and waiting costs. In this game, servers wish to minimize the expected delay of demand.[18] In [600], the authors consider a variation of the game where M/M/1 servers with a common *target response time* and given initial demand cooperate by shifting demand. The servers minimize energy costs equal to squared service rates μ_j^2.

Penmatsa and Chronopoulos [527] (2011) consider n M/M/1 servers with heterogeneous service rates μ_i and initial demand rates ϕ_i. Demand can be sent for processing from one server to another resulting in effective demand rates λ_i such that $\sum \phi_i = \sum \lambda_i$. The total rate of demand sent among servers $\lambda = 0.5 \sum |\phi_i - \lambda_i|$ is processed by an M/M/1 *communication subsystem*. The goal of servers and the communication subsystem is to minimize expected delays. The authors compute the bargaining point of the game with the disagreement point corresponding to $\lambda = 0$. They assess the quality of the solution according to a fairness index that measures the equality of response time at different servers: $\left[\sum D_i \right]^2 / \left[n \sum D_i^2 \right]$, where D_i is the expected delay at server i.

Experimental results indicate the solution is not much lower in efficiency but provides more fairness relative to the **SO** solution.[19]

---

[18]See [599] for a numerical study of an algorithm for computing an equilibrium in the non-cooperative version of this model.

[19]See §8.1.1 for a summary of a non-cooperative version of a routing model given in this paper.

5.8 Efficiency and price of anarchy

Bell and Stidham [85] (1983) (see [1] §3.7, [594] §1.5) were the first to solve the equilibrium and optimal joining strategies for customers selecting a server in an unobservable multiserver Markovian system with heterogeneous service rates, $\mu_1, \ldots, \mu_n$.[20] **Stidham [592] (1985)** uses these results to bound the **PoA** in heavy traffic, that is, in the limit as $\lambda \to \mu$, showing that the equilibrium choice can be worse than the **SO** assignment by as much as a factor of n, and that this bound is tight. The worst case is obtained when there is one very fast server and $n - 1$ very slow servers.

Friedman [238] (2004) constructs an example that, as in [592], has many slow servers and one fast server. The author does not assume heavy-traffic conditions, but the number of servers is arbitrarily large. Specifically, the example considers n servers with service rate 1 and a single fast server with service rate n. The arrival rate is $\lambda = n - 1$. In equilibrium, all customers select the fast server with expected waiting time of one unit and an aggregate expected waiting-time rate of $n - 1$. The optimal solution allocates an arrival rate of $n - \sqrt{n}$ to the fast server and $\frac{\sqrt{n}-1}{n}$ to each slow server resulting in an aggregate waiting-time rate of $\frac{n-\sqrt{n}}{n-(n-\sqrt{n})} + \frac{\sqrt{n}-1}{1-\frac{\sqrt{n}-1}{n}} = O(\sqrt{n})$. Therefore, **PoA**$\approx \sqrt{n}$.

Haviv and Roughgarden [330] (2007) modify the example of [238] to obtain **PoA** that asymptotically approaches the number of servers, n, and prove this is the worst possible example. **Wu and Starobinski [663] (2008)** independently obtain a similar result. They also extend their results to G/G/1 queues with high loads.

Johari, Mannor, and Tsitsiklis [374] (2005) analyze a general *network resource allocation game*. We simplify the description here and adapt it to their queueing application. Consider an M/M/1 facility with a fixed service rate μ and R customers. The utility customer r derives from submitting service requests at rate λ_r is a concave function $U_r(\lambda_r)$, and customers incur waiting costs c per unit time in the queue.

Given complete knowledge and centralized control of the system, the goal is to set quotas $\lambda_1, \ldots, \lambda_R$ maximizing $\sum U(\lambda_r) - c\lambda/(\mu - \lambda)$, where $\lambda = \sum \lambda_r$. The authors consider two models of decentralized systems where the utility functions are private information.

The first is a bounded-rationality model where quotas are set by the following auctioning mechanism: Each customer $j = 1, \ldots, R$ declares the total

---

[20]The routing formulas have often been rediscovered since.

amount w_j he is willing to pay. Given the bids $\mathbf{w} = (w_1, \ldots, w_R)$, the manager sets a price $p(\mathbf{w})$ per unit of allocated quota. All users are treated alike and are charged the same price. Hence the quota allocated to customer j is $\lambda_j(\mathbf{w}) = w_j/p(\mathbf{w})$. Assuming customer j optimizes expected net utility $U_j(\lambda_j) - w_j$ *ignoring the effect his choice has on price* and assuming the manager sets a price equal to the marginal cost, i.e., $p(\mathbf{w}) = c\mu/(\mu - \lambda(\mathbf{w}))^2$, there exists a unique equilibrium solution.

In the second model, customers are rational and anticipate the effect their strategy has on allocation. As in the first model, the manager sets the price equal to the marginal cost and $\lambda(\mathbf{w})$ is set as before. A unique equilibrium exists under these assumptions.

Denote by S^* the solution value of the full information centralized problem, and by S the equilibrium solution value in the second model. Then the **PoA**, S^*/S, is bounded by a constant (≈ 3).

Mazalov, Monien, Schoppmann, and Tiemann [479] (2006) consider a directed network with source s and sink t. A demand of rate r has to be routed from s to t along the $s - t$ paths. The cost of a path is the maximum delay of an edge along the path. Their main results are:

- A characterization of $s - t$ *series-parallel graphs* (assuming that every edge is contained in a simple $s - t$ path): *A directed graph is $s - t$ series parallel iff for any edge-latency function the equilibrium social cost is unique.*

- The price of stability is attained on parallel links.[21]

- The authors derive the price of stability in the case of M/M/1 latency as a function of the number of parallel links and demand rate. For a large r this converges to the number of parallel links, as in [330, 663].

Acemoglu and Ozdaglar [6] (2007) consider a general model, but we restrict the description to its queueing application. Users decide how much of their demand to allocate to each of a group of parallel servers.[22] The value of each served unit is constant and the delay cost at server i is a convex nondecreasing function $l_i(x_i)$ of the rate x_i of served demand.[23] Each of a set of service providers sets prices for a subset of servers it owns. Clearly, if all the servers are owned by a single owner this will extract all customer surplus and result in a **SO** solution. In general, the equilibrium solution is socially suboptimal, and the authors prove that **PoA**=$[2(\sqrt{2} - 1)]^{-1} \approx 1.21$.

---

[21] In this case the model reduces to that of [85].

[22] While describing the demand, the authors refer to small users but actually mean a single user with splittable demand, as seen in their Equation (1).

[23] We refer here to §6 in [6] that allows $l_i(0) > 0$, which is the case when the latency represents the sojourn time in a queue. The authors' tightness proof of this bound uses a linear non-queue latency function.

Czumaj, Krysta, and Vöcking [179] (2010) consider a finite number of customers routing demand to parallel servers. The demand of a customer is unsplittable and must be routed to a single server. The authors show that when the social objective is either the sum or the maximum of the delay costs across the queues, the **PoA** is unbounded. Therefore, they consider two variations of *bicriteria* **PoA**. One variation measures the factor by which the demand must be decreased so the equilibrium cost matches the original optimal cost. The other measures how much the servers' capacities must be decreased so that the optimal cost matches the original equilibrium cost. The authors provide bounds for both values. Other results concern many customers with small (but atomic) demand, heterogeneous customers, and loss systems. The authors conclude that rejection of demand requests in the case of an overload is necessary for the **PoA** to be bounded.

Altman, Ayesta, and Prabhu [44] (2011) assume Poisson demand routed to processor-sharing servers having heterogeneous service rates and customer waiting costs. Customers are homogeneous except for the size of service requirement, which is a random variable. The service requirement is known to the dispatcher in the centralized case but is private information in the decentralized case. The social objective is to minimize expected waiting cost in the system. The authors prove that, unlike the result in [330] where the waiting cost is the same at all servers, here the **PoA** is unbounded.

Ayesta, Brun, and Prabhu [76] (2011) investigate the dependence of the **PoA** on the number of customers K.[24] They consider unobservable single-server queues with heterogeneous service and waiting-cost rates. Customer i allocates a Poisson(λ_i) stream of jobs amongst the queues seeking to minimize waiting costs. Existence and uniqueness of the equilibrium is guaranteed by [506]. The main result is that the **PoA** is of order $\sqrt{K}$ independent of the number of queues. Interestingly, the worst performance is obtained when $\lambda_i = \lambda_j$ for all i and j.

Doncel, Ayesta, Brun, and Prabhu [206] (2014) refine the analysis of [76] to situations where there might be many servers, but only a few varying service rates, and homogeneous waiting costs. As observed in [85], self-interested customers overload the fast servers. The authors demonstrate that for given service rates, the **PoA** is non-monotone as a function of the system utilization with peaks corresponding to traffic intensities when the equilibrium solution requires more active servers.

Specific results obtained with two types of servers are:

- The equilibrium solution is close to optimal in most cases. Bad cases occur when the slow servers are numerous and very slow.

---

[24]This is in contrast to other papers, like [238, 330, 663], that focus on the dependence of the **PoA** on the number of servers.

- Let λ denote aggregate demand rate and assume service rates are fixed. There are (explicitly derived) thresholds, $\lambda^{\text{OPT}} < \lambda^{\text{NE}}$ such that when $\lambda \leqslant \lambda^{\text{OPT}}$ both equilibrium and **SO** solutions use only fast servers. Only the **SO** solution uses slow servers when $\lambda^{\text{OPT}} < \lambda < \lambda^{\text{NE}}$, and both solutions use fast and slow servers otherwise. **PoA** is 1 in the first interval, increases in the second, and decreases in the third. In particular, **PoA** reaches its maximum at λ^{NE}.

- With two types of servers and K customers, $\textbf{PoA} \leqslant \frac{K}{2\sqrt{K-1}}$. This bound can be achieved asymptotically when the number of servers is increased.

Anselmi, Ayesta, and Wierman [60] (2011) consider a system of N parallel profit-maximizing servers with heterogeneous service rates. The service value is assumed to be very high so price and lead times do not affect aggregate demand, but do affect on customers in their selection of a server. Customers joining servers with minimum full price. Even with homogeneous servers a price equilibrium need not exist. For example, no equilibrium exists when the system is heavily loaded such that all providers need to be used to keep congestion cost finite. The authors do however derive sufficient conditions for a unique price equilibrium to exist. They also show that the **PoA** is in general unbounded, but when an equilibrium exists and $N \to \infty$, **PoA**$\to 1$.

Knight and Harper [406] (2013) consider the routing of non-atomic customers at n distinct locations to service facilities. Service value and transportation costs depend on customer location, but waiting-cost rates are identical for all customers. Instead of net utility maximization, the authors consider cost minimization, where the service value is considered as a balking penalty. This modeling assumption greatly affects the definition and size of the **PoA**: When demand grows to infinity (such that most customers must balk) **PoA**$\to 1$, and when service value of any type grows to infinity **PoA** increases, but remains bounded. The authors demonstrate these results using data from a health service by modeling service facilities as $M/M/c$ queues.

Stidham [595] (2014) extends [330] in two ways. First, as in [44], servers have heterogeneous waiting-cost rates h_j. When $\lambda \to \mu \equiv \sum \mu_j$ the **PoA** approaches $\mu \sum h_j / (\sum \sqrt{h_j \mu_j})^2$. Secondly, the author extends the results to $GI/GI/1$ servers and shows that the same formula still holds, but with the modified waiting cost function, $\tilde{h}_j = h_j \left[\mu_j (C_a^2 - 1) + \lambda(C_{S_j}^2 + 1) \right] / 2\lambda$, where C_a^2 and $C_{S_j}^2$ are the squared coefficients of variation of the inter-arrival time and time of service at server j, respectively. A similar result is shown for a network of queues.

5.9 Trading positions

See [24, 315, 672] for related models where customer *reassignment* is carried out by a queue manager.

Gershkov and Schweinzer [249] (2010) consider mechanisms for rescheduling a clearing system (i.e., when there is no arrival process). Customers' heterogeneous waiting-cost rates are private information and service takes one unit of time. An efficient schedule serves the customers in decreasing order of waiting-cost rates. A trade mechanism specifies customer payments and schedule, given their waiting cost rate claims and the initial schedule. The mechanism must satisfy **IR**, **IC**, and must maintain a balanced budget.

The authors show that no such mechanism exists when the initial service schedule is deterministic (for example FCFS), and that to enable trade it is necessary that the players have weaker rights to initial positions in the schedule. In particular, in the extreme case where the initial schedule is random, an efficient mechanism exists, for example, an auction.

The authors mention that a stochastic queueing version is an interesting generalization of their deterministic model. Note that a bidding mechanism for an M/M/1 system which optimally regulates the arrival process and the service order is discussed in [303] (see [1] §4.5.2).

Yang, Debo, and Gupta [680] (2015) consider auction mechanisms for queue positions in an M/M/1 system with heterogeneous customer delay costs. The system is unobservable to the customers and joining customers have the option of maintaining their FCFS position. Alternatively, customers can pay a fee H and claim a waiting-cost rate (their bid). These fee-paying customers will sell positions to others with higher bids and buy from those with lower bids.[25] In each transaction the buyer pays the seller's price per expected waiting time exchanged. The main results are as follows:

- Suppose $H = 0$.

    - Customers with waiting-cost rates above the threshold balk. All joining customers participate in the auction.

    - Some trades are associated with a negative value to the buyer, but overall equilibrium trading makes all joining customers better off.

    - All customers overbid (i.e., bids are strictly greater than their true waiting-cost rate).

---

[25]Rosenblum (1992), see [1] §2.11, considered position trades in a queue where waiting costs are public information. As noted in [1] §2.11, a drawback of the model is that it ignores the potential value of future trades. The present paper amends this drawback.

- The authors characterize the equilibrium solutions when $H > 0$. Multiplicity of equilibria is possible. In particular, the solution where no customer participates in the auction defines an equilibrium. Customers with high waiting-cost rates balk. Among joining customers, those with low costs bid hoping to benefit from selling their positions while those with high costs bid hoping to shorten their delays. Customers with intermediate costs do not participate in the auction.

- Increased efficiency and profit to the broker can be achieved by limiting trade: Customers bidding in an interval $[\underline{R}, \bar{R}]$ do not trade position with one another (but do trade with customers bidding outside of the interval).

  - In equilibrium, all joining customers bid but there are no bids within the restricted interval except for at $\bar{R}$. Hence, these customers protect themselves from buying transactions associated with a loss.

  - The authors compute the optimal interval $[\underline{R}^*, \bar{R}^*]$ and bidding price H^*. $\underline{R}^* < \bar{R}^*$ and therefore this is not a special case of the unrestricted trade auction.

  - All joining customers participate in the optimal auction.

  - Customers bidding outside of the restricted interval are served in order of waiting-cost rates, but those with bids $\bar{R}^*$ are served in FCFS order.

El Haji and Onderstal [224] (2015) conduct laboratory experiments in a static environment (i.e., all customers arrive before service starts). Waiting cost rates are private information, and the authors consider two mechanisms for reallocating positions. In the *server-initiated* auction, all customers bid for the first position in the queue with the winner's bid being equally distributed among the others who now bid for the second position, and so on. In the *customer-initiated* auction, each new arrival observes the queue length and sequentially trades positions with queued customers. The authors compare the two approaches in terms of system efficiency and customer preferences and explain why the outcome deviates from theoretical predictions.

Chapter 6

Monopoly

Operations management literature on strategic queueing models naturally focuses on profit maximization. The part of that research that considers a monopolistic server is described in this chapter.

6.1 Profit maximization in Naor's model

Let S^* denote the maximum social welfare obtainable in Naor's model. Clearly, S^* is an upper bound on profit in the system. The firm can achieve this profit if and only if the following conditions hold: 1. Customers join according to the **SO** threshold n^*, and 2. all welfare generated in the system goes to the firm. These conditions are not satisfied by Naor's optimal static price when $n^* > 1$.[1]

Four mechanisms that achieve profit S^* are known, as described below. The first three are discussed in [229].

---

[1]It is numerically shown in [311] that optimal static pricing guarantees more than $0.8S^*$.

The simplest mechanism, suggested by Chen and Wan (2001), see [1] §2.8, uses dynamic pricing. The price leaves zero customer surplus at any state where customers are encouraged to join and is a higher price where they should balk. In Naor's model, this means charging price $p_n = R - C\frac{n+1}{\mu}$ from a customer arriving when the system occupancy is $n < n^*$, and charging a higher price that prevents joining when $n \geqslant n^*$. When customers are heterogeneous, achieving S^* by dynamic pricing is possible if each customer's type is observable (symmetric information) and dynamic price discrimination is permitted.

An alternative mechanism that achieves profit S^* follows from Hassin's [302] (1985) observation that customer behavior is socially optimal when the queue regime is LCFS-PR. In this case, the expected utility for any arriving customer is the same and is independent of the current queue length. Therefore, charging a fixed fee (almost) equal to this value will not deter the customer from joining and will preserve the socially optimal behavior while all gains go to the firm.[2]

A third possibility for achieving the profit S^* follows from work on priority sales by **Alperstein [39] (1988)** who showed that an LCFS-PR regime can be obtained through appropriate pricing of preemptive priorities while inducing the threshold n^* and leaving no customer surplus.

The fourth possibility is a limited form of dynamic pricing as suggested by **Hassin and Koshman [311] (2014)**. Customers are informed whether queue length is $< n^*$ (state L) or $\geqslant n^*$ (state H). The admission price is the highest possible price such that all customers join the queue when the state is L, i.e., $R - CW_L$ where W_L is the expected sojourn time in an M/M/1/n^* system. When the state is H, the price will be sufficiently large so that all arrivals will balk.

6.2   Price and capacity

In the short run the natural decision variable is price p, whereas in the long run the firm can also alter its capacity μ. In both cases, these two variables determine the delay W. It doesn't matter which pair of the three (p, μ, W) is known in the case of a monopoly, because the customer may deduce the missing value. However, the demand is naturally described in terms of (p, W) and authors often use W as a decision variable instead of μ. W is then often referred to as a *delay guarantee*. In the rational setting, customers of a monopoly can deduce the equilibrium W from the pair (p, μ), and need not rely on unreliable delay quotes.

---

[2]Note that with the alternative regulation mechanisms suggested in [637, 325] the server cannot fully extract customers' welfare by a single price.

6.2.1 Pricing

A notable feature of pricing in the E&H model is observed by **Chen and Frank [141] (2004)** (also see [1] §3.1.3). The short-run profit-maximizing admission fee *decreases* when demand increases and more customers are admitted to the queue. The reason being that a higher admission rate brings higher delay and a degradation in the quality of service supplied by the system. This phenomenon differs from common market behavior where an increase in demand leads to an increase in price. Chen and Frank also show that the optimal price increases in service value and capacity. Moreover, in the long run a sustained increase in the arrival rate will cause the firm to add processing capacity and raise the price. Some of the papers surveyed below reach similar conclusions concerning the sensitivity of prices to demand.

Ziya, Ayhan, and Foley [717] (2006) compute revenue-maximizing prices in queueing systems having multiple servers, a finite waiting room, price-sensitive but delay-insensitive customers, and demand satisfying increasing price elasticity. The main results are:

- **The $M/M/1/m$ queue:** The optimal price is monotonic in m. Whether it is increasing or decreasing depends on whether the system load at zero price is above or below a certain threshold.

- **The $M/GI/s/s$ queue:** The optimal price decreases in s.

- The authors report $M/GI/1/m$ and $M/M/s/m$ examples where the optimal prices are non-monotonic in the number of servers.

- In [718] the authors compute upper bounds on the optimal arrival rates for a multiple-class generalization of their model.

Maoui, Ayhan, and Foley [475] (2009) compute profit-maximizing prices and analyze their sensitivity to the system parameters in $M/G/1$ and $M/M/1/s$ queues. Customers are price sensitive and have heterogeneous service valuations leading to a demand function with an increasing generalized hazard rate (cf. [474]). The server sets the price of admission and incurs linear holding costs. The authors prove that an increase in potential arrival rate or a decrease in service rate results in a *higher* optimal price, which differs from the Chen-Frank observation,

Printezis and Burnetas [536] (2011) consider two customer classes with common service values but different delay sensitivities.[3] Customer types are observed by the server and price discrimination is allowed, but the queue regime is restricted to FCFS.

The authors compute the profit-maximizing prices. The solution is compared with that of the profit-maximizing solution when price discrimination is

---

[3]See [537] for a similar model with asymmetric information.

not possible. In both cases, customers from the more delay-sensitive class enter only if the other class joins at maximal rate. An interesting result, similar to the Chen-Frank observation, is that the optimal prices under price discrimination increase with capacity. However, this is not generally true when price discrimination is not possible. Another result states that when capacity is a decision variable, the benefits of price discrimination are higher with intermediate values of capacity or capacity costs.

Yeh and Lin [683] (2012) assume the expected frequency of facility breakdowns increases with demand. Demand itself is price elastic, meaning a price increase results in a revenue reduction. However, a price increase also results in reduced demand which decreases the frequency of breakdowns, which are immediately repaired, but at a cost. Therefore, the optimal price reflects balancing revenue and maintenance costs. Assuming increasing price elasticity and a strictly increasing failure rate of the facility, the authors prove that the expected number of breakdowns in a cycle is a concave and strictly increasing function of the demand rate. As a consequence, the profit function is strictly concave with a unique maximizing demand rate. The effect of the parameters on the optimal pricing policy is numerically demonstrated.

Zhou, Chao, and Gong [711] (2014) consider a market with finite potential arrival rates of two customer classes with heterogeneous service valuations and waiting-cost rates. The firm operates an FCFS regime with a single price. The main qualitative finding is that the optimal price is not monotone in the potential arrival rates. However, conforming with the Chen-Frank observation, as long as the firm does not change the customer classes it serves, the price (weakly) decreases when the potential arrival rate of either class increases.

Haung and Su [318] (2015) consider an $M/M/1/b$ system with homogeneous price-sensitive customers and two service classes, $k = 1, 2$, associated with service rates $\mu_1 > \mu_2$. Given prices P_1, P_2, it is assumed that class k demand rate is $\gamma_k = \frac{D_k}{D_1 + D_2}$ where $D_k = A_k P_k^{-E_k}$ for constants A_k and E_k (E_k is the price elasticity of class k's demand). The authors numerically study relations among the profit maximizing price ratio P_2/P_1, and the ratios A_2/A_1 and K_2/K_1.

6.2.2   Joint price and capacity optimization

Jahnke, Chwolka, and Simons [358] (2005) assume demand decreases linearly with the full price $\lambda(\pi) = \alpha - \beta\pi$. They define the full price π in a special way: $\pi(p, \mu) = p + \epsilon \max\{0, \rho - \Gamma\}$, where p is the price, Γ is a *market standard* for utilization, $\rho = \lambda(\pi(p, \mu))/\mu$ is the utilization level, and ϵ, α, β are constants. From these relations it follows that demand is a *kinked function* of p, meaning it is piecewise linearly decreasing and concave with a single breakpoint. The authors provide a closed-form solution for the profit-maximizing price and service rate subject to linear production and capacity costs.

Jiang [365] (2005) compares the **SO** and profit-maximizing solutions in an M/M/1 queue with heterogeneous service valuations. It turns out that the profit-maximizing solution has a higher price, a lower arrival rate, lower delay costs, and lower investment in capacity.[4]

Ray and Jewkes [545] (2004) consider a profit-maximizing M/M/1 firm operating in a market where price p, demand λ, and delay L are linearly related to each other: $p = d - eL$, and $\lambda = a - bL$. Delay is defined with respect to a given reliability level s, and capacity costs are linear.[5] In such a market the sign of b is important. If $b > 0$, then an increase in L comes with a decrease in demand despite the price reduction. Customers are *more lead-time sensitive than price sensitive* in this case. If $b < 0$, then a price reduction increases demand despite the increased delay, so customers are *more price sensitive than lead-time sensitive*.

The authors emphasize the importance of the manager knowing whether customers are more price sensitive or more lead-time sensitive. For example, when customers are more lead-time sensitive and the firm increases the delay, capacity will decline since both the delay requirement is relaxed and demand decreases. However, when customers are more price sensitive an increase in the delay causes the opposite effect since the firm now needs to satisfy a larger demand. In this case capacity may increase or decrease.

Serel and Erel [566] §4.4 (2008) prove conditions for uniqueness of the locally optimal solution in a pricing and staffing optimization problem. This is a profit maximization M/M/s model where the waiting costs are incurred by the server, customers are price sensitive, there is an exogenous upper bound constraint on expected waiting time, and customers are not delay sensitive as long as their expected delay remains below the bound.

Parra-Frutos [521] (2010) considers profit maximization through pricing and capacity setting in the E&H model with unbounded potential demand. The main finding is that when the capacity cost is concave (even linear), there is no optimal solution for the firm, and the higher the capacity the higher the expected profits it can obtain.

---

[4]Sufficient conditions for some of these properties are also provided by Mendelson and by Stenbacka and Tombak, see [1] §8.1-2.

[5]A very similar model is presented in [658]. This model has no capacity cost, but the firm incurs holding and tardiness costs.

6.2.3 Differentiation by quality and delay

Chayet, Kouvelis, and Yu [138] (2011) assume customer utility functions $\tilde{\theta}q - cw - p$, where $\tilde{\theta} \sim U[0,1]$ is the customer's quality-sensitivity parameter, w is the delay measure, q measures product quality, c is the customer waiting-cost rate, and p is the service price. When the firm offers quality q, a $\tilde{\theta}$-customer joins the system iff $\tilde{\theta} \geqslant \theta = (p + cw)/q$, resulting in an effective arrival rate $\lambda^e = (1 - \theta)\lambda$. The firm chooses q, p and μ and earns profit $[p - a(q)]\lambda^e - b(q)\mu = [\theta q - cw - a(q)](1 - \theta)\lambda - b(q)\mu$, where $b(q) = \beta q^2$ represents the unit investment in capacity and $a(q) = \alpha q^2$ is the unit production cost, when production quality is q.

The firm can segment the market by operating dedicated facilities offering quality levels $q_1 > q_2$, prices p_i, and lead times w_i, $i = 1, 2$. Alternatively, a single *flexible facility* can produce both quality types with a common FCFS queue (or possibly giving priority to customers choosing the high-quality product). In both cases, the market is segmented such that for some $0 \leqslant \theta_2 < \theta_1 < 1$, customers in $[\theta_1, 1]$ prefer product 1, those in $[\theta_2, \theta_1]$ prefer product 2, and the others balk. The authors assume the investment required for a flexible facility that can switch between products of qualities q_1 and q_2 is $b(q_1, q_2) = \beta[\delta q_1 + (1 - \delta)q_2]^2$, with $\delta \in (0, 1]$, and focus on the case $\delta = 1$.

The main results are:

- An appealing feature of this model is that the solution with dedicated facilities is expressed in terms of a single parameter $M = 2\sqrt{\beta c/\lambda}$. For a flexible facility, the optimal quality levels also depend on δ and β/α.

- The profit-maximizing quality levels and market segmentation under the dedicated and pooled solutions are quite different. Dedicated facilities can better optimize market coverage and product positioning, but lack the benefits of pooling capacity.

- Dedicated facilities yield some unexpected results: Capacity investment and quality may increase with higher marginal investment costs, and capacity investments are non-monotonic in market size.

Xu, Lian, Li, and Guo [673] (2016) consider a Hotelling model with customers uniformly distributed over $[0,1]$ and unobservable M/M/1 servers at the two ends of the interval. Customers have homogeneous travel and waiting costs and service valuation. The firm charges p_d from customers guaranteed to join their nearest queue and $p_p < p_d$ from customers assigned with probability 0.5 to each of the queues. In equilibrium, customers near the ends of the interval prefer the deterministic assignment, a fraction of the others chooses the probabilistic assignment, and the rest balk. This price discrimination enables the firm to increase profits.

The authors compare the profit-maximizing prices p_d and p_f in the case of FCFS and under prioritization of one of the two service types. The total revenue, however, is not affected by the regime.

6.2.4 Delay compensation

Any dynamic pricing scheme where higher congestion is associated with lower prices can be interpreted as delay compensation. We describe here different types of delay compensation. See §2.6.3 for delay compensation in observable queues. See [311] for a model where price is reduced when the system is congested.

Tuffin, Le Cadre, and Bouhtou [631] (2010) consider an M/M/1 system where customers whose *realized* system time exceeds threshold d receive delay compensation q. The expected compensation is therefore $q \cdot e^{-(\mu-\lambda)d}$. The firm's decision variables are price p, as well as q and d. Depending on q and d, customers may benefit from a higher arrival rate that increases chances of obtaining the delay compensation thus leading to FTC behavior and multiple equilibria. The authors assume that, because of long-run benefits, the firm benefits from inducing the equilibrium with the largest arrival rate even if it is not the equilibrium that maximizes profits. They demonstrate through examples that *a compensation scheme can increase revenues.*

Afèche, Baron, and Kerner [15] (2013) investigate profit maximization when customers are heterogeneous and risk averse and the server charges *lead-time-dependent tariffs*, that is, payments depend on *realized* lead times. Customers of the same *type* have the same utility function U and delay-cost function C, but differ in service valuations. The utility of a customer with value v who stays in the system for time w is $U(v - C(w) - P(w))$, where P is the price function set by the server. The main results are:

- Lead time-dependent pricing does not increase profits when customers are risk neutral. However, discounts based on the realized lead time can increase profits when customers are risk averse.

- If the server can distinguish among customer types it is optimal to eliminate delay-cost risk by fully compensating customer delay. If the provider serves indistinguishable customer types and charges a single tariff, then the more patient customers receive overcompensation and their utilities increase with lead time!

- When customer type is private information, pricing based on realized lead times allows for price discrimination. The first-best (i.e., when types are distinguishable) menu of tariffs may be **IC**, yielding higher revenues than with risk-neutral customers.

- For a single customer type the simplest refund policy, issuing a full refund for late delivery, performs well relative to the optimal lead-time-dependent tariff.

- Under joint pricing and capacity optimization, optimal pricing based on realized lead times yields higher profits with less capacity compared to flat-rate pricing. This profit gain can be significant, particularly if capacity is expensive.

Chen, Huang, Hassin and Zhang [144] (2015) assume that the price p of a product is exogenous and that an $M/M/1$ manufacturer with constant unit production and inventory holding costs applies a base-stock policy with base-stock level S. If customers arrive to a positive inventory level they immediately obtain a product and pay p. Otherwise, they choose between joining an FCFS queue or buying the product elsewhere at the same price p. While making decisions customers do not know queue length, though they do know if there is a shortage. Customer delay-cost rates are random variables from a given continuous distribution. When the product is out of stock it is offered under improved terms to encourage customers to wait until the product becomes available. The authors consider two compensation schemes, namely, uniform compensation and priority auctions:

- **Uniform compensation:** During stockout periods the product is sold for a reduced price $\alpha < p$. A fraction of the customers, consisting of those having lower waiting-cost rates, join the queue while the others balk.

- **Priority auctions:** During stockout periods priorities are determined through a highest-bid-first (HBF) auction. The firm's decision variables are the base-stock level S and the minimum allowed bid α.

- **Comparison:** Assuming linear delay costs, a priority auction yields higher profits,[6] uses lower S and α, and results in higher sales. Both mechanisms yield higher profits, lower S, and lower customer costs relative to the optimal base-stock solution without them. Hence both firm and customers profit from the institution of these mechanisms.

- **Speculation:** A customer with a low time cost that arrives during an in-stock period might reduce expected costs by waiting until a shortage occurs and buying at a reduced cost. To avoid this undesirable behavior the authors offer restrictions on price reduction.[7]

6.2.5 Advertising

Araman and Fridgeirsdottir [62] (2011) examine a web *publisher* that generates revenues by posting ads on its website and offers *advertising plans* to *advertisers*. A plan is basically defined by the expected number N of *viewers* that should see the ad during a period of expected length T, with price p charged per view. Advertisers and viewers arrive according to stochastic

---

[6]This is similar to [17].

[7]Similar restrictions are assumed in [79] excluding behavior where customers prefer a loaded server thereby increasing chances of being rejected and compensated.

processes with the arrival rate of advertisers a function of p and N. Ads are placed on webpages and when a viewer uploads a page all the ads posted on the page are delivered. The publisher's decision variables are the number κ of pages (sets of s ads each) and the price per view. Given that there are κ pages, every κth viewer sees the same ad. This affects the time until the contract is fulfilled and the ad is removed. When there is no vacant ad slot, new contracts must wait and *the publisher* incurs a linear waiting cost. The authors derive expected waiting-time equations and use them to compute the optimal plan. They also derive heavy-traffic approximations.

Jhang-Li and Chiang [364] (2015) consider a firm profiting from service fees while generating additional revenues from advertisements. There are two customer classes, premium and basic. Class i has waiting cost rates $\theta_i c$ and heterogeneous service valuations expressed by total value functions $\theta_i V(\lambda)$. The firm has the option of advertising and the price obtained $p(a)$ per customer depends on the *advertising level a*. However, advertising at level a also reduces i-customers' utility in proportion to $\theta_i a$. The firm decides on (i) capacity associated with a convex increasing cost function; (ii) subscription fee per customer class subject to class **IR** and **IC** constraints; (iii) advertising level; (iv) whether to prioritize premium customers; (v) whether to supply advertisement-free service to premium customers.

The authors solve six variations of the model with a specific form of the function $p(a)$, namely, advertising to both classes, only to basic customers, or not at all, and in each case with and without prioritization. Sensitivity analysis raises some interesting observations. For example, a reduction in customer disutility from service may increase or decrease optimal capacity and subscription fees. The authors also set conditions for the negative optimal subscription fees, in the form of rebates or non-monetary rewards, when gains from advertising are high.

6.2.6 Heavy-traffic approximation of price and capacity

Maglaras and Zeevi [463] (2003) consider heavy-traffic approximations to a Markovian system with large capacity C shared by the customers but having an upper bound on capacity allocated to a customer at any time. The demand function $\lambda(p)$, where p is full price, is assumed to be elastic, meaning a full-price increase results in a revenue reduction. The authors show that the optimal regime given this assumption involves high resource utilization with only minor degradation effects, as in the Halfin-Whitt **QED** regime, which the authors extend by incorporating sharing, pricing, and rational customer behavior. Heavy traffic is modeled by allowing the capacity C and potential demand Λ to grow in proportion to each other.

Let $\bar{p}$ be the price that matches demand to capacity. The authors consider both the short-run pricing problem and the long-run joint capacity and pricing decisions. In the short run, the revenue-maximizing price is $\bar{p}$ plus a second-order correction term proportional to $\frac{1}{\sqrt{C}}$. In the long run the problem is essentially decoupled, when C is a decision variable associated with a linear cost. In the first step, the stochastic effects of the system are neglected, it is assumed the system is fully utilized, and an optimal value of $\hat{C}$ is then computed. Given this value, the optimal price is computed as in the short-run problem.

Maglaras and Zeevi [464] (2005) examine an unobservable system with two customer types. Type **G** (guaranteed) users receive one unit of processing capacity and are admitted according to a control policy. Type **BE** (best effort) users equally share the residual capacity, with an upper bound of one unit of processing capacity per user, and are always admitted. Arrivals of both classes are Poisson where the arrival rate of **G**-users depends on price whereas the arrival rate of **BE**-users depends on the full price. The firm's goal is to determine revenue-maximizing class-dependent prices and the admission rule for **G**-customers.

The main model assumes **G**-users generate a higher revenue rate per unit capacity and a **G**-user is rejected only when service capacity is fully taken up by other **G**-users. A fluid approximation of the model is solved by setting prices so **G**-customers are never rejected and **BE**-customers have no delay. This solution is the basis for an asymptotic approximation where a sequence of systems are examined such that service capacity and demand grow by the same scale parameter n. The resulting system operates in the Halfin-Whitt regime. The utilization is of order $1 - \gamma/\sqrt{n}$ and the expected delay of **BE**-customers is of order $d/\sqrt{n}$, where the constants γ and d depend on the system's parameters. Furthermore, the blocking probability of **G**-customers is $o(e^{-cn})$ for some constant $c > 0$.

The authors consider a variation in which the server reveals the amount of capacity that **BE**-customers currently obtain. Under heavy-traffic assumptions, when service rates of both customer classes are equal, this revelation results in higher revenues and smaller expected delays.

Kumar and Randhawa [422] (2010) consider a heavy-traffic model where the arrival rate to an M/M/1 system depends on the full price and market size. Their goal is to study the impact of the delay-cost structure on the firm's pricing and investment in a growing market. Specifically, the arrival rate is $\lambda = n \cdot \Lambda(p + h \cdot \mathrm{E}[W_q^r])$ where n is market size, $\Lambda(\cdot)$ is a demand curve, p is price, W_q is queueing time, $r \geqslant 1$, and $h > 0$. The firm's decision variables are price p and capacity μ, which costs the firm $\kappa\mu$. Using the steady-state distribution $\Pr(W_q > t) = \rho e^{-(\mu-\lambda)t}$, where $\rho = \lambda/\mu$, the delay cost equals $h\rho\frac{\Gamma(r+1)}{(\mu-\lambda)^r}$, where Γ is the gamma function.

The authors show that optimal solutions lead the system to a heavy-traffic regime. The rate at which utilization approaches 100% when n increases depends on the form of the delay cost. A system with traffic intensity ρ_n is in a *k-heavy traffic regime* if $n^k(1 - \rho_n) \to C \in (0, \infty)$ as $n \to \infty$. Thus, the

Halfin-Whitt regime has $k = 0.5$. It is shown that when $r \geqslant 1$, i.e., delay costs are convex, $k = \frac{r}{1+r} \geqslant \frac{1}{2}$. Thus utilization approaches 100% at a rate faster than the conventional $O(\sqrt{n})$. The same result is obtained for $r < 1$ (concave costs) if the service order is FCFS. However, under concave costs the discipline that minimizes delay costs is LCFS and, in this case, the firm's optimal solution is to operate in $\frac{1}{2}$-heavy traffic.

The authors also consider a variation where capacity is fixed and the only decision variable is price, and another when arrival rate is fixed and the objective is to select a capacity level that maximizes social welfare.

Lee and Ward [430] (2014) derive an asymptotically optimal (static) price and capacity solution in a system with price-sensitive customers. Given price p, the demand rate is $\lambda(p)$, and the server incurs linear capacity and holding costs. The authors first solve a deterministic version with a unique optimal solution (p^*, μ^*). They then consider a sequence of systems where system n has demand $n\lambda(p)$ and show that the optimal solutions (p^n, μ^n) satisfy $p^n \to p^*$ and $\frac{\mu^n}{n} \to \mu^*$. Lastly, the authors deduce a refined policy which is asymptotically optimal on diffusion scale.

6.2.7 Intertemporal pricing

See [452] for multi-period pricing and delay guarantees where customers select a period according to a logit choice model.

Guo, Liu, and Wang [281] (2009) show that a unique subgame-perfect equilibrium exists in the following two-period pricing game. Customer service valuations are uniformly distributed over [0,1] and service values obtained in the second period are discounted. In the second period the server sets a profit-maximizing price. Customers anticipate this policy while deciding whether to join the first-period queue and pay the announced first period price. Customer behavior is characterized by two thresholds: those with high service valuations join the first-period queue, those with intermediate values join the second-period queue, and the rest balk. The authors show that revenue-maximizing prices are larger than welfare-maximizing prices.

The authors also present numerical results for the server's losses in a model where the server behaves as if the customers are myopic, though they are strategic.

Liu, Li, Xu, and Li [450] (2015) prove that an equilibrium exists in the following multi-period model. There are n full-price minimizing users of an M/M/m facility allocating demand among periods $h = 1, \ldots, H$. The queue owner maximizes profits by setting period-dependent prices. User waiting cost rates are heterogeneous and increase each period by a factor $\delta > 1$. It is assumed that the system is heavily loaded so the probability that an arriving request has to wait in queue is approximately 1.

6.2.8   Risk averse customers

See [678] on **SO** joining and profit maximizing pricing for risk averse customers with reference-point sensitivity, and see [15] risk-averse are compensated for their wait.

Başar and Srikant [81] (2002) consider an M/M/1 facility with $\mu = n \cdot c$. There are n users and the utility of user i from throughput rate x_i is $U_i(x_i) = w_i \log(1 + x_i)$. The server sets price p per unit of demand and the users follow by choosing demand rates such that user i maximizes $U_i(x_i) - px_i - 1/(nc - \sum x_j)$. [8] The authors provide explicit solutions for special cases of this model, particularly the cases of $c = 1$ and $n \rightarrow \infty$. They show that when n increases, the server's revenue per unit of capacity increases.

Shen and Başar [569] (2007) extend [81] for the case $c = 1$, allowing for nonlinear pricing $r_i(x_i)$. With symmetric information the monopoly can fully extract customer surplus and profit maximization coincides with welfare maximization. The authors derive an approximation to the optimal **IC** solution when there is a large number of users whose type cannot be identified by the server. The loss of profit due to incomplete information is numerically evaluated.

Hayel, Ouarraou, and Tuffin [331] (2007) consider *measurement-based pricing* in a multiclass M/M/1 FCFS queue where the arrival rates (throughputs) and average delay (response times) cannot be observed by the queue manager and need to be statistically estimated. Customers are risk averse and their cost consists of three components: Price, which depends on the measured throughput and delay, delay cost, which linearly depends on the actual experienced delay, and *cost of aversion*, which is proportional to the standard deviation of the price computation due to measurements. [9] The aggregate value of class j customers is an increasing concave function $V(\lambda_j)$ of their throughput λ_j. Equilibrium arrival rates are set so the cost applied to each class is equal to its marginal value.

Time is separated into periods of length T and measurements are performed in each period. The throughputs λ_j are estimated by sampling a fractile ϵ of the population. The delay (common to all classes) is measured by sending a Poisson(γ) stream of *probes* into the system. Probes are special jobs, served like standard jobs, but designed to measure response times. The probes do not contribute to the system's welfare but add to its congestion. The parameters ϵ and γ are decision variables which, like the prices p_j, depend on the measured throughputs and delay.

The authors design an algorithm for computing a pricing scheme and sampling parameters maximizing the total expected net value minus the cost of the measurements.

---

[8]Note that the expected waiting time $(nc - \sum x_j)^{-1}$ is not multiplied by x_i.

[9]The latter cost is not considered part of the aggregate social-welfare function.

6.2.9   Refurbished products

See [668] for a supply chain that re-manufactures defective items at no charge.

Vorasayan and Ryan [636] (2006) suggest a model for profit optimization when price-sensitive customers choose between a new and a refurbished product. **Ghosh, Ryan, Wang, and Weerasinghe [250] (2010)** derive an asymptotic optimal strategy for a similar model. Both of these models assume an M/M/1 production facility where a given fraction of purchased product is returned by customers, refurbished, and held in inventory. The price for a new product is exogenous and the firm sets the sale price for refurbished items.[10] Customer valuations are heterogeneous and a customer ready to pay p for a new product is also ready to pay δp for a refurbished item, where $\delta < 1$. Customers face three options: buy a new product, buy a refurbished product, or balk. The demand for refurbished products is lost if none is available. The firm incurs costs for lost sales of refurbished products and for backorders of new products.

6.2.10   Capacity auctions

Yolken and Bambos [685] (2011) investigate selling a fixed amount of processing capacity to N M/M/1 customers with heterogeneous rates of demand, job sizes, and waiting costs. Capacity is allocated through an auction such that user i submits a bid w_i and obtains a fraction $\theta_i = w_i / \sum w_j$ of the capacity. Customers minimize the sum of waiting and bidding costs. Results include:

- The game has a unique equilibrium.

- Total revenue $\sum w_i$ increases with customer demand and waiting costs.

- A closed-form expression for the symmetric equilibrium bid when customers are homogeneous.

- An example with one big customer and $N - 1$ small customers (similar to [238]) where the PoA asymptotically behaves like $\sqrt{N}$.

- When customers have homogeneous delay sensitivities, the PoA can be bounded by a function of the smallest, largest, and total demand rate of the N users.

---

[10] In [636], the proportion of returned products to be refurbished is also a decision variable. The authors show that with sufficient capacity to satisfy demand, the optimal proportion is either zero or close to 1.

6.2.11   Optimal buffer and batch size

Setting an optimal buffer size is equivalent to rejecting customers when the queue length exceeds a threshold. It is also a restricted version of dynamic pricing which sets a very high price when this queue length is reached. The marginal advantage of increasing the buffer size in Naor's model is also investigated in [311].

See [475] §4.3 for a discussion on optimal buffer size in an M/M/1/s queue with price-sensitive customers and linear holding costs incurred by the server.

Masarani and Gokturk [476] (1987) consider an M/M/1/N queue where the server incurs a cost $C(N)$ and the buffer size N is a decision variable. Demand is assumed to have constant elasticity $\lambda(p) = \mu p^\epsilon$ where p is the price and $0 < \epsilon < 1$. The authors prove that for any given N, if ϵ is less (more) than 0.5, the profit-maximizing prices lead to $\rho > 1$ ($\rho < 1$). In the special case of linear costs $C(N) = cN$ and $\epsilon = 0.5$, the optimal buffer size is $N = \sqrt{\mu/c} - 1$.

The authors also consider competition among identical firms assuming $C(N) = cN$, entry to the market is free, the service price is exogenously determined, and customer joining probability is the same for all servers *regardless of their waiting room size*. This bounded rationality assumption leads to an equilibrium where all firms choose a unit-size buffer. The authors compute the equilibrium number of firms and show that when $\epsilon = 0.5$ and at the monopolistic price, competition increases the total number of buffer slots.

Wang and Choi [652] (2013) consider a multiclass G/G/1 model. Jobs of type i are accumulated until they number Q_i and are then placed in a queue for bulk service. The authors approximate expected waiting times W_i as functions of arrival and service parameters and batch sizes Q_i. For class-dependent prices p_i, arrival rates are determined through usual equilibrium conditions of the form $R_i - p_i - C_i W_i = 0$. The firm's problem is to set prices and batch sizes that maximize a long-term objective function. The authors numerically demonstrate that optimal batch sizes can be highly sensitive to the objective function. The authors present a single-class variation of the model in [653].

6.3   Expert systems

In expert service markets, customers cannot fully assess the type and amount of service they need.[11] Experts performing both the diagnosis and

---

[11]Some authors use the term *discretionary* in describing systems where duration of service is determined by the server, as distinguished from *non-discretionary* service of which duration is determined by objective standards.

the service may supply too much or too little service to optimize their own utility. One way to avoid the problem is by employing independent agents for handling the diagnosis. Another would be by setting an adequate price structure that would eliminate or diminish the motivation for fraud.

Expert systems inherently assume asymmetric information, but in contrast to common models where customers have private information, it is the expert who has an informational advantage over the customer.

Another distinguishing property of expert systems is that the service duration can be the basis for the price, unlike in other models with service rate decisions.

We distinguish between models where a longer service benefits the customer from those in which it doesn't. See §9.2.2 for value-creation models in which longer service benefits the server.

6.3.1 Duration-independent service value

This section assumes the service value to be independent of the service duration. Customers obviously prefer shorter service which reduces price and waiting costs. A notorious example is the taxi market, having some distinguishing features: (i) customers are heterogeneous with respect to the amount of service they need; (ii) it is not possible to undertreat a customer because the desired service is the shortest possible, but it is possible to refuse service;[12] (iii) it is unlikely for a customer to encounter the same driver and therefore cheating involves no loss of future business; (iv) the charging scheme is exogenous and is often a two-part tariff with the variable part being proportional to the service duration.

Glazer and Hassin [257] (1983) considered a non-queueing model of the taxi market with a two-part tariff $F + pt$, where F is a fixed price, t is the duration of the ride, and p is the charge per unit time of service. The authors recommend setting $F = pW$, where W is the expected customer interarrival time, i.e., the time it takes the driver to find a new customer. Otherwise, if $F < pW$, a driver would find it profitable to serve fewer customers and cheat them by taking long routes. If $F > pW$, the driver would find it profitable to refuse long rides.

Price structures that eliminate service refusal by taxi drivers who maximize discounted profit are offered by **Janssen and Parakhonyak [360] (2011)**.

We note that when taxis line up and wait for customers drivers may decline short rides knowing they would quickly return to the back of the line. In such a case the customer should be allowed to go down the line until a driver agrees to serve him. Drivers along the line will have a decreasing threshold distance they would agree to serve and if no driver accepts a service request

---

[12]A similar assumption is made in other models described in this section.

the customer must balk and find other means of transportation. This setting is somewhat similar to models of kidney allocation (see §2.3).

Debo, Toktay, and Wassenhove [186] (2008) analyze profit-maximizing two-part tariffs in an expert system under the following assumptions:

- The system is an almost-unobservable $M/M/1$ queue. Customers cannot observe queue length upon arrival, but know whether the server is idle (state 0) or busy (state 1). Customer strategy is characterized by a pair of joining probabilities, one for each state.

- Customers are homogeneous, having a common service value and waiting-cost rate.

- The server admits all customers who choose to be served and chooses the amount of service to supply depending on the state (0 or 1) at the time of the customer's *arrival*. Undertreatment is not possible.

- There are two service procedures with different service rates. The strategy of the server is to choose the service rate for each of the two system states. Choosing the lower rate is considered a *service inducement*.

- The price consists of a fixed fee charged to each customer obtaining service plus a variable rate per unit of service time.

- In equilibrium, both the customers and the server are aware of the other's strategy.

The authors focus on the conditions that lead to service inducement. The waiting-cost rate should be small and, most interestingly, the system load should be between upper and lower thresholds. When these conditions hold, service inducement allows the expert to extract all surplus from customers without introducing large inefficiencies. That is, the surplus customers would enjoy with a fixed service rate in Naor's model is now extracted by the expert via a variable rate and slow service.

6.3.2 Duration-dependent service value

This section deals with *customer intensive services*, where *quality of service* increases as the time spent by the service provider serving a particular customer increases. A customer's decision to join the queue depends on quality of service, expected delay, and price. The server thus faces quality-speed tradeoffs. Longer service increases service quality but may cause longer delays. Models of *co-production*, where customers choose service rates, are also related to this subject, see §4.7.

Note that the research reported here mostly assumes service value to be proportional to service *rate*. In contrast, [188] assumes instead that the service value is proportional to service *duration* and [626] assumes dependence on the square root of the service duration. Two papers [654, 655] use a Brownian motion modeling of the diagnostic accuracy.

See [175, 188, 442] for other models where customers benefit from long service durations.

Wang, Debo, Scheller-Wolf, and Smith [654] (2010) assume two customer types of known proportions in the population, with customers not knowing their own type. No cost is incurred when a customer obtains the correct type of service/treatment, but there are type-dependent costs incurred by both customer and service facility when the wrong treatment is given. It is assumed that the customer's decision is "difficult" in the sense that the parameter values are such that, without any additional information, the customer is indifferent between the two treatment options.

Before seeking treatment each customer decides whether to go through a diagnostic process by weighing the value of information to be gained against the *inconvenience* of waiting to be diagnosed. Customer type is defined as $+1$ with prior probability π, or -1 with the complementary probability. The diagnostic process is modeled as a Brownian motion with drift r representing the *skill* of the diagnosing agent. Belief updating starts from π at time zero and evolves until it hits one of two boundaries $y \leqslant 0 \leqslant x$ set by the service facility. Once the process hits the boundary customer type is diagnosed and the customer is provided the corresponding service. A nice feature of this model is that the choice of the boundaries determines both the accuracy (error probabilities) and the distribution of the process duration.

The customer's decision variable is the probability of requesting the diagnostic procedure. The manager's decision variables are the number of diagnosing servers and the boundaries x, y. The equilibrium solution is computed using a heavy-traffic approximation for the expected waiting time. The authors reach many interesting conclusions including the following:

- Increasing asymmetry in error costs may affect the error rates either in the same direction or in opposite directions.

- For the symmetric case where $\pi = 0.5$ and the costs for erroneous diagnosis are type-independent (and therefore $x = -y$):

  - When the diagnostic skill level increases, callers experience longer waits on average.

  - Adding staff (servers) may increase congestion or error, but not both.

Anand, Paç and Veeraraghavan [54] (2011) consider an M/M/1 system with potential demand rate Λ and an initial service rate μ_b. The server costlessly sets a service rate $0 < \mu \leqslant \mu_b$, and the resulting value of service is linearly decreasing with service rate (faster service comes with lower quality): $V(\mu) = V_b + \alpha(\mu_b - \mu)^+$. The *intensity parameter* $\alpha \geqslant 0$ determines the sensitivity of service value to service speed. The server's decision variables are price and service rate. As in the E&H model, the equilibrium arrival rate is Λ if this leaves customers with nonnegative expected utility, or else it is the value that sets the expected utility to 0 so potential customers are indifferent between joining or balking.

- The interesting characterization of the solution is that the service cannot be too slow because, in this case, the expected delay is too high. Also, service cannot be too fast as this results in low quality. Both extremes force the server to lower price causing a profit loss. Therefore, there is an intermediate profit-maximizing service rate.

- As can be expected, when the intensity parameter α increases the optimal rate decreases which results in better service, but with longer delays. The optimal price and associated equilibrium demand are both unimodal functions of α, first decreasing and then increasing.

- The paper also considers capacity competition when the price is restricted to be uniform, but quality can vary among servers. It turns out that when the number of competing servers increases, service value *and also price* rise while the expected delay remains constant. This is in contrast to traditional models of service rate competition.

Ni, Xu, and Dong [502] (2013) extend [54] by assuming two classes of customers differing in the intensity parameter α. The server's decision variables are the service rate and class-independent prices. The authors demonstrate that revenue is not, in general, a unimodal function of the service rate. They solve the firm's problem by computing maximal revenues in four regions and then choosing the largest one. They also demonstrate that price-discrimination could increase profits.

Dai, Sycara, and Lewis [181] (2011) assume the value of service to be linearly decreasing in μ. Customers incur linear waiting costs and have a reservation utility r. Potential demand is large so for a given service rate the arrival rate equates the expected net utility of an arriving customer to r. The authors derive the service rate that maximizes the net utility *of the arriving customers*.[13]

Wang, Debo, Scheller-Wolf, and Smith [655] (2012) consider a variation of their model [654]. They assume the staffing level is fixed and is not a decision variable. A fixed *inconvenience cost* is incurred by any customer undergoing diagnostic service, independent of waiting time. The authors consider

---

[13]This is different from social optimization which considers aggregate added value relative to the reservation value r.

mainly the symmetric case (numerically demonstrating that the major results hold in the asymmetric case). A *mismatch cost*, independent of customer type, of an incorrect diagnosis is split in a given proportion between customer and server (acting as an insurer). For a given boundary value x, customer equilibrium is characterized by the joining probability $p_e(x)$. Joining the diagnostic system may save the mismatch cost for both server and customer, but the customer also considers the inconvenience and waiting costs which affect the equilibrium joining probability. The server sets the boundary value x which minimizes its share of the mismatch costs.

The authors show that the equilibrium joining rate is non-monotone in x because of the opposing effects of increasing accuracy and congestion. The main result of the paper is that the optimal solution leads to an equilibrium where the server captures the whole potential demand only when its skill, as reflected by r, is sufficiently high, or the fixed inconvenience cost is sufficiently low, or the customer's share in the mismatch cost is sufficiently high. In this case, the server will choose the largest x such that $p_e(x) = 1$.

Dai, Akan, and Tayur [180] (2012) consider a healthcare system where patient costs are partially covered by insurance. The system is modeled as an M/M/1 queue.

The value (quality) of service $Q(\mu)$ is assumed to decrease linearly with the service rate. For constants $0 \leqslant \beta \leqslant 1$ and π, if price $p \geqslant \pi$, then the part covered by the customer is a convex combination $\beta p + (1 - \beta)\pi \leqslant p$. Customers also incur linear waiting costs. The equilibrium arrival rate $\lambda(\mu, p)$ is determined by equating customer net utility to 0, i.e., $Q(\mu) = cW + \beta p + (1 - \beta)\pi = \frac{c}{\mu - \lambda(\mu,p)} + \beta p + (1 - \beta)\pi$ where cW denotes expected waiting cost. The price p^* and service rate μ^* that maximize $p \cdot \lambda(\mu, p)$ can be explicitly computed. The authors prove that p^* decreases with both π and β while μ^* increases in β and decreases with π.

The authors show that the profit-maximizing solution provides longer (better) service but serves fewer customers than socially desired. Interestingly, the expected waiting time is the same under both solutions. The authors show that setting a reimbursement ceiling or penalizing the server for slow service is not sufficient to coordinate the system. Also discussed are heterogeneity in customers insurance parameters and time values, allocation of time between diagnosis and analysis, misdiagnosis costs incurred by the server, and uncertainty about the skill level of the provider.

Kostami and Rajagopalan [413] (2014) present a multi-period variation of [54]. Longer service increases quality, and this increase is reflected in the *potential* demand rate which at period $j + 1$ satisfies $\Lambda_{j+1} = \Lambda_j - \delta\lambda_j(\mu_j - \hat{\mu})$. In this relation $\lambda_j = \Lambda_j - \alpha p_j$ ($\alpha > 0$) is the demand rate,[14] μ_j is the service rate, and p_j is the price, at period j, and $\hat{\mu}$ is a *benchmark service rate*. It

---

[14]Compare this assumption with equation (4) in [54] where the demand relation to price is endogenously derived.

is assumed the *quality sensitivity* parameter δ is small and that λ_j is not too small. Customers are quality and price sensitive but not delay sensitive as all delay costs are incurred by the firm. The authors show that when the firm can dynamically control price but can only set the service rate once, it is optimal to keep demand constant over time. In general, the firm benefits from a high benchmark speed.

Tong and Rajagopalan [626] (2014) propose a model of profit maximization in which both service and demand rates are endogenous. Customers are indexed by *type* α, a priori unknown to both customers and firm, which is a random variable with pdf f. Arriving customers go through a *diagnostic phase* which reveals their type. The value of service of total duration τ (including the diagnostic phase time τ_0) to an α-customer is $v(\alpha, \tau) = \sqrt{\frac{\tau - \tau_0}{\alpha}}$. Given α, the customer is provided with service of duration $\tau(\alpha)$ for a fee $p(\alpha)$, independent of the system state.

Service duration $\tau(\alpha)$ is deterministic for a given customer but becomes a random variable given the customers' heterogeneity. Therefore, this is an M/G/1 queue.[15] In equilibrium, the expected net utility of customers is zero. The firm sets the functions $\tau(\alpha)$ and $p(\alpha)$ to maximize $\lambda \mathrm{E}(p) = \int \lambda p(\alpha) f(\alpha) d\alpha$.

The profit-maximizing solution determines the expected price $\mathrm{E}(p)$, but the corresponding price for each customer type can be implemented in several ways. At this stage a second condition is added, namely that each customer gets zero utility *ex-post*, and with this condition the solution is uniquely determined.

The authors compare the optimal solution with two second-best single-parameter solutions: fixed price and service value guarantee (through type-dependent service duration); and fixed price per unit time of service, letting customers choose their service duration after they learn their type. Both schemes are reported to do well relative to the optimal solution.

Paç and Veeraraghavan [511] (2015) focus on the signals an expert provides to customers and the effect of congestion on the expert's cheating and system efficiency. They show how cheating can emerge even when customers can verify whether their request was treated (but not whether they obtained excessive service). Customers' problems are either *major* or *minor* with known probabilities. The resolution of a minor problem provides the customer value v_L and requires expected service time τ_L, whereas treating a major problem gives value $v_H > v_L$ and requires expected time $\tau_H > \tau_L$. Waiting cost rate c is common to all customers. The server first announces the diagnosis of the customer type L or H, and the customer can then either join or costlessly balk based on the customer's updated belief concerning his type. The expert's

---

[15]Expected waiting time is approximated by assuming that the service coefficient of variation is independent of the firm's decisions.

decision variables are service fees, p_L and p_H, and the state-independent diagnostic strategy.

Quantitative results are derived assuming *diminishing marginal returns* $v_H/\tau_H \leqslant v_L/\tau_L$, but the qualitative conclusions also hold without this assumption. The main results include:

- Committing to an honest diagnosis is costly for the expert, especially in small markets.

- As v_H increases, over provision is more likely and more customers are deterred from joining. To make his actions credible, an honest expert charges high prices for both treatment types. Thus, *low prices may serve as a warning for excessive service.*

- As potential demand decreases, signaling an honest diagnosis becomes more costly since the excess capacity makes overtreatment easier. As a result, overtreatment is unavoidable in small markets.

- Social welfare optimization may require either specializing and serving one class of customers or prioritizing a class. Whether the server specializes in or prioritizes minor or major problems depends on the relation between service values and service times.

6.3.3 Optimal screening of strategic applicants

Wang and Zhuang [656] (2011) consider an M/M/1 screening process where strategic individuals apply for entry. An *approver* can immediately approve an application or decide to *screen* the applicant. Screening numerous applicants reduces the chance of admitting a bad applicant but increases waiting times, which could reduce the willingness of good applicants to apply.

Applicants are either *good* or *bad*. Their type is private information but they also posses an observable attribute $t \in \{1, 2\}$ and the approver knows the conditional probabilities relating applicant type and observable attribute. The approver gains R when admitting a good applicant and loses C when admitting a bad applicant. An applicant of type θ obtains reward r_θ when admitted and incurs waiting cost rate c_θ when screened. A bad applicant will incur a penalty if screened.

The approver maximizes the reward from admitted good applicants net of the penalty from admitted bad applicants. The approver's strategy is the probability of screening a t-applicant while the applicants' strategies are the probabilities that a (θ, t)-potential applicant decides to apply, $\theta \in \{\text{good,bad}\}$ and $t = 1, 2$.

The authors compute the equilibrium strategies when discriminatory screening based on the customer's attribute is and is not allowed. They then investigate conditions on the model's parameters such that the benefit from discriminating is sufficiently large as to justify the discrimination.

6.4   Subscriptions and nonlinear pricing

Firms offer long-term service contracts (subscriptions) which guarantee subscribers a certain price and delay. The common price structure is a two-part tariff consisting of a fixed subscription fee and variable usage cost. At the same time, the firm can also exploit capacity by serving additional demand at a comparatively high price and without a delay guarantee. It is well known in the general economics literature that such differentiation may increase both profits and social welfare.[16] This section describes the implementation of similar principles to queueing systems.

See [569] for nonlinear pricing when customers are risk averse.

Cheng and Koehler [161] (2003) consider a Markovian model with a finite population of customers. The distribution of the time from end of service to the customer's next demand is $\exp(\lambda)$. Customers face two options for obtaining service during a period T and both have the same distribution. One is self-service, which costs a fixed exogenous *ownership cost OC*; the other is subscribing to an $M/M/s$ firm that charges a two-part tariff, consisting of a fixed subscription fee F and a fee p per service and offers *compensation* of y for each time unit of queueing. Customers have heterogeneous service valuations.

The average time between consecutive services for a customer is $\lambda^{-1} + \mu^{-1}$ in the case of self-service and $\lambda^{-1} + \mu^{-1} + W_q$ at the firm, where W_q is the expected queueing time. Therefore, a customer who values service at R will subscribe if $\frac{T}{\lambda^{-1}+\mu^{-1}+W_q}(R - p + yW_q) - F \geqslant \frac{T}{\lambda^{-1}+\mu^{-1}}R - OC$. Given the distribution of R, this relation determines the number of subscribers M.

The authors prove that under an "economies-of-scale" assumption that restricts F and p from being too large there exists a unique equilibrium solution.[17] Moreover, they prove that the firm's profit can be written as a unimodal function $V(M)$ which is independent of the way M is achieved through the decision variables F, p and y. Therefore, M can be viewed as the firm's decision variable and there is a wide range of optimal pricing and reimbursement policies. Numerical examples are used to obtain insights into both the short-run and long-run solutions.

Caldentey and Wein [123] (2006) consider a firm offering long-term contracts with a fixed price r per use. Customers arrive at rate Λ and have heterogeneous reservation prices. The market splits into two customer types: A fraction of the customers sign long-term contracts if their reservation price

---

[16]See, for example, A. Glazer and R. Hassin, "On the economics of subscriptions," *European Economic Review* **19** (1982), 343-356.

[17]An FTC feature arises from the firm's compensation policy, namely, the longer the wait the higher the compensation. Therefore, the uniqueness of an equilibrium does not follow directly from the ATC argument.

is larger than r; the remainder are *speculators* who wait to see the *spot price* $R(t)$ upon arrival. Speculating customers consist of a mixture of customers operating exclusively in the spot market and of others whose reservation price happens to be below the price r. Using the above information it is possible to compute $E[\nu_S]$, the mean reservation price for a spot-market customer. It is assumed that spot market prices behave like a variant of geometric Brownian motion and their average is equal to $E[\nu_S]$ plus a normal random variable with mean 0.

The firm sets the contract price and dynamically controls production and admission of speculative demand. The firm incurs linear costs per unit of inventory or backorder and wishes to maximize an exponential utility function. The proposed solution consists of a base-stock production policy and an admission policy for speculative demand that rejects an order if the inventory level is below a threshold that is linearly related to the spot price.

Masuda and Whang [477] (2006) analyze a model with two customer types, $i = 1, 2$, and a nonlinear pricing scheme called *fixed-up-to* (FUT). With this scheme, the server offers options (π_i, λ_i, p_i), one intended for each customer type. A customer selecting the ith option and using the service at rate λ obtains benefit $V_i(\lambda)$ and is charged $P_i(\lambda) = \pi_i + p_i(\lambda - \lambda_i)^+$, where π_i is a fixed cost for any usage rate below λ_i, and p_i is a penalty per unit of over-limit use.

Let f_i be the population size of i-customers. If each i-customer selects demand rate λ_i^o then the total demand rate is $\Lambda = \sum f_i \lambda_i^o$, and the expected waiting time in the system is an increasing convex function, $W(\Lambda)$.

The benefit to an i-customer is a concave function $V_i(\lambda)$ and it is assumed that the marginal values satisfy $V'_{i+1}(x) > V'_i(x)$ for every $x \geqslant 0$ so these functions intersect only at 0 (this is a crucial assumption). The customer's net utility is $V_i(\lambda) - P_i(\lambda) - c\lambda W(\Lambda)$, where the waiting-cost rate c is common to all customer types. Customer type is private information not known to the service provider.

The main result is that there exists an **IC** and **IR** FUT menu that achieves maximum profit among all nonlinear pricing schemes. Interestingly, the resulting equilibrium rates do not exceed the allowances λ_i, and the specific values of over-limit penalties p_i do not matter as long as they are large enough to guarantee this property.

Randhawa and Kumar [541] (2008) compare charging customers with heterogeneous service valuations by means of a subscription or on a per-use basis. Demand arriving when the server is busy is lost, subscribers pay a fee p per unit time, service is $\exp(\mu)$, and the time from the end of service or from a failed attempt to the subscriber's next demand is $\exp(\lambda)$. A demand request is denied (a failed attempt) with probability γ depending on the number N of subscribers and the capacity k of the facility. Only the potential subscribers with service valuations exceeding the expected fee to be paid for the time

interval between successful attempts actually subscribe. This yields a relation
between γ and $N(p, \gamma)$ leading to an equilibrium probability of denial. The
decision variables of the firm are capacity k, and fee p. Given these values,
an equilibrium N and γ can be computed. However, the computations are
difficult and the authors resort to asymptotic analysis in which the number of
potential subscribers n grows without bound.

The authors compare the solution with an alternative model in which customers pay per use rather than subscribe. In both cases customers have a
desired level of use and they take into consideration the probability of failed
attempts when choosing an attempt rate so that the rate of successful attempts conforms with their desired level. The authors prove that $O(n)$ scaled
profits in the two models are equal. In some interesting cases it is proved, and
demonstrated numerically in others, that scaled $O(\sqrt{n})$ profit with subscriptions is higher. In contrast, neither solution dominates the other with respect
to quality of service as reflected by the denial probability, consumer surplus,
or social welfare. A final result is that a firm may increase profits by offering
both subscription and pay-per-use.

Hall, Kopalle, and Pyke [298] (2009) consider a Markovian FCFS
queue with a single server committed to supplying service to *core customers*
for a fixed price and within a given expected waiting time W_0. The arrival
rate of these customers is fixed. The server can utilize excess capacity and
admit occasional *fill-in customers* as long as all commitments are kept. Fill-
in customers are price sensitive but not delay sensitive and their demand is
$\lambda_f(p_f)$, where p_f is the charged price.

The authors compare three options: (i) a constant price p_f independent of
the state of the queue, (ii) a constant price up to a threshold while blocking
fill-in customers above the threshold (both price and threshold are decision
variables), (iii) and state-dependent dynamic pricing. Clearly, option (iii) is
more flexible than (ii) which is more flexible than (i), and hence profits associated with the optimal choice in each option should increase from (i) to
(iii). The authors conclude from a numerical study that the main gain relative to option (i) (almost 80%) comes from implementing option (ii), and the
additional gain from dynamic pricing is much smaller (about 4%).[18]

Cachon and Feldman [117] (2011) consider a service provider in a
market with a finite number of homogenous customers. Each customer needs
service on multiple occasions, referred to as *service opportunities*, which occur
at rate τ.[19] The value of a service opportunity is a uniform random variable.
When a service opportunity occurs, a customer decides whether to submit a

---

[18]Note the similarity of these results to those in [311] where customers are full-price
sensitive and the queue is unobservable but that customers know if its length exceeds a
cutoff value.

[19]It is assumed that average sojourn time is much smaller than the average time between
service opportunities and therefore arrival rate to the system is almost independent of queue
length.

service request to the server based on three factors: service value, expected delay, and price. It is observed that a two-part tariff consisting of both a *per-use fee* and a *subscription price* can maximize the service provider's profits. The per-use price achieves the **SO** congestion and the subscription fee extracts all customer welfare. The authors look, however, for a simpler pricing scheme and compare profits that could be obtained by using one of these price structures alone.

- The authors consider both the short-term model with a fixed service capacity where the firm only chooses the pricing policy, and the long-term version where the firm also sets capacity. In both versions, subscription pricing generates more revenue than a per-use fee unless the system is highly congested (and capacity is sufficiently expensive in the long-run model).

- The loss of revenue relative to the optimal two-part tariff solution is small when congestion is low or high and is more significant in intermediate levels.

- If customers are heterogeneous with usage rates $\tau - \delta$ or $\tau + \delta$ and know their type, subscription pricing is preferred if capacity is fixed, utilization is high, and δ is sufficiently large.

Afèche, Baron, Milner, and Roet-Green [16] (2015) consider an $M/M/1$ monopoly facing two customer types, $i = 1, 2$. The market consists of N_i i-customers, receiving value r_i per service and each generating demand at rate $\gamma_i \ll \mu$. The waiting cost for both types is c per time unit. The firm designs a menu of lead times and two-part tariffs. The two-part tariff contract for each customer type consists of a subscription fee F_i and price per use p_i. The firm also controls the number of class i customers it serves.

The authors show that under asymmetric information, although all customers have the same delay sensitivity, the optimal solution prioritizes customers with higher demand rates if they have lower marginal valuations per use. Moreover, the added profit relative to the best FCFS contracts may be highly significant.

6.4.1 Hyperbolic discounting

Hyperbolic discounting explains the observed behavior of individuals preferring an immediate smaller reward to a larger one in the near future, but having the opposite preference when the choice is between rewards given in the farther future. These preferences are represented by the hyperbolic discount function $(1 + \alpha t)^{-\gamma/\alpha}$ which corresponds to a discount rate $\gamma/(1 + \alpha t)$ that decreases with time t.[20] Thus, customer time preferences are not consistent and the choice between two rewards in different future times depends not

---

[20]Common discounting with time consistent preferences is obtained when $\alpha \to 0$. The undiscounted model is obtained when also $\gamma \to 0$.

only on when these options will be realized but also on the time in which the decision is made.

Plambeck and Wang [529] (2013) consider a queue where M homogeneous customers generate independent Poisson streams of demand for a single $\exp(\mu)$ server. The server offers a contract containing a subscription fee at a constant rate $s \geqslant 0$ and a fixed usage fee u. Completion of service W units of time after arrival generates a rate r/L of benefit during the interval $[W, W+L]$ for some $L \geqslant 0$. Applying hyperbolic discounting, the present value of service is

$$
\mathrm{E}\left[\int_W^{W+L} (r/L)(1+\alpha t)^{-\gamma/\alpha}dt\right] - u - \mathrm{E}\left[\int_0^W c(1+\alpha t)^{-\gamma/\alpha}dt\right],
$$

where c is the waiting-cost rate. The expected discounted value for a customer is obtained by summing the present value of all future services and subtracting the subscription cost $\int_0^\infty s(1+\alpha t)^{-\gamma/\alpha}dt$.

The authors consider an asymptotic model where $\mu \to \infty$ while M/μ and c/μ remain constant. In this asymptotic regime, the service value has the form $\beta r - \bar{P}$ where $\bar{P} = u + c\mathrm{E}[W]$ is the full price and the condition for accepting the subscribing offer has the form $\hat{\lambda}(\eta r - \bar{P}) \geqslant s$, where $\hat{\lambda}$ is the rate at which the customer *believes* he will opt for service.[21] The parameter β ($\beta > \eta$) is interpreted as the customer's "self-control" and the customer's time preferences are consistent when $\beta = \eta$.

Some of the main results are:

- The **SO** usage fee can be positive (customers pay) or negative (the manager pays the customers). The explanation for this is that hyperbolic discounting discourages customers from seeking immediate service, and social welfare can be improved by encouraging more arrivals.

- Under a break-even budget constraint, the manager maximizes social welfare by charging for subscriptions in addition to the negative usage fee. Revenues from subscriptions will be used to cover the cost of paying customers to go for service.

- When customers are heterogeneous with respect to discounting parameters, priority scheduling can be used to dramatically increase system performance.

- The authors also investigate the impact of *naivety* when customers overestimate β, and therefore also overestimate their frequency of use. They show that in this case, under revenue-maximizing pricing, revenue is higher and utilization is lower.

---

[21]Customers may realize they will not always go for service when needs arise in the future and the anticipated frequency of getting service determines their willingness to pay for subscriptions.

Plambeck and Wang [530] (2012) ask whether a queue manager would choose to reveal queue length when customers apply hyperbolic discounting. They show that when the manager sets price and capacity to maximize profits, and β is small, revealing queue length decreases social welfare but increases profits. Therefore, it may be desirable to force the server to hide the length of the queue. This contrasts the case without hyperbolic discounting (Hassin (1986), see [1] §3.2).

6.5 Providing substitute services

The main goal of the research described in this section is to investigate the advantages of market segmentation by supplying options with differing delay or quality, and price. Some of these models are similar to material surveyed in Chapter 8. The main difference being that the focus here is on a monopolist operating a multiserver system to segment the market. See [517] for a comparison of dedicated vs. pooling servers when customers reside in a linear city, and [667] for a supply chain offering substitutable services.

6.5.1 Differentiation by price and delay

Boyaci and Ray [104] (2003) compute profit-maximizing prices and capacities for a monopoly that differentiates the market by offering two service classes. Each class is served by a dedicated M/M/1 server. Demand at server i linearly depends on price p_i and delay L_i, and on the substitution effects resulting from the price and delay of the alternative server j:

$$\lambda_i = a - \beta_p p_i + \theta_p(p_j - p_i) - \beta_L L_i + \theta_L(L_j - L_i).$$

The capacity cost is assumed to be linear, the delay L_2 of the *regular class* is fixed according to market standards, the delay L_1 of the *express class* is constrained to being smaller than L_2. The authors investigate the solution's sensitivity with respect to capacity costs and observe the importance of the ratio $\frac{\beta_p/\theta_p}{\beta_L/\theta_L}$ in characterizing the solution.

Kim [402] (2007) investigates a queueing system with two customer classes and a *movable-boundary* regime: Loss-sensitive c-customers are served in an $M/M/N_c/N_c$ subsystem and delay-sensitive p-customers are served in an $M/M/N_p$ subsystem. When all N_p p-servers are busy, free c-servers can serve p-customers, but such services are preempted when the servers are needed by new c-customer arrivals. Customers have heterogeneous service valuations and linear costs associated with loss probability (for c-customers) and delay (for p-customers). The firm's goal is to set profit-maximizing prices P_c and P_p. The author illustrates the problem by solving an example.

Printezis, Burnetas and Mohan [537] (2009) consider a service provider operating two identical $M/M/1$ queues. There are two customer classes with finite potential demand rates and waiting-cost rates $c_1 > c_2$. Customer type is private information but all customers share the same service value and service rate. The service provider sets prices $p_1 \geqslant p_2$ for servers 1 and 2, and customers react by joining a queue or balking. The solution of this model is simplified by observing the following property: The equilibrium effective joining rate of the more time sensitive class 1 can be positive in equilibrium only if class 2 fully joins. The authors give an explicit expression for a threshold M of the service rate on which the solution depends:

- $\mu \leqslant M$. In this case only 2-customers join.

- $\mu > M$. In this case class 2 is fully captured with positive net utility, while class 1 customers have zero net utility.[22]

Jayaswal, Jewkes, and Ray [362] (2011) extend [104] by considering class-dependent prices β_p^i and delay sensitivities β_L^i, such that $\beta_p^1 < \beta_p^2$ and $\beta_L^1 > \beta_L^2$. The decision variables are price and service rate.

The authors also solve the shared capacity variation of the model where all customers are served by the same server with preemptive priority given to express customers (class 1). Notable conclusions include:

- When capacity is expensive, express customers obtain faster and more expensive service under the sharing option while regular customers obtain slower and less expensive service, as compared to the dedicated servers option.

- The introduction of substitutability increases the cost of service for the regular class and decreases it for the express class, i.e., it results in a more homogeneous pricing scheme.

- When capacity becomes more expensive, a dedicated firm reduces both price and delay differentiation while a shared capacity firm reduces its delay differentiation (and may increase or decrease its price differentiation).

- **Teimoury, Modarres, Monfared, and Fathi [619] (2011)** consider a similar model and numerically demonstrate that dedicated queues and substitution effects can lead to less price and delay differentiation.

Zhao, Stecke, and Prasad [707] (2012) consider a Markovian system with two customer types, l and p (lead-time and price sensitive), which cannot be distinguished by the server. Arrival rates are λ_l and λ_p, service valuations are v_l and v_p, and waiting-cost rates are $\beta_l > \beta_p$. The utility of an i-customer

---

[22]Note that since customers have positive surplus the profit and social-welfare objectives do not coincide and thus the results of the paper do not apply to social-welfare optimization.

is $U_i(L, P) = v_i - \beta_i L - P$, where P is price and L is lead time, $i \in \{l, p\}$. Regular and express services are fulfilled by separate servers with rates $\mu_l > \mu_p$ targeted for l- and p-customers. The firm's decision variables are prices P_l and P_p, and service capacities μ_l and μ_p associated with the same linear costs.

The goal of the research is to determine conditions under which the firm can increase profits by maintaining dedicated servers rather than pooling the two customer types and operating a single server with a common price.[23] It is shown that two queues are less attractive to the firm than a single queue when $\frac{\beta_l}{\beta_p} < \frac{v_l}{v_p}$, or $\frac{v_l}{v_p} < 1$. However, the opposite outcome is possible when $\frac{\beta_l}{\beta_p} > \frac{v_l}{v_p} > 1$.

Teimoury and Fathi [618] (2013) offer a two-stage production Markovian model. In the first MTS stage the firm completes a portion θ of the manufacturing process using a base-stock policy with level s.[24] In the second MTO stage the firm completes the production process. The fraction of semi-finished products suitable for completion decreases with θ. There are costs associated with operating the system and the retailer sets θ, s, and price to maximize profits. The authors formulate this optimization problem and propose a multi-product model with linear demand, substitution effects,[25] and shared storage capacity.

6.5.2 Capacity allocation

Ros and Tuffin [550] (2004) consider a general network model with a queueing-related result (Theorem 2). Customers have heterogeneous service values and only those with a positive expected full price join. The system has fixed capacity which it can allocate to two M/M/1 facilities and charge different prices. The authors prove that revenue is maximized by pooling capacity and operating a single facility with a single price.

Kostami and Ward [414] (2009) construct a heavy-traffic approximation to a model motivated by theme park applications. The system consists of two queues with dedicated service, an *inline queue* $\mathbf{Q_I}$ and an *offline queue* $\mathbf{Q_O}$. $\mathbf{Q_I}$ is observable and reneging is not permitted. Customers renege at a constant rate from $\mathbf{Q_O}$ and although the number of customers joining this queue is observable, reneging is only recognized when the customer doesn't show up to obtain service.[26] Customers have different waiting-cost rates at the two queues and an arriving customer enters the queue with a smaller expected

---

[23]The authors mention as a possible extension the comparison to a single server with differentiated services.

[24]It is common to refer to θ as the order *penetration point* or *decoupling point*, expressing the manufacturing stage where the product becomes linked to a specific customer order.

[25]Customers are not delay sensitive and differentiation is by price only.

[26]This assumption results in an information setting similar to that of the ticket queue model in [671].

waiting cost while ignoring the reneging process from $\mathbf{Q_O}$.[27] The server incurs holding costs proportional to the length of $\mathbf{Q_I}$. The server's motivation for maintaining $\mathbf{Q_I}$ is in reducing the possibility of abandonment from $\mathbf{Q_O}$, which is associated with a penalty for each abandonment. The server's decision is how to best allocate capacity between the two queues and minimize costs. The authors use simulation to demonstrate the accuracy of their approximations.

Chau, Wang, and Chiu [136] (2010) consider a monopoly serving customers with homogeneous service valuations and heterogeneous waiting-cost rates. The firm has a fixed capacity to be allocated to several M/G/1 servers. The authors prove a general theorem which implies that for the M/G/1 special case if the firm is restricted to a single price, then profits and social welfare are maximized when capacity is pooled (i.e., allocated to a single server).

6.6   Priorities

This section on priority regimes is closely related to §6.5 on providing substitutable services. Here we deal with a single-server system that allocates priority rights to customers ready to pay the required amount to reduce their waiting costs while the alternative considered in §6.5 is maintaining multiple servers for the same purpose.

A main characteristic of a priority model with strategic customers is that, in most cases, customers choose the priority class. This is in contrast to systems with *predetermined* priority classes. These alternatives are also sometimes classified as *open class* vs. *closed class* models. Closed class models are often associated with symmetric information, whereas open classes are typical of models with asymmetric information where customer class or type is private information.

Some of the models explicitly refer to priority classes while in others the use of priorities is implied implicitly by a requirement that the solution is achievable or follows the $c\mu$-rule.

A priority regime can naturally be used to increase welfare or profits when customers are heterogeneous. It is interesting to note that priorities can also be used for these purposes when customers are homogeneous, as was first observed in [302]. Similar conclusions can be found in [70, 467, 560].

Static-priority regimes assign customers to priority classes and apply absolute-priorities with or without preemption. Relative priorities extend the achievable space, and their use for profit-maximization and social-welfare maximization is the subject of §5.3.2. The present section also includes more

---

[27]The existence of reneging from $\mathbf{Q_O}$ reduces the expected wait there but also affects the probability the customers will eventually enjoy the service. It is assumed that customers ignore both effects.

general scheduling policies that select the next customer to be served in a state-dependent manner and are not required to be work conserving (see, for example, [12] for relevant definitions).

Priority models aimed at social-welfare maximization are surveyed in §5.3.

Mandjes [467] (2003) considers an M/M/1 queue with nonlinear waiting costs represented through the value $u(w) = w^{-\alpha}$ of service completed after a delay w.[28] There are two customer classes: *Data-users* with $\alpha = \alpha_d$, and *voice-users* with $\alpha = \alpha_v$, where $0 < \alpha_d < \alpha_v$ (time is normalized so the two functions coincide at $w = 1$.) This means data-users are more sensitive to shorter delays but less sensitive to longer delays. The important qualitative property here is that the functions intersect, whereas this doesn't happen under the common assumption of linear waiting costs. The main results are:

- When an identical non-differentiating entry price is imposed only one type of user will arrive in equilibrium. This is the *class dominance* property (see [1] §3.4.1, [594] §4.5). The author computes the profit-maximizing price.

- Suppose the service rate is low such that only data-users arrive under the non-differentiating solution.

  - A server who can distinguish customer types may increase profits by directing the voice-users to a (preemptive) high-priority class with price p_H, while the data-users are directed to a low-priority class with price p_L. The author computes the profit-maximizing prices.

  - If customers are free to choose a priority class then a different equilibrium may result, i.e., these prices are not **IC**. The author computes the profit-maximizing **IC** prices.

- An example with homogeneous customers demonstrates that a two-priority system can increase profits even in this case.

- **Hayel, Ros, and Tuffin [333] (2004)** compare *generalized processor sharing* (GPS) with FCFS and priority queueing in a heavy-traffic version of [467].

Katta and Sethuraman [393] (2005) consider an M/M/1 queue with N customer types. Type i is characterized by potential demand Λ_i, service value R_i, and time value C_i. The server cannot recognize customer type and offers a profit-maximizing **IC** menu of prices and priorities. The authors prove that if ratios R_i/C_i are increasing in C_i, then the solution is work conserving

---

[28] A similar utility function is assumed in [265]. **Nel and Zhu [501] (2011)** numerically solve examples of duopoly price competition in an M/M/1 system with homogeneous atomic customers assuming utility is reciprocal to expected waiting time, i.e., $\alpha = 1$.

(includes no strategic delays), and is characterized by an index n such that the arrival rates of classes $i = 1, \ldots, n-1$ equal Λ_i, and those of $i = n+1, \ldots, N$ are 0. The main result is that profit can be further increased by *pooling* some of the customer types, provided that the delay cost distribution has a non-monotone hazard rate.[29]

Zhang, Dey and Tan [704] (2007) assume that waiting-cost rate h and service value v are perfectly correlated and uniformly distributed and customer type is private information.[30] The number of priority classes is exogenous. Priority i is associated with an expected delay guarantee d_i and a price p_i. In equilibrium there will be thresholds distinguishing the types of customers who buy different priority levels, or balk.

The authors compute profit-maximizing prices in several variations of the model according to: whether the number of priority classes is one or two; whether delay guarantees are exogenous or decision variables;[31] and if capacity is fixed or can be modified at a linear cost.

The authors also treat the single-class problem with a constraint on the *variance* of the delay.

Gilland and Warsing [255] (2009) consider equilibrium behavior in an M/M/1 queue with two priority classes. For the base model with homogeneous customers, a fixed price for high priority, costless low priority, and no balking, it is shown in [1] §4.2. that due to FTC behavior there can be three equilibrium solutions, where either all buy priority, nobody buys, or there is a mixed equilibrium. The authors continue this line of research allowing for heterogeneous waiting-cost rates uniformly distributed over [0,1] and considering a monopolistic server that maximizes profits from priority sales.

The solution is characterized by a threshold such that jobs with waiting costs above threshold buy priority. The main result of the paper is that the profit-maximizing price induces the **SO** threshold. This result differs from E&H in that the server does not obtain all customer surplus and therefore the server's objective differs from the social objective. Yet, the same solution is reached because, as follows from the authors' derivation (in their Appendix B), when service rate is normalized to 1, we arrive at $2C + \Pi = \lambda + \frac{\lambda^2}{1-\lambda}$, where C is social cost and Π is profit. Therefore, maximizing profit is equivalent to minimizing social cost. This result is extended to general service distributions and to more than two priority classes. The authors show that most of the gains (social and revenue) obtained by segregating the market can be achieved with few priority classes.

The authors also consider a variation where all service requests belong to a single customer. When the customer buys priority for urgent jobs, less urgent

---

[29] In [18] pooling can be optimal for any delay cost distribution.

[30] The case where h and v are independent is also considered and analyzed; details can be found in the online supplement of the paper.

[31] Strategic delay is not optimal in the cases considered here.

jobs will wait longer. This reduces the incentive of the owner to buy priority and indeed the authors prove that more priority is sold in equilibrium when jobs belong to nonatomic customers.

Yu, Zhao, and Sun [687] (2013) assume a waiting-cost rate $c \sim U[0,1]$, a service value R common to all customers, and the firm's capacity cost to be proportional to μ^2. The benchmark model has a single-server shared queue. The firm's decision variables are price and capacity. In the main model the firm segments the market by offering preemptive priority classes with delays $L_1 < L_2$ and prices $p_1 > p_2$. Customer equilibrium is characterized by two thresholds such that customers with low, intermediate, or high waiting costs join class 1, class 2, or balk, respectively. The profit-maximizing solution depends on a single parameter $R\Lambda$, where Λ is the potential arrival rate. The authors prove that the two-class solution outperforms the benchmark single-class solution and that the lead time in the latter solution is between optimal L_1 and L_2.

He and Chen [336] (2014) consider a two-server system where customers differ in their waiting-cost rates which can be C_H or $C_L < C_H$ and in *flexibility*. Some customers are *dedicated* to a specific server while the others are *flexible* and can be served by either server. The firm offers six contracts to differentiate the six customer types. Each contract specifies price and expected waiting time, and the contracts designed for flexible customers also specify the probability of being routed to each server. The contracts are designed to maximize the firm's profit rate subject to **IC** and **IR** constraints, and guaranteed expected waiting times must be achievable. In addition, it is assumed that dedicated customers never pretend to be flexible.

The authors prove that it suffices for the firm to observe customer delay costs to achieve the first-best solution, but knowing their flexibility alone is not sufficient. They also consider a simpler mechanism with only four contracts and the same conditions offered to both flexible customers and dedicated customers in the same queue.

Güler, Bilgiç, and Güllü [266] (2014) consider profit maximization in an M/M/1 queueing-inventory system.[32] Customers belong to a finite set of classes with heterogeneous reservation prices and waiting-cost rates. The firm follows a base-stock policy and arriving demand is immediately satisfied when the inventory level is positive.[33] Otherwise, customers join a non-preemptive priority queue. The authors first prove that the optimal priority rule for profit maximization is the $c\mu$-rule and therefore the remaining decision variables are base-stock level and class-dependent prices. Given these values, the arrival

---

[32] Extensions to M/M/m and M/G/1 systems are also considered.

[33] A higher profit might be achieved with a policy that satisfies the demand of class i only when the inventory level exceeds a threshold c_i. However, this is more difficult to solve and the policy can be harder to implement.

rates of the classes are determined as in the E&H model. The authors compute the optimal base-stock level for given arrival rates and show that:

- Prices given by the first-order conditions are **IC**.

- A continuous approximation of the state probabilities yields an explicit solution for the single-class case.

Deng, Chen, and Shen [194] (2015) formulate the design of a profit-maximizing set of contracts in the presence of valuation and waiting-cost heterogeneity as a *second-order cone programming problem*. The main qualitative part of the analysis assumes two levels of delay sensitivity with waiting-cost rates $c_H > c_L$, and two levels of service valuations $v_H > v_L$. Thus there are four customer types, LH, HH, LL, and HL, which are private information.

The authors note that their model, in contrast to previous ones, does not assume strict order among customer types. They find that under the optimal solution, conforming with the $c\mu$-rule, customers with higher waiting costs enjoy higher priority, but this is not necessarily so for customers with higher valuations.

Liu and Berry [449] (2014) consider an unobservable M/G/1 queue with two priority classes and homogeneous full-price-minimizing customers. High-priority service is controlled by a profit-maximizing *primary* service provider (SP) and regular priority service is shared by $N > 2$ competing *secondary* SPs. Competition leads the secondary SPs to charge zero price and the question is what price will the primary SP charge.[34] The main insights include:

- A characterization of conditions on the first and second moments of the service distribution leading to a profit-maximizing solution where all buy priority.

- The primary SP may earn less profits when capacity increases and the system is less congested.

- Social welfare may increase or decrease as a result of the cost-free low-priority option.

Nazerzadeh and Randhawa [500] (2015) assume customer waiting-cost rates and service valuations are random variables perfectly correlated in a sublinear way, i.e., the waiting-cost rate of a customer with valuation v is $w(v)$, and $w(v)/v$ is a decreasing function. A profit-maximizing firm designs a menu of K feasible price and delay values, where K is exogenous and the preemptive $c\mu$-rule is used. The menu induces a partition of the customer population

---

[34]It is shown in [1] §4.2 that the M/M/1 version of this model has FTC behavior and at most three equilibrium probabilities for buying priority. All buy priority can be an equilibrium associated with *rent dissipation*.

according to waiting-cost rate thresholds. The authors consider a heavy traffic approximation when the potential arrival rate is $n\Lambda$ and service rate is n, where $n \to \infty$. They show that for $K \geqslant 2$ the revenue is of the form $An + B\sqrt{n} + O(n^{1/(2K)})$, meaning the solution with two classes ($K = 2$) captures almost the entire benefits of differentiation and that offering additional classes can only increase revenues by $O(n^{1/(2K)})$.

Chen, Cui, Deng, and Shen [155] (2015) add a new dimension to contracts used to differentiate customers in an M/M/1 system under asymmetric information. Their model assumes two possible waiting-cost rates $c_H > c_L$ and two possible service valuations $v_H > v_L$, defining four customer types, LH, HH, LL, and HL with arrival rates λ_{ij}, $i, j \in \{L, H\}$. Customers are offered a menu with a choice of four contracts of the form (q_{ij}, w_{ij}, p_{ij}), $i, j \in \{L, H\}$, where q_{ij} is probability of admittance, w_{ij} expected delay if admitted, and p_{ij} price charged if admitted. A customer with service value v and waiting-cost rate c obtains utility $q_{ij}(v - cw_{ij} - p_{ij})$ when accepting the contract. The contracts and associated priority scheduling are designed to maximize revenue subject to **IR**, **IC**, and feasibility.

The use of probabilistic admission control may increase revenues since customers with high values are willing to pay more for a higher probability of admittance. The authors solve the server's problem by decomposing it into 16 subproblems. They numerically verify that adding probabilistic admission control can increase the server's revenues. While in more than 95% of the scenarios the relative gain was less than 5%, in some cases the revenue gain reached 20%.

6.6.1   Strategic delays

Afèche [9, 12] (2004, 2013) designs a revenue-maximizing menu of price and lead time as well as scheduling policy for an M/M/1 queue with two customer types differing in delay-cost rates $c_1 > c_2$ and service value distributions $F_i(v)$, $i = 1, 2$. If customer types can be distinguishable, then revenue is maximized by applying the $c\mu$-rule and by controlling arrival rates with appropriate pricing. As shown in [487], if the server cannot distinguish customer types the $c\mu$ priority policy also yields an **SO** and **IC** menu of price and lead time. However, this is not true when the server maximizes revenues, and the author derives a solution method that uses the achievable-region approach. The objective is to maximize the revenue rate over arrival rates $\boldsymbol{\lambda} = (\lambda_1, \lambda_2)$ and lead times $\boldsymbol{W} = (W_1, W_2)$ subject to the constraints that lead times be *operationally achievable* (**OA**) and **IC**. The problem is solved by first determining the optimal lead times and the corresponding scheduling policy for fixed $\boldsymbol{\lambda}$ and then optimizing the resulting revenue function over $\boldsymbol{\lambda}$.

Given a fixed $\boldsymbol{\lambda}$, the $c\mu$-rule optimizes revenues among all **OA** policies, but it need not be **IC**: 2-customers may prefer the high-priority 1-class to the low-priority 2-class or 1-customers may prefer 2-class to 1-class. In the latter case, incentive compatibility can be restored by reducing the 2-class price and artificially increasing the 2-class lead time. The author coins the term *strategic delay* for this artificial delay policy which manipulates customers' strategic service class choices, and its operational impact is that scheduling is no longer work conserving. Strategic delay involves a trade-off between revenue gain from high-priority customers and revenue loss from low-priority customers. The author identifies necessary and sufficient conditions for the revenue-maximizing **IC** solution to include strategic delay.[35]

The author also shows in [9] that for types with heterogeneous service requirements, delay tactics other than strategic delay may be optimal; namely, if patient customers have the higher $c\mu$ index it may be optimal to alter priorities relative to the $c\mu$ policy, in some cases prioritizing customers in the *reverse $c\mu$* order.

Yahalom, Harrison and Kumar [676] (2006) consider a variation of [12] in which the delay-cost function is nonlinear. The cost of a class i customer associated with delay d is $a_i C(d)$, $i = 1, 2$, where C is a convex function, and $a_1 > a_2$.

The main results are:

- The feasible region is the intersection of the achievable region and the required **IC** constraints. Therefore the solution need not be on the efficient frontier. The optimal solution in this case is either Pareto efficient, or class 1 is given priority while class 2 is *intentionally delayed*.

- For the quadratic cost function $C(w) = w^2$, it is possible to give an explicit description of the Pareto efficient frontier. Additional assumptions guarantee that a variant of the $Gc\mu$-rule will be optimal.

Maglaras, Yao, and Zeevi [465] (2015) complement the analysis of [12] using an approximate analysis to investigate the importance of strategic delays in large-scale multiserver systems characterized by large capacity and market potential. The service provider operates an $M/M/s$ system and faces N customer types. Each customer type is characterized by a finite market potential, a willingness-to-pay distribution, and a delay cost. Customer types are private information. The service provider offers $k \leqslant N$ classes distinguished by price and delay and customers select a class according to **IC** and **IR** conditions.

The authors define a simple *deterministic relaxation* (DR) which ignores queueing and is readily translated into a price-delay menu and scheduling policy that are near optimal for the stochastic problem, provided the system is sufficiently large. Their main results are:

---

[35]The policy of inflating lead-time quotes that appears in [528] resembles the notion of strategic delay. In the asymptotic analysis of [528], patient customers become infinitely patient and strategic delay does not lead to lost revenue from these customers. In contrast, the analysis here considers the tradeoff between revenue gain from high-priority customers and revenue loss from low-priority customers.

- The significance of strategic delay for systems with sufficiently large capacity and market potential depends on the number of service classes and system utilization: (a) If the DR solution prescribes $k = 2$ service classes, then strategic delay is significant only if the system is also underutilized in the DR solution. However, for two-class systems where the asymptotic effects are not in force, strategic delay can be significant even with optimized capacity. (b) If the DR solution prescribes $k \geqslant 3$ service classes, then strategic delay is significant regardless of the system utilization prescribed by the DR solution.

- If the DR solution prescribes a fully utilized system and $k \geqslant 2$ service classes, then in a large-scale system that implements the corresponding near-optimal stochastic solution all priority classes except the lowest class operate in the **QD** regime; the lowest priority class operates in the **ED** regime (in contrast to [463] where delay costs are homogeneous and there is a single service class).

Afèche and Pavlin [18] (2015) study a multi-type version of [12]. Customer types are described by a continuous distribution F of delay costs c which are perfectly correlated with the service value $V(c) = v + c \cdot d$ ($v, d > 0$). This property leads to an interesting outcome, where it may be optimal to exclude the customers with the least, the most, or with moderate delay sensitivities and service valuations. This behavior can be explained by noting that the net utility $v + c(d - w)$ to a c-customer from service with lead time w increases in c if $w < d$ and decreases in c if $w > d$.

A menu of price and lead-time contracts satisfying **IR** and **IC** conditions is such that lead times are nonincreasing in c, prices are nondecreasing in c, and the set $\mathcal{C}_a$ of admitted customers has the following structure: the set of customer types that buy low lead-time qualities, $C_l := \{c \in \mathcal{C}_a : w(c) > d\}$, and the set of types that buy high lead-time qualities, $C_h := \{c \in \mathcal{C}_a : w(c) < d\}$, are (possibly empty) intervals that include the least time-sensitive type ($c_{\min}$) and the most time-sensitive type ($c_{\max}$). Customer types (if any) that buy the intermediate lead time, $C_m := \{c \in \mathcal{C}_a : w(c) = d\}$, randomize between buying a contract and balking.

Based on these properties, the decision variables of the server can equivalently be the rates λ_h, λ_m, and λ_l of joining customers from C_h, C_m, and C_l, respectively, and the lead times $w(c)$ targeted to $c \in \mathcal{C}_a$. Revenue maximization calls for prioritizing customers according to their *virtual delay costs* whereas, **IC** requires the delays $w(c)$ to be non-increasing in customer types c. This may cause conflicts which are resolved by *pooling* adjacent types of customers into a single service class.

The authors provide necessary and sufficient conditions for three nonstandard features of the optimal solution: (i) pooling without strategic delay may be optimal for *any* delay-cost distribution, and the paper specifies whether such pooling occurs at the high, medium, or low end of the delay-cost spectrum; (ii) pricing the middle of the delay-cost spectrum out of the market; and (iii) pooling with strategic delay at the low end of the delay-cost spectrum.

6.6.2 Delay-dependent dynamic priorities

Sinha, Rangaraj, and Hemachandra [580] (2010) consider Poisson streams of *primary* and *secondary* customers with arrival rates λ_p and λ_s, identical service distributions, and the following *delay-dependent dynamic priority scheme*: At time t, the *instantaneous priority* of a customer arriving at time T is $q = b_i(t - T)$, where $b_i = b_p$ ($b_i = b_s$) for primary (secondary) customers. After completing service the server selects the customer with highest instantaneous priority. Let $\beta = b_s/b_p$, then $\beta = 0$ means that primary customers obtain absolute priority, $\beta = \infty$ means that secondary customers obtain absolute priority, and $\beta = 1$ induces the FCFS regime.

The firm has an agreement with primary customers guaranteeing them an expected waiting time of at most S_p. The firm can admit secondary customers as long as it fulfills this commitment. Secondary-class demand rate is a linear function $\lambda_s = a - b\theta - cS_s$ where θ is the admission fee and S_s the expected delay for this class.

The authors characterize the profit-maximizing price θ and priority parameter β. When S_p is large the solution gives priority to the secondary customers. For intermediate values $0 < \beta < \infty$, if S_p is too small there is no feasible solution.

An analytic proof for the exact characterization of the interval of S_p where $\beta < \infty$ is given by **Gupta, Hemachandra, and Venkateswaran [293] (2015)**. These authors obtain in [292] analogous results for a variation of [580] that allows for service preemption.

Gupta, Hemachandra, Raghav, and Venkateswaran [291] (2014) obtain complementary results to the analysis of [580]. They also consider the *switching frequency*, which is the number of times the server switches classes per number of customers served. Computational experience indicates that this performance measure is highest under FCFS, i.e., when $\beta = 1$.

Hemachandra and Gupta [339] (2015) provide a game theoretic interpretation for the optimal solution identified in [580]. The solution corresponds to the unique equilibrium of a two player non-cooperative game: Player 1 represents the queue and Player 2 represents the secondary class of customers. Player 1 maximizes the revenue rate $\theta\lambda_s$ subject to the expected waiting-time constraints by setting the admission fee θ, the expected secondary class delay S_s, and dynamic-priority parameter β. Player 2 sets the offered arrival rate $\lambda_s = a - b\theta - cS_s$. This is a constrained game as strategy sets available to each player depend on the strategy picked by the other player.

6.6.3 Priority auctions

See §6.2.10 for revenue and PoA analysis of a capacity auction, [144] for priority auctions in an MTS system with delay compensation, and [468] for a discrete-time model where auctions for service are repeated in every period.

Afèche and Mendelson [17] (2004) assume a customer's net value from service completed t time units after arrival to be $v \cdot D(t) - C(t)$, where $D(t)$ is a decreasing function with $D(0) = 1$ and C is an increasing delay-cost function with $C(0) = 0$.

The first part of [17] assumes uniform pricing for FCFS service. Let $\bar{D}(\lambda)$ and $\bar{C}(\lambda)$ denote the expected values of D and C upon service completion of a random customer. If a uniform price P is imposed, then the inverse demand function is $P(\lambda) = V'(\lambda) \cdot \bar{D}(\lambda) - \bar{C}(\lambda)$, where $V(\lambda)$ is the expected rate of value generated by the system when the arrival rate is λ. The model's assumptions ensure a unique λ^* satisfies the first-order conditions for maximizing the social-welfare objective $V(\lambda) \cdot \bar{D}(\lambda) - \lambda \cdot \bar{C}(\lambda)$. Similarly, a unique λ^M satisfies the first-order conditions for maximizing the revenue $\lambda \cdot P(\lambda)$. The authors show that λ^M may exceed λ^*, that is, a monopolist serves more customers and charges a lower price than is **SO**.

The second part of [17] considers an M/M/1 highest-bid-first model. This model assumes D and C are linear. The model with preemptive priorities is a variation of Hassin's model [303] (see [1] §4.5) with two main changes: It uses the generalized delay-cost structure, and the server requires an entry fee (equivalently, a minimum bid). The following results are obtained:

- With preemptive priorities, as shown in [303], bidding without entry fees leads to social optimality. In contrast, revenue can be increased by requiring a positive minimal bid.

- With nonpreemptive priorities the results are similar to those under uniform pricing and FCFS: the revenue-maximizing entry fee may be larger or smaller than the **SO** fee.

- Compared to uniform pricing, priority auctions give both higher social welfare and higher revenue. The percentage gains are much larger under preemptive priorities.

- The delay-cost structure greatly affects performance. The value of priority auctions significantly increases in the value-delay cost correlation.

- The scheduling policy also affects the priority auction benefits. For example, the most impatient customers always benefit under preemptive priorities as compared to uniform pricing, but may be worse off under nonpreemptive priorities.

- **Zhou and Huang [712] (2015)** consider an M/M/1 queue with heterogeneous service valuations $v \sim \mathrm{U}[0,1]$ and perfectly correlated delay cost rates $D(v) = \alpha v^\beta$. The main result is, again, that bidding will guarantee higher revenues and social welfare than will a flat fee.

Zhang, He, Ma, Cheng, and Yang [697] (2005) assume an M/M/1 queue with a highest-bidder-first (HBF) auction for preemptive priority. Balking is allowed, service value is V, and waiting-cost rates $c \sim U[0, c_{max}]$. Let $p_{min} = V - c_{max}/(\mu - \lambda)$. The authors show that the minimal bid (entry fee) which maximizes the firm's profits is $\max\{V/2, p_{min}\}$.

Abhishek, Kash, and Key [3] (2012) study the pricing schemes of a cloud service provider. Service valuations take one of two values, v_1 or v_2, and customer waiting-cost rates are continuous random variables.

Under the *pay as you go* (PAYG) scheme there is no waiting, each customer obtains a dedicated server, and pays a fixed price per unit time of service.

The firm can also sell excess capacity at a reduced price in a *spot market* using an auction. The spot market is modeled as a GI/GI/k system with preemption and bidding for job priority. An equilibrium in the spot market is associated with cutoff values (c_1, c_2) such that customers with value v_i participate in the auction if their waiting-cost rate is at most c_i, $i = 1, 2$. The authors compute the equilibrium cutoff values and the associated revenue.

When a hybrid system consisting of both a PAYG scheme and a spot market is implemented, the PAYG price set by the firm will affect cutoff values in the spot market and also the associated profit. The authors conclude (combining analysis and simulation) that in many cases revenue raised by the PAYG system in isolation dominates that of the hybrid system.

6.6.4 Bribery

The following relationship has been claimed by Gunnar Myrdal: corrupt bureaucrats may slow service to attract more bribes. This behavior conforms with analytical results obtained in a decentralized bidding model investigated by Hassin, see [1] §8.3.

Some papers in this survey reach similar conclusions with respect to a firm's incentive to invest in capacity. See [95, 159] and [361].

Jayaraman and de Véricourt [361] (2013) analyze data on bribery and find that bribery increases with delays. To explain this finding the authors develop a simple M/M/1 model with linear waiting costs and no balking. Each time a customer waits t time units in the system (even during service), the server requests fee b (a *bribe*). If the customer declines, the server sends him to the end of the queue. A bribe transaction is also associated with a fixed loss K to the server reflecting the risk of being caught. The authors compute the profit-maximizing pair (t, b). The interesting outcome is that optimal t is strictly positive, so bribes are only requested during periods of congestion.[36]

---

[36]Note that this is an FTC model. A customer's incentive to pay the fee increases when

6.6.5   Advance reservations

Oh and Su [505] (2012) consider a profit-maximizing restaurant allocating a portion of its capacity for reservations while keeping the rest for walk-in customers. Unfilled capacity from reservation no-shows cannot be reallocated to walk-in customers. The base model assumes homogeneous customers with random service values that are realized only after they decide whether to make a reservation. If the restaurant allocates a fraction of its capacity to reservations it must take into account **IC** relations guaranteeing that customers prefer to make a reservation before their service value is realized. Customers who cannot make reservations, due to the limited capacity allocated, wait for their service values to be realized and then choose between walking in or balking. A reservation comes with a no-wait guarantee while walking in is associated with a waiting time depending on the capacity allocated to walk-ins and their arrival rate. The restaurant's decision variables are the capacity allocation, the prices for customers who make reservations and for walk-ins, and the no-show penalty (often called a "nonrefundable reservation fee"). The main results are:

- Profit can be increased by levying a no-show penalty and by giving discounts to customers who make reservations. The optimal penalty equals the price of the meal.

- When market size increases, the optimal policy changes from allocating the entire capacity for reservations to a hybrid system that allocates less capacity, and finally to allocating the entire capacity for walk-ins.

Simhon and Starobinski [578] (2014) consider an N-server system with Poisson arrivals and instantaneous service granted at given discrete instants. Customers have homogeneous service valuations, know the time to the next service but cannot observe the queue. A customer's *lead time* is the time from arrival to the next service instant. The customer's decision is whether to pay a fee and make a reservation in advance. Customers with reservations obtain priority and are accepted to service if there is a free unassigned server. The remaining servers, if any, are randomly allocated to customers without advance reservations.[37]

The authors prove that in equilibrium either all customers make reservations, or none do, or there exists a threshold such that the customers who make reservations are those whose lead times are above a threshold. Moreover, there is always at least one threshold equilibrium. Assuming that the service provider can choose the equilibrium when it is not unique, the authors show that profit from reservations is maximized by charging a reservation fee only from those whose reservation is granted.

---

more customers are doing so. Therefore the equilibrium, if one exists, need not be unique. The authors observe there can also be an equilibrium where all customers reject the bribe, but don't discuss the possibility of mixed equilibria.

[37]Note the similarity to models presented in §4.1.1, but in those the arrival time of a customer is a decision variable.

6.7   Hotelling-type location models

Dobson and Stavrulaki [204] (2007) consider a profit-maximizing monopolist offering a product at price p. The firm serves customers located on a
line that contains the origin, where a single facility is located. Customers are
homogeneous except for their location. The aggregate arrival process for inter-
val I is Poisson with intensity $l|I|$, service value is $\hat{p}$, and the waiting-cost rate
is α. The service rate μ is a decision variable costing $c\mu$ per unit time. Shipping
the product to a customer at distance s takes $g(s)$ units of time. In equilibrium,
customers buying the product are those for which $\hat{p} \geqslant p + \alpha[W(\mu, \lambda) + g(s)]$
where $W(\mu, \lambda)$ denotes expected delay. This means that if the effective served
demand is λ it includes those customers located in $I(\lambda) = [-S(\lambda)/2, S(\lambda)/2]$,
where $S(\lambda) = \lambda/l$. The price $p(\mu, \lambda)$ is determined by a marginal customer
at the end of $I(\lambda)$, giving $p(\mu, \lambda) = \hat{p} - \alpha[W(\mu, \lambda) + g(S(\lambda)/2)]$. The two decision variables are now λ and μ. Assuming exponential service, and given a
value λ, the optimal service rate $\mu(\lambda)$ is shown to satisfy $\mu(\lambda) = \lambda + \sqrt{\alpha\lambda/c}$.
This square root type law resembles that of the Halfin-Whitt regime, but is
obtained here in a different non-asymptotic model.

The optimal price consists of two separable additive terms, one congestion related, and the other transportation related. An interesting outcome
is that when transportation costs are increasing convex in traveled distance,
an increase in the served demand accompanied by the appropriate profit-
maximizing service rate may be associated with an *increased* price. This result contrasts common intuition that faster facilities serve customers at lower
prices due to economies of scale.

Lastly, the authors deal with maximizing profit per unit distance of served
demand and use the solution to approximate the case where the market size
(length of demand interval) is very large and facilities can be established at a
fixed cost.

Pangburn and Stavrulaki [517] (2008) assume customers situated on a
line containing the origin where a profit-maximizing service facility is located.
There are two customer classes differing in service valuations $\hat{p}_1 > \hat{p}_2$, waiting-
cost rates, and constant densities (per unit distance). The paper considers
several models:

1. **Pooled services:** The firm sets a single FCFS server and a common
   price for both customer classes. The resulting demand of class i is an
   interval $[-S_i, S_i]$, so total demand is $\lambda = 2\lambda_1 S_1 + 2\lambda_2 S_2$. Under M/M/1
   assumptions, the authors obtain a closed-form solution for this case.

2. **Dedicated services:** The firm sets a dedicated server for each class
   with rates μ_i and price p_i, $i = 1, 2$, and the firm bears operations cost
   $c(\mu_1 + \mu_2)$. The optimal solutions are separately computed for each class.

3. **Comparison:**

    (a) The optimal pooled price is not necessarily between the two prices in the dedicated model; it is greater than the minimal but can also be higher than both.

    (b) In both pooled and dedicated models an increase in time value of either class causes a decrease in demand and in optimal capacity.

    (c) The profit ratio $\pi_{\text{dedicated}}/\pi_{\text{pooled}}$ increases, in general from a value below 1 to a value above 1 as $\hat{p}_1/\hat{p}_2$ increases from 1. Thus choosing whether to pool or to dedicate depends on the parameter values.

4. **Self-selection:** Clearly, asymmetric information, where customer types are private information and customers are free to choose their server, reduces the firm's profit. However, this reduction turns out to be minor (less than 10% on average in the computational experiments).

5. The authors also consider other queueing models. For example, they show that a pooled model with class priorities and self-selection can be a profitable alternative.

Alptekinoğlu and Corbett [40] (2010) investigate a multi-server (or multi-product) Hotelling-type model. Products are characterized by location on [0,1]. Customers are homogeneous except for their locations (ideal-product preferences) which are independent and identically distributed on $[0, 1]$. The expected utility of a θ-customer buying product z and guaranteed delay t is $\bar{p} - p - rt - d|\theta - z|$, where $\bar{p}$ denotes the reservation price and p is the price. Complete market coverage must be achieved, meaning every customer obtains nonnegative utility. The firm makes three simultaneous decisions: product location, an inventory (base-stock) policy, and price. The firm partitions the interval into segments and allocates an M/M/1 server to each.

The authors characterize some essential properties of the profit-maximizing strategy:

- MTS (i.e., positive base-stock level) is optimal for a given market segment when the segment is sufficiently narrow and densely populated.

- For a unimodal customer preference-location distribution, the optimal product line has a hybrid MTS-MTO design with MTS products clustered around the mode and MTO products at the tails.

Tan, Li, Zhang, and Yang [615] (2015) consider a directed Hotelling-type model motivated by inland waterway transportation. Customers are uniformly scattered over [0,1] and wish to reach a common destination at point 0. For a special point x_1 (the location of a river port) road-system travel from x to x_1 costs $C_R^0|x - x_1| + C_R^1$, and waterway transportation from x to the destination 0 costs $C_W(x)$. It is assumed $C_W(x)$ includes a fixed cost $C_W(0)$

and is convex on $(0, 1]$. Constraints are imposed on the cost functions, in particular $C_W(0) > C_R^0$ and $C_W'(1) < C_R^0$. Customers choose the less expensive alternative, either directly traveling from their location to the destination or travelings through x_1. Those using the port also incur M/M/1 (linear) waiting costs. The problem is to choose location x_1, capacity μ (costing $I(\mu)$ where I is convex), and price τ that maximize net profit of the port.

The authors prove that for a given location x_1, either all choose land transportation, or for $x^* < x_1$ the customers using the port are those located on $[x^*, 1]$. The authors characterize the profit-maximizing price when location and capacity are fixed, the optimal price and capacity when location is fixed, and the combined problem of selecting price, capacity and location. The solution has the port located at a point x_1^*, and the customers it serves are those from $[x_1^*, 1]$.

6.8 Searching for customers

See §9.2.1 for models where firms affect demand by setting a marketing effort level, and [703] where the server searches for the first customer in the orbit queue in a loss system with breakdowns.

Son and Ikuta [589] (2007) consider a discrete-time Geo/Geo/1/1 loss system where a profit-maximizing server incurs a fixed cost when searching for customers (i.e., the arrival process at the beginning of the next period is activated). There are also *sideline profits* obtained when the server is idle. Customers have heterogeneous service valuations. In the *admission-control* model an idle server either accepts or rejects customers according to their service valuations. In the *price-control* model, the server sets a price and customers whose valuations exceed the set price and arrive when the server is idle, join. In both cases, the server faces three options: never search for customers, always search, or search only when idle. The authors describe recursive equations for this Markov decision process and characterize the solution.

Son [586] (2007) extends [589] to a Geo/Geo/1/n queue. In this model optimal strategy may dictate searching for customers only when the queue is sufficiently *long* and the server prefers to serve more customers rather than wait for the queue to empty and enjoy the sideline income. The author derives conditions such that never searching or always searching is optimal. Clearly, without the sideline profit, both the optimal admission threshold for service valuation in the admission control version, and the optimal price in the price control version, are monotone increasing with queue length. However, here these functions first decrease with queue length and then increase. Thus, in general, optimal admission policy admits a customer with given service value only if the queue is neither too short nor too long (i.e., there are two threshold values such that the customer is admitted only in intermediate states).

Son [587] (2008) studies a variation of [586] with deterministic service time and an unbounded queue. In the admission-control case there is a state-dependent threshold such that only a customer whose service value exceeds this threshold is admitted. Under profit maximization there is a state-dependent price. When the sideline profit is high and the queue is short, the queue manager becomes more selective, knowing that if no customers are admitted, after a while the queue will empty and the server will start earning the sideline price. In such a case, both threshold and price decrease with queue length up to a point before increasing. In an extreme case, both functions can even be monotone decreasing.

Son and Ghamari [588] (2008) further extend [586] to a Geo/Geo/n/N system. Both admission thresholds on service valuation and prices set by the firm need not be monotonic in the number of customers in the system. When the sideline profit is large and the queue short, the firm may wish to have fewer customers unless they are willing to pay highly for service. As queue length increases, this effect diminishes and thresholds may decrease. However thresholds will again increase when the queue is almost saturated.

Economou and Kanta [220] (2011) consider an M/M/1/1 model where customers finding a busy server upon arrival choose between balking or joining an orbit queue. The reward from service is fixed and the waiting-cost rate is constant. The authors consider two cases. In the observable case, a customer encountering a busy server is informed of the number of customers in the orbit queue. This information is not available in the unobservable case.

In the observable model the orbit queue is FCFS. When the server becomes idle, a search for the first customer who joined the orbit queue is started. The search-time distribution is exponential with parameter α independent of queue length. Service starts once the customer is found. However, if a new arrival occurs before the search is completed, the search stops and the new arrival enters service. The same mechanism is assumed in the unobservable case. Here, a discipline of the orbit queue need not be explicitly assumed since customers are risk neutral and base their decision to join on expected values.

The unobservable case is ATC, and there exists a unique equilibrium probability that a customer facing an active server will join the orbiting queue. The authors give explicit formulas for the equilibrium, the **SO**, and the profit-maximizing probabilities. Of these three probabilities, the profit-maximizing probability is the smallest and the equilibrium probability is the largest.[38]

---

[38]Unlike the E&H model, profit and social-welfare maximization generally differ here. In the present model, customers observe the state of the server and a customer arriving when the server is idle obtains a higher surplus than one arriving when the server is busy. This means that, unless all customers balk when the server is busy, the server cannot fully extract customer surplus, in contrast to the unobservable E&H model.

In the observable case, the authors derive formulas similar to those of Naor [497] for the equilibrium, **SO**, and profit-maximizing threshold strategies. The equilibrium threshold is the greatest of the three and the profit-maximizing threshold is the smallest. This is the same order as in Naor's model, and also conforms to the order among the respective probabilities in the unobservable case. We note that though the orbit queue is FCFS the general order of service isn't FCFS since new arrivals overtake customers waiting in orbit.

The authors compare the unobservable and observable cases. Depending on the parameters, social benefit in equilibrium may be smaller when the queue is observable.

6.8.1   Attracting demand

Zhou, Lian, and Wu [713] (2014) consider an $M/M/1$ profit-maximizing firm with potential demand rate Λ of *informed* customers (I-customers). The firm can attract more demand by offering a free (shorter) *experience service* to *uninformed* customers (U-customers). It is assumed that the number of U-customers is unlimited and the firm could initiate an experience service whenever the server is free. However, service cannot be interrupted once started and therefore serving U-customers increases I-customers' waiting times. The advantage of offering the free service is that a given proportion of U-customers will continue to obtain the regular service at the regular price.

The authors derive the expected waiting time of I-customers when the firm serves a given demand rate of U-customers with experience service, and characterize the optimal price and admission rate of U-customers. In particular, only in a scarce demand market, i.e., when Λ is below a given threshold, will the firm provide experience service.

Nair, Wierman, and Zwart [495] (2015) consider a two-stage (Stackelberg) game between a profit-maximizing firm providing for free service and its *user base*. Revenues are proportional to demand (such as revenue from advertising), and the firm incurs a fixed cost per server. Users incur linear waiting costs. Service value per user is an increasing function of arrival rate, thus expressing positive *network effects*. The firm first declares the number of servers it will employ, each with an exogenous fixed service rate, and demand is determined consequently. The authors consider two cases. In the first, users cooperate and set λ so as to maximize aggregate welfare. In the other case, users are noncooperative and equilibrium demand is achieved either at its maximal potential size or when customers are left with no surplus.

As market size becomes large, profit maximization leads to heavy traffic.[39] When customers do not cooperate, the firm essentially provides the minimum number of servers required to serve the full potential user base. Stronger positive network effects and lack of customer cooperation lead to increased de-

---

[39]Similar results are obtained in [422, 528] and [532].

mand and profit for the service provider. A main finding here is that lack of cooperation has a more significant impact than the positive network effects.

Chapter 7

Competition

Competition in queueing systems takes different forms. Firms compete by setting price, capacity, delay guarantee, information, and queue discipline. When capacities are fixed, firms can compete by quoting prices, and delays then result from equilibrium demands. Alternatively, firms can quote delay guarantees and let prices adjust accordingly. The latter option has been named *time-based competition*. In a long-run model, any two out of price, lead time, or capacity, can be set by competing firms while the third variable will be determined by equilibrium conditions. See, for example, [32].

Relevant literature is surveyed in [1] §7. Summaries on competition research are also scattered in other chapters. See the following sections for competition models focusing on various topics: Observable queues §2.6.4; control of systems with private and public servers §5.6; PoA §5.8; providers of complementary services §8.3; breakdowns §10.2.3; demand allocation that induces competing servers to invest in capacity §9.3; supply chains §9.4; firms selected according to an attraction model §11.2.2; and competition over location §11.2.3. See also [439] and [276] for lead-time and ordering-fee competition, respectively, where customers place duplicate orders, [476] for competition when firms choose buffer size, [54] for capacity competition when utility increases in service duration, [342, 567] for delay-quote competition, [679] for competition with benchmark effects, and [127] for competition between servers facing a single customer.

Models of routing customers to servers often relate to competition among servers. Several such models related to competition are described in Chapter 8.

7.1   Competition when customers maximize utility

Sattinger [561] (2002) considers competition in a closed unobservable system with a finite population of customers. The time from service termination to the instant a customer returns to the queue is exponentially distributed. The firms are homogeneous, incur a fixed cost per customer served, and set prices to maximize expected profit. Arriving customers select a firm that minimizes discounted expected costs.

The effective arrival rate induced by this system is approximated by the waiting time in an M/M/1 system. The author proves that the equilibrium is unique and symmetric, and provides an explicit formula for the equilibrium price which can be used to determine the long-run number of firms in the market when firms also incur a fixed operating cost rate.

Chen and Wan [142] (2003) study simultaneous price competition between two firms in a market with homogeneous customers. Firm i has fixed capacity μ_i; the potential total arrival rate to both firms is Λ. Customers choosing service from firm i incur a waiting cost h_i per unit time, pay price pi, and receive service value R_i.

With identical firms, the authors define thresholds $\underline{\Lambda} < \bar{\Lambda}$ and show that:

- A pure Nash equilibrium always exists.

- For the *scarce demand case* where $\Lambda \leqslant \underline{\Lambda}$, there exists a unique equilibrium given by $p_1 = p_2 = 4h\Lambda/(2\mu - \Lambda)^2$ and corresponding arrival rate $\lambda_1 = \lambda_2 = \Lambda/2$.

- In the *moderate demand case* where $\underline{\Lambda} < \Lambda < \bar{\Lambda}$, there may exist a continuum of equilibria. In particular, a unique symmetric equilibrium always exists such that $\lambda_1 = \lambda_2 = \Lambda/2$ and $p_1 = p_2 = R - h/(\mu - \Lambda/2)$.

- In the *ample market case* with $\Lambda \geqslant \bar{\Lambda}$, each server charges the monopoly price and obtains the arrival rate of a monopoly.

- As opposed to the other two cases where consumer surplus is zero, in the scarce demand case consumer surplus is positive.

With heterogeneous firms the authors show that:

- When demand is scarce there exists a unique equilibrium in which only one firm operates, but the price it sets is lower than the monopoly price and consumer surplus is positive.

- For higher values of Λ there is a unique equilibrium where both firms operate and consumer surplus remains positive.

- When Λ further increases, an equilibrium does not exist or a continuum of equilibria exist in which consumer surplus is zero.

- When Λ exceeds a certain threshold there exists a unique equilibrium in which both firms charge monopoly prices.

- When a unique equilibrium exists, the firm with the higher service rate or lower waiting-cost rate can charge a higher price and capture a larger market share.

Cheng, Demirkan, and Koehler [160] (2003) independently consider the model of [142]. They do not carry out a complete case analysis as [142], however, they also consider the long-run version where capacity is a decision variable. The authors present proven results and conjectures supported by a numerical study. The main results are:

- As in [142], in the short run the firm with higher capacity charges a higher price and enjoys a larger market share.

- In the long run, assuming fixed but heterogeneous marginal capacity costs, the firm with a lower marginal capacity cost charges a higher price and realizes a higher profit.

- An increase in customer delay cost increases profits in the short run but reduces profits in the long run!

Touati, Dube, and Wynter [627] (2004) assume customers are characterized by a parameter, $\alpha \sim U[0,1]$ such that the utility of an α-customer paying fee p and waiting w time units is $\alpha p + (1 - \alpha)\gamma w$. The market has two M/M/1 service suppliers with capacities μ_1 and μ_2 and prices $p_1 > p_2$. If both suppliers are active, then clearly w_1 must be less than w_2. Consequently, there is a threshold value such that only customers having α below the threshold join supplier 1. Waiting times depend on arrival rates and hence on the threshold value, giving a system of fixed-point equations. The main result is that a solution to this system exists iff $|\lambda - \mu_2| \leqslant \mu_1$. Moreover, the solution is unique. The authors also analyze the sensitivity of the solution with respect to capacity costs, price difference, and total arrival rate.

Chen and Wan [143] (2005) extend their short-run model of two homogeneous firms [142] by allowing the firms to choose their capacities. The model assumes firms incur cost c per unit capacity, operating cost r per customer, and fixed cost σ, all per time unit. Denote the optimal profit rate of a monopoly firm as $F(\Lambda)$. The authors find that:

- If $F(\Lambda) < \sigma$, no firm operates.

- If $F(\Lambda/2) < \sigma \leqslant F(\Lambda)$, there are two equilibria in which one firm operates as a monopoly.

- If $F(\Lambda/2) = \sigma$, three equilibria exist: either of the firms operates as a monopoly or both firms operate with identical market share ($\lambda_1 = \lambda_2 = \Lambda/2$).

- If $F(\Lambda/2) > \sigma$, there exists a continuum of equilibria in which the firm installing the higher capacity also sets a higher price, captures the larger market share, and enjoys higher revenues.

- Consumer surplus is zero in all cases.

- Equilibria where both firms operate together are not **SO**, and each firm sets lower capacity and higher price than when operating as a monopoly (which is **SO**).[1]

Zhang, Dey and Tan [705] (2008)[2] consider two identical M/G/1 servers competing in a market where customer service valuations $v \sim \mathrm{U}[0,1]$ are perfectly correlated with the waiting-time cost rate γv.

A delay guarantee d is exogenously dictated by market standards and decision variables are the prices p_j, $j = 1, 2$. The utility of a v-customer selecting server j is $v(1 - \gamma d) - p_j$, and arriving customers either select one of the servers or balk.

Let $\bar{p}$ be the price for which demand induces delay d when there is just a single server. Let $\underline{p}$ be the price with the same property when demand is equally divided between two servers. The authors argue that a price p defines a symmetric equilibrium if (and only if) $\underline{p} \leqslant p \leqslant \bar{p}$. The reason being that an increase in price would reduce the firm's demand to zero whereas decreasing the price would attract the entire demand and increase the delay at that firm above the standard d. The results are extended by allowing for two priority classes and letting the delay guarantee be a decision variable.

Variations of [705] are treated by **Zhang, Tan, and Dey [706] (2009)**. In

---

[1]It is interesting to compare the results of [142, 143] with those obtained by **de Palma and Leruth [515]** for a *static* model where each member of a population of customers chooses between joining one of two heterogeneous servers or balking, and expected waiting time at a server is proportional to their workload. A unique equilibrium exists when capacities are fixed. When firms choose capacities at a linear cost and customers are homogeneous, a unique and symmetric equilibrium exists. When customers have heterogeneous waiting-cost rates, firms tend to offer differentiated capacities.

[2]Another part of this paper is described in §11.2.2.

one such variation, each competitor chooses one of two service levels $d_L > d_H$. It is shown that for thresholds $\rho_1 < \rho_2 < \rho_3$, when $\rho_1 < \rho < \rho_2$ both providers choose d_H, when $\rho > \rho_3$ both choose d_L, otherwise they opt for differentiated service. Another variation considers sequential (Stackelberg) price competition where the leader chooses delay guarantee d_1 and the follower chooses delay guarantee d_2. The authors show that for thresholds $\rho_4 < \rho_5 < \rho_6$ the solution is (d_H, d_L) if $\rho \leqslant \rho_4$, (d_H, d_H) if $\rho_4 < \rho < \rho_5$, (d_L, d_H) if $\rho_5 \leqslant \rho \leqslant \rho_6$, and (d_L, d_L) if $\rho > \rho_6$.

Fan, Kumar, and Whinston [231] (2009) consider competition between two profit-maximizing sellers with products having different *implementation costs* $c_1 > c_2$. Customers have heterogeneous sensitivities $\theta \sim U[0,1]$ to these costs, and the utility of a θ-customer obtaining the product from seller i for price p_i is $V - p_i - \theta c_i$. Assuming $p_1 < p_2$, customers buying from seller 2 are those with $\theta > (p_2 - p_1)/(c_1 - c_2)$.

It is further assumed that while seller 1 supplies the product instantaneously upon demand, seller 2 is an MTO M/M/1 system. The sequence of events is such that first the two sellers set prices and then seller 2 sets capacity μ which guarantees an expected delay of no more than a given exogenous standard, incurring capacity costs of $\gamma_0 + \gamma_1 \mu$ per unit time.

The authors solve the equilibrium of this game. They also solve a variation where γ_1 is a Bernoulli random variable whose realization is known to seller 2 but not to seller 1 when they set prices.

Melo [483] (2014) derives sufficient conditions for a pure price equilibrium to exist in a directed acyclic network. The network has multiple origins, each with a given demand rate and a common destination d. Links are owned by profit-maximizing servers that set usage fees. Users are heterogeneous such that the utility associated with using a link for a random customer is composed of a random valuation minus the full cost associated with traversing this link. The utility of traversing a path is the sum of link utilities and in equilibrium users select utility-maximizing paths.

The author specifically treats a load-balancing special case with M/M/1 parallel servers. In this case, a pure strategy price equilibrium may fail to exist in highly congested networks. When there is free entry of new firms (links of the network), the number of firms exceeds the social optimum.

Sadat, Abouee-Mehrizi, and Carter [553] (2015) derive a utility function $U_i = C_i \left(1 - \frac{r\theta}{\mu_i - \lambda_i + r\theta}\right) - P_i$, based on patients' perceived quality of life when selecting hospital i, $i = 1, 2$. In this expression r is a discount rate, P_i is the admission price, and C_i and θ are constants. The hospitals are modeled as M/M/1 queues.

The authors derive sufficient conditions for four equilibrium types: (i) each hospital behaves as a monopoly; (ii) one hospital dominates the market; (iii) high price competition and positive customer surplus; (iv) moderate price competition and zero customer utility. An example demonstrates the possibility that no equilibrium exists.

Do, Tran, Tran, Pham, Golam, Alam, and Hong [202] (2015)
consider competition between an M/M/1 *public* firm and an M/M/∞ *cloud broker*.[3] All servers have the same service rate. Customers minimize full costs and firms maximize revenues. Numerical results indicate that the broker would enjoy higher revenues if it plays the leader's role in a Stackelberg game.

7.1.1   Cournot competition

See [576] for Cournot competition and partial cooperation.

Nam [496] (1997) examines duopoly Cournot competition between identical M/M/1 servers in a market where customers have heterogeneous service valuations and homogeneous waiting-cost rates. Servers compete by setting the rate of demand they are willing to serve which in turn determines prices satisfying equilibrium conditions. The author provides sufficient conditions that exclude the possibility of an asymmetric equilibrium.

Musacchio and Wu [489] (2008) consider Cournot competition between M/M/1 servers and with two types of price-sensitive customers: *voice traffic* and *web traffic*. If service provider i offers voice customers service rate y_i and web customers service rate x_i such that $x_i + y_i \leqslant \mu_i$, then the resulting prices are $p_v(y) = ak - by$ and $p_w(x) = k - x$, where $y = \sum y_i$ and $x = \sum x_i$. Social welfare is the sum of providers' profits and consumer surplus, which is simply $kx - 0.5x^2 + aky - 0.5by^2$. Two service architectures are analyzed:

- **Priority architecture:** Voice traffic arrives as a constant fluid at rate y, while web traffic arrivals are Poisson and served as in an M/M/1 system with capacity $\mu - y$. In this case there exists a unique equilibrium and **PoA**$\leqslant \frac{4}{3}$.

- **Shared architecture:** Both customer types are queued together and the expected delay of voice customers is constrained not to exceed an upper bound $D_{\max}$. If $y_i = 0$ the constraint is simply $x_i \leqslant \mu_i$. The latter exception induces a non-convex strategy space, and it remains unclear whether an equilibrium will always exist. However, provided one does exist, the authors prove that **PoA**$\leqslant 2$ and the bound is tight.

Afanasyev and Mendelson [8] (2010) consider a Cournot game between two M/G/1 servers. Customers are homogeneous except for service valuations. Let $V(\lambda)$ be the expected total value created for users per unit

---

[3]See [199, 628] for similar models.

time when the effective arrival rate is λ.[4] Servers announce capacities μ_j and arrival rates λ_j, $j = 1, 2$. Customers then decide which server (if any) to join and equilibrium prices adjust accordingly.[5] The server incurs an operations cost c per customer and a unit capacity cost g_j.

A joining customer with service value U selects a server that maximizes $U - dW_j - P_j$, where W_j is expected waiting time at server j, d is waiting-cost rate, and P_j the price. For given λ_j and μ_j, $j = 1, 2$, there will be a threshold $\bar{U}$ such that only U-customers with $U \geqslant \bar{U}$ join. If only one server, say server j, is active, then P_j is obtained from $V'(\lambda_j) = P_j + dW_j$. If both servers are active customers must be indifferent between them and $P_j = V'(\lambda_1 + \lambda_2) - dW_j$ for $j = 1, 2$.

The equilibrium may involve both, either, or neither servers being active and multiple equilibria, possibly of different types, may exist.

In a more general setting the authors assume heterogeneous waiting costs perfectly correlated with service valuations. In general, the qualitative difference between this case and the previous one is that the equilibrium segments the market such that one server supplies fast service and serves customers with higher waiting costs while the other serves customers with lower valuations.[6]

7.1.2 Investment incentives

Johari, Weintraub, and van Roy [375] (2010) assume each of N service providers chooses price p_j and investment level I_j. The congestion cost incurred by a customer joining the jth server is a function $l_j(\lambda_j, I_j)$ of the demand served λ_j and investment I_j. Customers have heterogeneous service valuations and in equilibrium the marginal utility obtained by an additional infinitesimal customer is equal to the full price $p_j + l_j(\lambda_j, I_j)$ at every active server.

The authors concentrate on models exhibiting *nonincreasing returns on investment*. Let $K_j(\lambda_j, I_j) = \lambda_j l_j(\lambda_j, I_j)$ denote the *total congestion cost* at j. Then for all $\alpha > 1$, $K_j(\alpha\lambda, \alpha I) \geqslant \alpha K_j(\lambda, I)$. This means that for a fixed total investment the congestion cost associated with a single firm serving the whole market is *greater* than the total cost of several firms equally splitting the demand and the investment expenditures. Some of the main results are:

- Sufficient conditions for the existence of equilibrium. The simplest condition is: If all l_j functions are concave in λ, and the inverse demand function $P(\lambda)$ is concave, then an equilibrium exists.

- Suppose the congestion functions exhibit constant returns on investment

---

[4]The inverse demand function $V'(\lambda_j)$ is assumed to satisfy a condition that generalizes linear ($V' = \alpha - \beta\lambda$, $\alpha, \beta > 0$) and constant elasticity ($V' = k\lambda^{-\beta}$, $k > 0$, $0 < \beta < 1$) demand functions.

[5]The authors note that when *prices* are announced and demand reacts accordingly an equilibrium may not exist, as shown in [118].

[6]This is similar to classic results by Levhari and Luski (1976, 1978), see [1] §7.1.1.

(for example, the loss probability of an $M/M/1/s$ system). If an equilibrium exists, then it is unique and the number of customers served will be less than the **SO** level.

- Assume all firms have the same congestion cost functions $l(\lambda, I)$. Define the *marginal rate of substitution* $\mathrm{MRS}(\lambda, I) = -\frac{\partial l(\lambda, I)/\partial \lambda}{\partial l(\lambda, l)/\partial I}$ and assume that $\frac{\partial}{\partial I}\mathrm{MRS}(\lambda, I) \geqslant \frac{1}{\lambda}$. Then, prices and investment levels in equilibrium are uniquely determined, demand is equally divided among active servers, and total demand served is less than **SO**.

- When new firms can enter the market, and under additional assumptions on the model's functions, the number of entering firms will exceed the **SO** number. However, if the fixed cost of entry decreases to 0 the resulting free-entry equilibrium will be asymptotically efficient.

DiPalantino, Johari, and Weintraub [197] (2011) consider a two-stage setting. In the first stage, each firm chooses whether to offer it customers a contract of *price and investment level* (P-IL) or *price and service* (i.e., delay cost guarantee) *level* (P-SL). In the second stage, firm j chooses price p_j, and either investment I_j or delay-cost guarantee h_j that conform with its commitment. It is assumed that the delay-cost function exhibits *constant returns on investment*, i.e., is a function $h_j(\lambda_j/I_j)$.

The equilibrium full price experienced by a customer served by firm j, $f_j = p_j + h_j(\lambda/I_j)$, is equal for all active firms. The authors prove the uniqueness of the equilibrium market full price.

The main results are:

- The market full price in the second-stage is lower when all firms choose P-SL than when all choose P-IL (the latter case is investigated in [375]). This result is interpreted as expressing more intense competition in the P-SL game.

- If firms are identical, offering P-IL contracts is weakly dominant and the only subgame-perfect equilibrium where all firms make positive profits is when all offer P-IL contracts.

- In the asymmetric case, either all firms offer P-IL contracts or there is a unique active firm that offers a P-SL contract.

7.2   Competition with exogenous demand functions

Cachon and Harker [118] (2002) consider two $M/M/1$ servers, $i = 1, 2$, competing for homogeneous full-price sensitive customers. Full price amounts to $f_i = p_i + g_i$, where p_i is service fee and $g_i = (\mu_i - \lambda_i)^{-1}$ is expected waiting time. It is assumed the firm can reliably commit to full price level, which it achieves by selecting the necessary service rate μ_i at a linear cost $k_i\mu_i$. Given full prices f_1 and f_2, the demand rate for server i is $\lambda_i(f_1, f_2)$. It follows that optimal values p_i and μ_i that achieve f_i satisfy $g_i(f_i) = \sqrt{k_i/\lambda_i(f_i, f_j)}$, where f_j is the full price at the other server. This leads to the following expression demonstrating scale economies of profit for server i:

$$\pi_i(f_i, f_j) = (f_i - k_i)\lambda_i - 2\sqrt{k_i\lambda_i(f_i, f_j)}.$$

As a result, it is possible that operating is profitable for a firm only if a minimum size of demand is secured. This increases price competition and equilibrium does not always exist. The authors analyze different possibilities, such as an equilibrium with both firms participating, an equilibrium where only one firm participates, or no equilibrium. The authors also demonstrate how outsourcing can mitigate competition, see §9.4.1.

Ilmakunnas [354] (2002) investigates investment decisions with two competing profit-maximizing service providers. The competition has two stages. In the first stage, firms choose their capacities; in the second they select prices. Demand is a linear function of full price. Firm i incurs costs $\lambda_i c_1(\mu_i) + c_2(\mu_i)$. In equilibrium both firms offer the same full price. A firm's investment in increasing capacity has two effects on its rival. For the given de-mand level, the market share of firm i increases and that of its rival decreases until the full prices re-balance. However, this decrease in the market full price causes an increase in total demand which both firms enjoy.

The author proves that the resulting increase in the rival's demand does not compensate for the loss of market share. The main result is that firms underinvest in capacity in this two-stage setting in comparison to the solution obtained when they set price and capacity simultaneously. The explanation provided for this result is that underinvestment commits firms to longer queues and relaxes the second-stage price competition.

Allon and Federgruen [32] (2007) consider competition under the *sep-arable demand* model

$$\lambda_i = \left[a_i(\theta_i) - \sum_{j \neq i} \alpha_{ij}(\theta_j) - b_i p_i + \sum_{j \neq i} \beta_{ij} p_j \right]^+,$$

where θ_i denotes *service level*, specifically $\theta_i = \bar{w} - w_i$ where w_i is expected delay and $\bar{w}$ a benchmark upper bound, p_i denotes price, a_i is increasing concave, and α_{ij} are nondecreasing functions.[7,8,9] This demand model includes two additional assumptions:

1. A uniform price increase by all firms cannot result in an increase in demand volume for any firm, i.e., $b_i > \sum_{j \neq i} \beta_{ij}$, $i = 1, \ldots, N$.

2. A price increase by a firm cannot result in an increase in industry aggregate demand, i.e., $b_i > \sum_{j \neq i} \beta_{ji}$, $i = 1, \ldots, N$.

Firm i selects service rate μ at a linear cost $c_i(\mu) = \gamma \mu$ so as to guarantee any given waiting-time standard between 0 and $\bar{w}$. The authors consider three types of competition differing in the way firms make strategic choices: *simultaneous competition* (SC), where firms make all choices simultaneously, *service-level-first competition* (SF), where firms initially choose waiting time standards and then select prices in a second stage, and *price-first competition* (PF), where these choices are made in reverse order.[10] In all three cases an equilibrium pair of price and service-level vectors exists, and a numerical study gives strong indications that the equilibrium is unique. It turns out that equilibrium solutions are identical for the (SC) and (PF) cases and if prices are fixed, then each firm's equilibrium service level is independent of the service levels adopted by its competitors. Prices and service levels under the (SF) setting are higher than those under (SC) and (PF) competitions.

Allon and Federgruen [33] (2008) generalize their model [32] of M/M/1 servers with linear capacity costs. The capacity cost required for satisfying the delay goal $\theta_i = 1/w_i$ for a given demand λ_i is denoted $C_i(\lambda_i, \theta_i)$. Firm i maximizes $\pi_i = p_i \lambda_i - C_i(\lambda_i, \theta_i)$.

The cost function has the form:

$$C_i(\lambda_i, \theta_i) = B_1 \lambda_i + B_2 \theta_i + \sqrt{B_3 \lambda_i^2 + B_4 \lambda_i \theta_i + B_2^2 \theta_i^2}.$$

The authors show that a wide range of queueing models can be captured, exactly or approximately, by this cost function. For the M/M/1 model $\mu_i = \lambda_i + (1/w_i)$ and therefore C_i is affine. The authors show that the same property holds when capacity is optimally allocated over the nodes of an open Jackson network so as to maintain given θ_i and λ_i. Other examples include the M/G/1 queue and the well-known Kingman's bound for the GI/GI/1 queue.

The authors present sufficient conditions for the existence of an equilibrium in the three models considered.

Allon and Federgruen [34] (2009) examine a Markovian model of service providers competing in a market with customer classes by setting prices,

---

[7]The authors describe an economic model that leads to this demand function. They also show that it generalizes the full-price model, see p. 41 in [32].

[8]Part of [32], but with different notation, can also be found in [513].

[9]A similar model is independently discussed in [669].

[10]The second stage of "price only" and of "service only" are interesting on their own account.

capacities, and priority disciplines. Each class has its own demand function according to the separable demand model of [32]. Information is symmetric, so customer types are observable to the server.

The authors first solve the problem of a server facing demand rates λ^l, $l = 1, \ldots, J$ and choosing the service discipline to fulfill the expected waiting time guarantees w^l, $l = 1, \ldots, J$ while using the minimum possible service rate.

An equilibrium always exists when waiting-time standards are exogenous and competition is only over price (price competition). The authors also give sufficient conditions for the existence of an equilibrium when firms compete over delay guarantees while prices are exogenously fixed (waiting-time competition), and when competing over price and delay guarantees (simultaneous competition).

Hong, Hsu, Wu, and Yeh [343] (2012) consider competition between two M/M/1 servers in a market where customers have alternative ways of obtaining service at given values of price p_M and lead-time t_M. The arrival rate to server 1 depends linearly on the price and lead-time differences $t_1 - t_M$, $t_1 - t_2$, $p_1 - p_M$, and $p_1 - p_2$, and similarly for server 2. The authors provide sufficient conditions for the existence and uniqueness of price equilibrium allowing for heterogeneous service rates. In particular, a unique equilibrium always exists when service rates are equal.

Kavurmacioglu, Alanyali, and Starobinski [394] (2015) consider competition between $M(n)/G/C_i/C_i$ service providers, $i = 1, 2$. Provider i serves demand of *primary users* (PUs) arriving at rate λ_i and earns K_i per served PU. The servers also offer service to *secondary users* (SUs) where price is a decision variable. SUs demand rate is a decreasing function $\sigma(p)$ of $p = \min\{p_1, p_2\}$. If $p_1 \neq p_2$, the provider with the lowest price captures the entire SU market.[11] Otherwise, if $p_1 = p_2$, demand is divided between service providers according to fixed exogenous proportions.

The authors compare the outcome of the competition under two scenarios:

- **Coordinated access:** The providers implement admission control on SUs (assumed to have no effect on demand). The authors define the *break-even* price p^{BE} where immediate revenue balances the opportunity cost of a secondary request. If $p < p_i^{\mathrm{BE}}$, it is not profitable to admit SUs. If $p_1^{\mathrm{BE}} < p_2^{\mathrm{BE}}$, competition results in provider 1 capturing the entire SU market. If $p_1^{\mathrm{BE}} = p_2^{\mathrm{BE}}$, both providers will set prices to this value.

- **Uncoordinated access:** The providers cannot apply admission control to SUs. In this case there is a continuum of equilibria that induce the providers to share the SU market at a price above the break-even price.

---

[11]Cf. [705] where a similar assumption leads to a continuum of equilibria as obtained here in the uncoordinated access case.

7.3 Competition with limited cooperation

Competing firms often recognize that some degree of indirect cooperation can serve their interests. In the context of queueing systems, servers may compete for customers while at the same time cooperating to reduce overall balking, share fixed costs, or even outsource to each other when lacking sufficient capacity.

Tan, Chiang, and Mookerjee [614] (2006) consider two profit-maximizing M/M/1/1 servers with demand rates λ_i and service rates μ_i, $i = 1, 2$. An arriving customer can be *redirected* from a busy server to one that is idle.[12] In the main model, servers gain a fixed amount (normalized to 1) when a customer from their customer base is served. The decision variables are the price p_i to charge the other server for redirected demand and the probability q_i of accepting demand redirected from the other server.

If q_2 decreases, server 1 will respond by reducing q_1 and reserving increased capacity to satisfy its own demand. Therefore, we have an FTC case. The authors prove that for given prices p_1 and p_2, the best response is a threshold policy: $q_i = 0$ if $q_j \leqslant \theta_j^*$, and $q_i = 1$ when $q_j \geqslant \theta_j^*$, for $i \neq j$. Therefore, there can be three equilibrium solutions in general: $\{(0, 0), (\theta_1^*, \theta_2^*), (1, 1)\}$. In other cases there is a unique pure equilibrium, and the authors give a complete characterization of the equilibria in the (p_1, p_2) space.

Given this characterization and assuming the dominant strategy $(1, 1)$ prevails when the equilibrium is not unique, the authors compute the best price response. They conclude that $(p_1, p_2) = (1, 1)$ is always an equilibrium (each server charges the full price for serving demand from the other server), and in some cases there is also an internal equilibrium. In particular, such an equilibrium always exists if $\lambda_1 = \lambda_2$ and $\mu_1 = \mu_2$.

Shrimali [576] (2010) discusses a model with queueing examples where delay costs are incurred by the service providers and represent queueing time or loss probabilities. The author considers a two-stage Cournot competition between two asymmetric service providers over price-sensitive customers. In the first stage, each provider sets the size of its customer base and the amount it charges customers belonging to the competitor's customer base. The market retail price is determined through an inverse demand function. In the second stage, each server decides the fraction of demand to be outsourced to the competing server. The providers are subject to nonlinear variable operating costs and thus it may be profitable to enter the resource-sharing agreement. The model results in a subgame-perfect equilibrium with both providers serv-

---

[12]See [195] for a related model.

ing positively sized customer base, but with only one using the other server's resources.[13]

7.3.1 Marketplace competition

Allon, Bassamboo, and Çil [29] (2012) consider price competition in a marketplace with many identical service providers. Customers are homogeneous, obtain a fixed reward R upon service completion, incur linear waiting costs, and renege after an exponentially distributed time.

In the base model each server is modeled as an unobservable M/M/1 system with exponential reneging. Servers announce prices and customers select a server. A unique customer equilibrium choice exists for any price vector and a unique symmetric price equilibrium exists if the reneging rate is high.

The base model assumes customers choose *preferred prices*, and that a *moderating firm* informs them of the lowest price offered by an idle server. The customer will join that server unless the offered price exceeds his preferred price, in which case the customer joins a queue and waits until admitted to a server with a price below the preferred price.[14] The authors resort to asymptotic analysis when the number of servers is very large. The queue regime is not specified in the model and does not affect the asymptotic results.

The outcome depends on the size of the demand:

- **A buyer's market:** Arrival rate is below aggregate service rate. In this case, when the number of servers grows to infinity the symmetric equilibrium price is zero. Thus, providing information on available servers increases competition and decreases revenues.

- **A seller's market:** Arrival rate exceeds service rate. In this case many equilibrium prices may exist. The authors show that if nonbinding communication among the servers is enabled, their equilibrium profit is asymptotically maximal, they charge the maximum price $p = R$, and customers incur no wait.

Allon, Bassamboo, and Çil [30] (2013) add skill and capacity management to their model [29]. There are two customer classes, A and B, and k *candidate servers*. The values S_A and S_B that a server generates when serving an A or a B-customer are random variables with a given joint pdf. The moderating firm runs two skill tests, exam A and exam B, and servers found to have $S_I \geqslant \omega_I$ are *eligible* to serve I-customers, $I \in \{A, B\}$. Servers that pass both exams are *flexible* and can choose the customer type they want to serve. Customers know the server's type and the expected skill level given this information and obtain their expected *net reward* by subtracting the quoted

---

[13]This outcome can be contrasted with the inefficiency in [384], where both providers end up serving customers from the competitor's customer base.

[14]See [300] for similar behavior in a single-server system.

price. Customers announce their *preferred net rewards* from the available selection and, if no server offers this net reward, or a higher one, they wait in queue.

By setting the exam thresholds ω_I, the moderating firm controls the number and quality of eligible servers. The firm obtains a fixed share of revenue generated in the marketplace and therefore sets the threshold pair (ω_A, ω_B) to maximize the total system revenue. In particular, the higher the thresholds the higher the service value but with less eligible servers.

The authors solve a fluid approximation of the model. The optimal solution depends on the correlation between S_A and S_B. When skill levels are perfectly correlated, capacity allocated to one class mainly affects the firm's profit from this class, and the firm sets low thresholds to use the potential capacity. In contrast, when skill levels are independent variables, capacity allocated to one class strongly affects the profit from the other class and the firm may need to fail candidates in both exams to maximize profit.

7.3.2   Competition between firms sharing a server

Note the difference between the subject of this section and the models of [75, 168] where two competing firms share a queue but have separate servers.

The first model of **Le Cadre, Bouhtou, and Tuffin [429] (2009)** considers Stackelberg competition between two firms sharing an M/M/1 FCFS service facility while operating in separate markets. Thus, the firms interact through the use of a common facility, but they do not directly compete against each other. Customer service valuations are uniformly distributed over [0,1]. The game has three stages. First, server 2 sets a price and chooses a level of advertising investment used to decrease the perceived full cost in its market. Then, server 1 sets a price. Finally, demand is determined by the equilibrium conditions so customers with valuations greater than the perceived full cost arrive. The authors prove that the third stage has a unique equilibrium and offer to solve the whole game by means of backward induction.

The authors also characterize the equilibrium solution in a second model where the servers compete for market share by operating separate facilities. The *entire capacity is initially owned by server 1* who leads the game by setting a wholesale price for capacity to server 2. The servers set retail prices in the second stage, and in the last stage customers select a server or balk.

Guijarro, Pla, and Tuffin [265] (2013) consider competition between two operators using the same unobservable M/M/1 facility. The *primary operator* serves priority (class 1) customers. When there are no class 1 customers, the *secondary operator* uses the same facility to serve non-priority (class 2) customers. If the expected waiting time for class i is T_i and a i-customer pays for the service p_i then his utility is $cT_i^{-\alpha} - p_i$, for constants $c > 0$ and $0 < \alpha < 1$.[15]

The model assumes service of a 2-customer cannot be preempted and therefore the primary operator's profit declines following entry by the secondary operator. The authors derive an explicit solution for the sequential game where operators first determine their prices and the arrival rates of the two classes are then determined by the equilibrium resulting from **IO** behavior of the customers.

7.4 Multi-period competition

See §6.2.7 for multi-period pricing by a monopoly.

Chayet and Hopp [137] (2007) consider a two-period model: In the first period, firm 1, *the incumbent*, acts as a monopoly setting capacity μ_1 (at a linear cost) and price p_0. Then, firm 2, *the potential entrant*, installs capacity μ_2 after observing μ_1. Once installed, firms cannot alter capacity. In the second period the firms set prices p_1 and p_2 according to a profit-maximizing simultaneous subgame. Firm 1 maximizes total profits during the two periods, whereas firm 2 operates only in the second period. Customer demand rates in both periods are linearly decreasing in full price, which in the second period equilibrium should be equal at both servers if both servers are active.

In general, for a given capacity cost to any one firm the other firm is profitable if its capacity cost is below a threshold. An interesting case occurs when the capacity cost of the incumbent is 0 and thus installs infinite capacity. The entrant can still be profitable if its capacity cost is below a threshold, and then any price equilibrium will have $p_1 > p_2$.

To isolate the first-mover advantage, the firms are assumed to be identical, in particular, having the same unit capacity cost c. The authors demonstrate that the first-mover advantage, though partial, is strong. Specifically, there are thresholds such that:

- When $c \geqslant c^u$ no firm is profitable. For $c^b \leqslant c \leqslant c^u$, the market is small and can only bear one firm, and the incumbent acts as a monopoly.

- When $c^d \leqslant c \leqslant c^b$, the incumbent is still able to prevent entry by installing additional capacity beyond the monopoly level as a deterrent. The threat of entry in this region benefits customers and reduces the incumbent's profits.

---

[15]A similar utility function is assumed in [467].

- When $c < c^d$ the incumbent cannot deter entry. The incumbent still installs more capacity than a monopolist and substantially more than the entrant. The entrant offers lower price and longer lead time while capturing less demand and lower profits.

Guo and Hassin [274] (2013) consider a Stackelberg game between two servers, which we describe in terms of servers operating at two different times. Each server conducts an M/M/1 queue and waiting-cost rates are constant. Customers in the first service period observe the price of server 1 and decide whether to obtain service or wait until the second service period. In the second period, server 2 announces a price and customers join the queue or balk. When customers are homogeneous, the leader obtains, in equilibrium, a larger profit than the follower and customer utility is zero. However, suppose server 2 can also announce a price in the first period and commit to it. In this case, depending on system utilization, server 2 may be able to obtain a higher profit than server 1 and customer utility may be positive.

7.5   Hotelling-type models

Kwasnica and Stavrulaki [423] (2008) add competition to [204] assuming now that price is exogenously given and the shipping delay incurred by a customer located a distance s from the server is a linear function $g(s) = G_0 + G \cdot s$ where $G_0 \geqslant 0$ and $G > 0$. With these simplifications, the authors obtain a closed-form solution for the optimal capacities and service semirange S^M in the monopolistic case, i.e., the server locates itself at a point x and serves all customers in $x \pm S^M$.

The main part of the paper considers a two-stage duopoly competition where in the first stage firms choose capacity and in the second they choose location.[16] The authors show, in general, that each firm serves a smaller interval than the monopoly and selects a lower capacity. They characterize several types of subgame-perfect equilibria depending on the relation of the capacity unit cost c and the other parameters (demand rate per unit distance, price, shipping cost, fixed and variable coefficients, waiting cost rate, and service value):

- Assume the monopolist does not serve the entire interval.

  - For large c, the servers will act as *local monopolists*, locating themselves sufficiently apart from each other and applying the monopolistic solution.

---

[16]Alternative models are described by **Pangburn and Stavrulaki [516] (2005)**.

- For medium c, the solution is a *constrained local monopoly*. The servers serve the entire interval but install exactly the minimum capacity so no customer can obtain positive utility from both.

- For small c, a *constrained competition* arises and despite the reduced benefit of capacity investment associated with it, both firms invest in additional capacity. In equilibrium each firm still serves only half of the interval, but the increased capacity discourages the competitor from expanding any further.

• Assume the monopolist serves the entire interval.

- For $p \leqslant 2c$, capacity costs are sufficiently high so a symmetric subgame-perfect equilibrium will involve co-location (both firms are located at the middle of the interval). Firms then have no incentive to continue investing in capacity to gain a larger market share.

- For $p > 2c$, each firm has a unilateral incentive to purchase additional capacity and no symmetric pure-strategy subgame-perfect equilibrium exists.

Gallay and Hongler [241] (2008) assume *observable* M/G/1 queues with average service rates μ_1 and μ_2 located at points $x_1, x_2 \in \Omega = [-\Delta, +\Delta]$. Customers arrive at rate Λ from locations uniformly distributed over Ω. An arriving customer located at x chooses which server to join by comparing the expected utilities $U_i(x)$ from joining the ith queue, $i = 1, 2$:

$$U_i(x) = a - p_i - c_t|x - x_i| - c_w \mathrm{E}(W_i|N_i),$$

where a is service value, N_i is queue length and p_i is service fee at server i, c_t is transportation cost per unit distance (though travel is instantaneous), c_w is waiting-cost rate, and W_i is expected sojourn time at i.

Given the fixed parameters $x_1 < x_2, a, c_t, c_w$, and the state variables N_1 and N_2, there exists a break-off point Y such that arrivals at $x < Y$ join server 1, while those at $x > Y$ join server 2. Because the state (N_1, N_2) is a random variable, Y is a random variable as well. The main focus of the paper is to characterize the probability distribution of Y. The authors restrict the analysis to heavy traffic, that is, Λ is only slightly smaller than $\mu_1 + \mu_2$. They first consider the *symmetric version* where $x_1 = -x_2 = L/2$, $p_1 = p_2 = p$, and $\mu_1 = \mu_2$. An important dimensionless parameter in this case is $\gamma = \frac{c_w}{\mu L c_t}$, which quantifies the relative importance of the costs. The authors demonstrate through an example that when $\gamma \to \infty$ the density of Y has two peaks at $\pm\Delta$, whereas when $\gamma \to 0$ there is just one peak at 0. Thus, for high values of γ (associated with high waiting costs) the main concern of customers is queue length, whereas for low values (associated with high transportation costs) the main concern is the distance from the server. The authors emphasize the existence of the *phase transition* between these two modes. They analyze types of asymmetry emerging from heterogeneous servers or from non-symmetric locations and prices of the two servers.

7.5.1   Product assortment

An assortment problem involves determining which of the possible variations of a product should be stocked when it is not possible or desirable to stock them all. See §6.7 for similar models in monopolistic markets.

Mendelson and Parlaktürk [486] (2008) consider a Hotelling-type model where customer preferences over product space are expressed by parameter $\theta \in [0, 1]$ and there is a linear misfit disutility relative to the ideal choice. Two modes of operation are considered. A *mass-customizing* (**MC**) firm is modeled as an unobservable M/M/1 system that supplies exactly the desired type θ. This firm sets a uniform unit price and incurs a unit production cost. A *traditional firm* holds an assorted inventory of products, outsources production to a supplier with a given deterministic lead time, and applies continuous inventory review. The traditional firm decides on unit pricing, positioning of available products, and on an inventory replenishment policy. It incurs a fixed cost per order, unit purchase costs, and linear inventory holding costs. The inventory positions of the traditional firm are unobservable to customers who base their choices on the expected waiting time at the **MC**, the expected waiting time for each product type at the traditional firm, and the relevant prices.

The equilibrium solution determines for each product supplied by the traditional firm a market segment of customers who prefer buying this item over other standard products and over waiting at the **MC** firm. The authors obtain several interesting and unexpected results. Among them:

- A traditional monopolistic firm will reduce product variety when the **MC** firm enters the market.

- If one firm buys the other, thus becoming a monopoly, the variety of product may increase or decrease.

- The effect of market size on the **MC** firm's profit is not monotone.

- Speeding up service by the **MC** firm may lead to a *decrease* in profits as a result of the competitor's response.

- The traditional firm does not necessarily benefit from improving replenishment lead times and decreasing unit holding costs.

7.6   Customer loyalty

This section covers models in which current demand is a function of past service. Within that theme, some papers consider the monopoly case whereas others consider competition.

See [582] where customers are sensitive to promised delay, and customer loyalty depends on a weighted average of past tardiness values.

Sobel [585] (1973) models customer loyalty in a market with homogeneous customers and heterogeneous service providers. Let $W_1, \ldots, W_t$ be successive waiting times experienced by a customer at a given server. It is assumed that successive service requests from a given customer are sufficiently dispersed so that successive waiting times are considered independent. The weighted average waiting time is $U_{t+1} = \alpha U_t + (1 - \alpha) W_{t+1}$, with $U_0 = 0$, and $0 < \alpha < 1$. For some constant m, if $U_t \leqslant m$ customers remain loyal to their server, but if this condition is violated they will randomly select another server. Assuming waiting time at server j has a cdf of the form $G_j(x) = 1 - a_j e^{-\theta_j x}$ for $x \geqslant 0$ (for example, M/M/c or G/M/1 queues), the author derives market share approximations for the cases of perfect competition (i.e., many servers) and duopoly.

Hall and Porteus [299] (2000) consider a finite time horizon dynamic multi-period model of two competing M/G/1/1 firms. Service price and unit capacity costs are fixed, and time periods are long enough so queues act close to their steady-state behavior. At the end of the time horizon the firms obtain a reward proportional to market share. A blocked customer switches to the other firm in the next period with an exogenous probability (the firm's *loyalty coefficients*) independent of any other information or past experience the customer may have accumulated. This is a form of bounded rationality. The firm's goal is to maximize expected present value by means of setting the capacity level in each period. The authors obtain an explicit solution to the unique subgame-perfect equilibrium of this game.

Sankaranarayanan, Larsen, van Ackere, and Delgado [558, 559] (2009, 2010) describe a behavioral multi-period model of a queueing system with identical parallel servers and N customers. In each period, each customer chooses a queue based on waiting-time expectations. Customers are connected according to a neighborhood graph and neighboring customers share information on their experiences in these queues in each period. Customers apply a specific weighing scheme to their own and their neighbors' experiences in an attempt to identify the queue with the minimal waiting time to join in their next visit. A numerical example seems to show convergence to a uniform selection of server. A variation of the model in which customers are risk averse and update the variance of their estimates in a similar way is treated by **Delgado, van Ackere, and Larsen [191] (2011)**.

Filliger and Hongler [237] (2005) and **Gallay and Hongler [242] (2009)** consider simple queueing networks with *feedback loops* and *intelligent customers*. These customers are not strategic in the sense of acting to achieve a specific goal, but use queue information in making decisions. The authors characterize the evolution of the queue using deterministic approximations and hydrodynamic analogues. They consider several variations and show interesting stable temporal oscillations corresponding to *delayed responses* to queue length in such systems.

In the simplest model, analyzed in [237], there is a single queue and each customer possesses an *impatience factor* P. As long as customer waiting time falls below P customers remain *loyal* and will line up to be served once more. If the waiting time exceeds P customers leave the system.

Two extensions are provided in [242]. The first considers two parallel servers with separate queues. Several variations are discussed, one of which assumes only one queue is observable.[17] New arrivals join the observable queue if its length is below a threshold, or otherwise join the unobservable queue. At the end of a service customers either leave or return to the same server for an additional service. In another variation both queues are observable.

The second extension considers a closed two-server network.

If the time it takes for service completion is at most P the customer will return to the same queue. Otherwise, the customer will switch to the other queue.

Afèche, Araghi, and Baron [13] (2015) do not explicitly model competition, but their model deals with customer loyalty towards a given firm, and the existence of competition is implicit. An $M/M/s$ firm must invest a convex increasing amount $S(\lambda_n)$ to achieve a flow of *new customers* with rate λ_n. The firm also serves multiple types of *base customers* characterized by their calling (visit) rates, service rates, call-dependent and call-independent profits, and their loyalty parameters. The base customers revisit the server an exponentially distributed time after previous service (the expected length of this time interval is of a higher order of magnitude than the expected service time). Customers renege after an exponentially distributed time. A fixed fraction of new customers receiving service join the customer base while a fixed fraction of the base who reneged will leave. In addition to promotion (advertising) expenses the firm incurs a fixed cost per time unit per server. The firm maximizes profit, setting the number of servers, promotion expenditures, and the priority policy.

The authors characterize optimal controls based on a deterministic fluid approximation and validate these analytic prescriptions through a simulation study for medium and large call centers. They define new performance metrics to guide the firm's decisions. For example, the optimal policy prioritizes base customers in decreasing order of their weighted *one-time service value* (the value of serving a new customer's current request but not any of his future requests) and gives new customers the priority level that maximizes their value per unit of processing time.

---

[17]This scenario resembles that of [304] except here the feedback loop is added.

Chapter 8

Routing in queueing networks

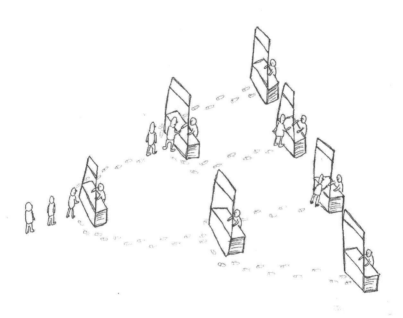

Queues can be formed at the nodes of a network or on its arcs. The simplest form of a queueing network is a set of parallel servers supplying *substitutable* services. An arriving customer selects one of these queues, or balks. In the case of a central planner, customers are *routed* to the servers. Another simple topology is that of *serial* or *tandem* queues. In this case services are *complementary* and customers typically need all of them. More complicated topologies involve selection of substitutable *routes* with each supplying complementary services. Further complications arise when routes have different source and destination nodes. For example, it is possible for customers to leave a serial network at any node. In the case of complementary services some variations dictate a given order while others don't.

In the case of a very large number of users, the common equilibrium concept is the *Wardrop equilibrium*: all routes used between a given source-destination pair have the same delay, which is not larger than the delay across any unused route between these nodes. When the number of users is not very large, the regular Nash equilibrium applies. These differences are explained in [47] §2.4.

The literature on routing in telecommunication networks is extensive and we cover here only papers explicitly referring to delay functions resulting from queueing models. For a broader survey on the subject see [47].

As pointed out in [45] (see §V there), expected delay under light traffic can often be approximated by a linear demand function. This approximation means that results on routing when delay is a polynomial or a linear function of demand also have some relevance to queueing systems. However, we do not include such models in this survey.

See §2.6.4 for throughput maximization by dynamic pricing in a network, §9.3 for related models on demand allocation to suppliers in a supply chain, [60, 483] for price competition in networks, and [557] for demand allocation in a network under bounded rationality.

Models of competition often include a subgame where customers choose among parallel servers. These models are mainly surveyed in Chapter 7, while Chapter 5.7.3 contains cooperative routing games.

8.1 Parallel servers

The problem of optimally routing demand to (unobservable/observable) parallel servers providing substitutable services is often referred to as *load balancing*.

Research on routing customers to parallel servers is mostly devoted to unobservable queues. The pioneering work was done by Bell and Stidham [85] (1983) where the equilibrium and **SO** routing in an unobservable model with non-atomic demand is computed assuming $M/M/1$ latencies. This material and extensions of it can be found in [1] §3.7 and [594]. Specifically, Chapter §6 of [594] includes models of equilibrium, social-welfare optimization, and competition for extensions where arrival rates are decision variables, utility is a nonlinear function of throughput, waiting costs are nonlinear, total demand is fixed or variable, service rates are fixed or variable with their sum being fixed (capacity allocation models) or variable.

8.1.1 Equilibrium routing

See [46] where customers route demand to parallel servers under the power criterion, and [572] on equilibrium routing decisions in a system with breakdowns.

Economides and Silvester [213] (1991) consider two users, α and β, who probabilistically route Poisson demands of rates λ^α and λ^β to two unobservable $M/M/1$ servers. Requests of user α that find the server busy join a queue, but those of user β are blocked and lost. The two users have different goals concerning their requests. User α's goal is to minimize expected waiting time while user β's is to minimize the probability of being blocked. The authors prove the existence of a unique equilibrium and give a detailed case analysis of the solution types.

Orda, Rom, and Shimkin [506] (1993) extend Lee and Cohen (1985) (see [1] §3.7.3) allowing for heterogeneous service facilities and general convex delay-cost functions and prove sufficient conditions for uniqueness of equilibrium routing of demand to parallel queues by a finite number of users. **Richman and Shimkin [549] (2007)** extend the uniqueness result to a family of *nearly parallel networks*.

Libman and Orda [444] (2001) study routing of unsplittable demand by n customers (each user directs demand to a single server) and prove the existence of an equilibrium under the M/M/1 delay function. The solution can be achieved through a simple algorithm in which users sequentially update their routing choice (in arbitrary order) and the total number of updates is bounded by $n(n+1)/2$. The authors also provide an algorithm that checks whether the solution is unique.

El Azouzi and Altman [223] (2003) study a constrained game of customer routing decisions subject to coupling constraints, like, for example, a bound on the total demand sent to a server. They show that an equilibrium still exists in such cases but multiple solutions are also possible.

Sahin and Simaan [554] (2006) consider equilibrium routing of demand to parallel M/M/1 servers with heterogeneous service rates μ_j. Waiting-cost rates c_{ij} and service value R_{ij} depend on both user i and server j. Routing as well as total user demand are decision variables. The authors show that a unique internal equilibrium (i.e., all users send positive flows to all servers) exists if $R_{ij} \geqslant \frac{c_{ij}}{\mu_j}$ for every user i and server j.

Kumar and Krishnamurthy [421] (2008) report the results of a series of experimental studies on the way risk-averse customers choose a server in an unobservable multiserver system. The system has two service providers having similar expected service durations and levels of congestion, but different variability of service durations. Customer sojourn time is the sum of service duration (supply-side) and queueing time (demand-side).

The study first verifies that with exogenous data on expectation and variability, the participants consider the variability at least as important as the expected values. Then, the following conclusions are reached in the main model:

- The presence of the demand-side risk associated with congestion reduces the risk aversion associated with the supply-side uncertainty. A managerial implication of this finding is that gains from reduction in service time variability may be limited.

- When customers are informed that the overall level of congestion in the market is expected to differ from normal levels, they focus more on the service process and shift to low-uncertainty service providers.

Penmatsa and Chronopoulos [527] (2011) consider n M/M/1 servers and m users. Initially, demand at rate ϕ_{ij} that belongs to user j reaches server i. Users can redirect their demand so that eventually server i serves rate β_{ij} of user j's demand. All redirected demand is processed by an M/M/1 communication subsystem, and users minimize the total delay of their demand (including communication delays). The authors use the following fairness index for measuring the equality of the equilibrium solution: $\left[\sum C_j\right]^2 / \left[m \sum C_j^2\right]$, where C_j is the expected delay of user j's jobs. This measure is 1 when delay is equal for all servers and decreases as the differences increase.

Experimental results indicate that the equilibrium solution is not much lower in efficiency but provides more fairness relative to the **SO** solution.

Cardellini, de Nitto Personé, Di Valerio, Facchinei, Grassi, Lo Presti, and Piccialli [129] (2015) consider delay-sensitive atomic customers allocating demand in a three-tier network consisting of a local tier, a middle tier with M/G/1 EPS queues, and a remote tier with an M/G/∞ queue. The processing rate of a job and the energy required depend on the customer and on the chosen tier. Customers minimize their expected response time subject to energy and utilization constraints. Thus both the objective and the strategy sets of a customer depend on the strategies of other players. The authors prove the existence of a *generalized Nash equilibrium* and present a distributed algorithm for its computation.

8.1.2   Cooperation

See §5.7.3 for cooperative routing games.

Customers' *perceived utility* is a weighted sum of their utilities and those of their peers. The *degree of cooperation* is measured according to these weights. In particular, *non-cooperative* customers assign zero weight to others, *altruists* assign zero weight to their own delay costs, and *equally cooperative* customers allocate weights uniformly.

Azad, Altman, and El Azouzi [77] (2010) consider routing decisions under partial degrees of cooperation. The authors consider two simple examples. One has two customers and two parallel servers. The other example has two customers, each associated with a server, and if one customer routes demand to the other server this customer incurs M/M/1 *communication costs* (similar to [384]). The authors numerically demonstrate the following possibilities:

- The equilibrium need not be unique. This result is of interest because a unique equilibrium exists when there is no cooperation, as proved in [506].

- A unilateral increase in a user's degree of cooperation (which affects other users' strategies) may induce a new equilibrium associated with a *smaller* actual cost to that user.

Blocq and Orda [93] (2014) consider routing games with *malicious players*. The following example is illustrative. There are two servers with unit service rate and two users, a selfish user **S** with demand rate $\lambda_S = 1$ and a malicious user **M**. The goal of **S** is to minimize the expected delay W_S of its demand, while **M**'s objective is to *maximize* W_S. It is easy to see that there is no pure equilibrium in this game: **S** must send demand at rate ≥ 0.5 to at least one of the servers, and by sending all of λ_M to this server **M** can induce $W_S = \infty$. However, given a routing choice by **M**, it is always possible for **S** to reroute its demand and guarantee $W_S < \infty$.

The authors analyze a cooperative version of the game. Atomic customers wish to minimize expected waiting costs and route demand to parallel M/M/1 servers. The cost of a coalition is defined from a worst case perspective assuming the other users act as a malicious leader **M** in a Stackelberg game and distribute their demand to *maximize* costs to the coalition. In response, the members of the coalition route their demands so as to minimize aggregate average cost.

- The authors show that in this game **M** acts as a continuum of self-optimizing nonatomic users. They use this finding to investigate solution concepts such as the Nash bargaining point, the inner core and the nucleolus.

- In [92] the authors show that if the disagreement point is the (unique) Nash equilibrium point and customers have homogeneous costs, then the Nash bargaining point is also **SO**.

Koutsopoulos, Tassiulas, and Gkatzikis [415] (2014) investigate a *peer-to-peer network* of parallel servers where each server (peer) also acts as a customer and routes demand to other servers. In addition, each server determines an absolute priority ordering for the other demands. Peers cannot directly affect delay of their own requests, but they can do it indirectly by affecting the behavior of those they serve, and consequently the equilibrium behavior.

The authors derive conditions for equilibrium under fixed priority rules (including FCFS). They suggest best-response heuristics for updating priority orders and provide numerical results for various degrees of cooperation.

8.1.3 Welfare maximization

See §5.8 for **PoA** analysis of competition associated with routing to parallel servers.

Chen, Zhang, and Huang [146] (2008) (also [145]) consider a Markovian multiserver model with heterogeneous quality of service and nonlinear waiting costs. Specifically, $v_1 < v_2 < \cdots < v_n$ denote service values obtained from servers $1, \ldots, n$, respectively, and the cost of waiting W time units is $\theta T(W)$. $T(W)$ is convex and common to all customers and customers' waiting-cost rate is $\theta \sim U[\theta_{\min}, \theta_{\max}]$. An arriving customer joins the queue that provides maximum expected net utility if it is positive, or otherwise balks. The system incurs a service cost c_v per service and a lost-sale penalty c_p for each balking customer. Customer type is private information.

- The **SO** solution has thresholds $\theta_{\min} \equiv \theta_{n+1} \leqslant \theta_n < \cdots < \theta_1 \leqslant \theta_{\max}$ such that customers with $\theta > \theta_1$ balk, and those with $\theta_{i+1} < \theta < \theta_i$ choose server i.[1] These threshold values have the property of the intervals (θ_{i+1}, θ_i) having increasing lengths. Thus, better servers obtain higher portions of the demand with rates $\lambda_i = \Lambda(\theta_i - \theta_{i+1})$, impatient customers are routed to lower quality servers, and the most impatient balk.

- A similar structure of the solution results when the system is controlled by a profit-maximizing firm. The firm applies **IC** prices that achieve the desired assignment.

- Prices inducing the **SO** solution are numerically compared with the monopoly's prices for $n = 2$. When the market potential Λ is low the two solutions are identical. However when Λ is high, the **SO** price of the high-value server is lower than that set by the profit-maximizing firm and is *higher* for the low-value server.

Sun, Tian, and Li [610] (2010) solve equilibrium and **SO** routing in the parallel servers model of [85] when time is discrete and service duration is geometrically distributed. The qualitative results are preserved under this variation.

Gupta, Jukic, Stahl, and Whinston, [289] (2011) apply simulation to a model with a finite number of users routing service requests to parallel M/G/1 service providers each offering a subset of services. Customers are heterogeneous in service valuations and waiting-cost rates. Servers have heterogeneous unit capacity costs. The authors consider the long-run problem of setting **SO** capacities. They conclude from numerical experiments that congestion-based pricing can result in more capacity investment relative to the no-price equilibrium if the relative cost of capacity is high compared to service value.

Bodas and Manjunath [97] (2011) compute **SO** routing in a system with two customer classes having heterogeneous waiting-cost rates and two

---

[1] The solution resembles Ghanem's (1975) partition into priority classes according to the impatience factor, see [1] §4.4.1.

servers having heterogeneous service rates and admission fees. They then consider the equilibrium outcomes in three models: (i) both classes consist of nonatomic noncooperative customers, (ii) both classes act as atomic customers (equivalently, they consist of nonatomic customers who cooperate to maximize the aggregate class welfare), and (iii) a mixed model with one atomic customer and one class of nonatomic customers.

Filippini, Cesana, and Malanchini [236] (2013) demonstrate how to compute **SO** and equilibrium routing of customers to N M/G/1 servers with heterogeneous service rates.[2] The performance measure is the average sojourn time of a customer. Numerical results are given for the **PoA** for $N = 2$.

Bodas, Ganesh, and Manjunath [96] (2014) consider customer classes $i = 1, \ldots, M$ with heterogeneous waiting-cost rates $\beta_1 > \cdots > \beta_M$. The waiting time at server j is a convex increasing function $D_j(\gamma_j)$ where $\gamma_j = \sum_i \lambda_{ij}$ is the total arrival rate to server j.

The **SO** solution is characterized by numbers $n_1 \leqslant \cdots \leqslant n_M$ such that the routing functions satisfy $\lambda_{ij} = 0$ for $j \notin \{n_{i-1}, \ldots, n_i\}$ and $\lambda_{ij} > 0$ for the internal indices of this interval. Thus, each customer class uses a nearly dedicated set of queues with overlap possible only at the ends of the intervals. The equilibrium routing is similarly characterized but arrival rates at the internally indexed queues are not necessarily positive. Lastly, the authors prove that the system can be coordinated by charging admission fees corresponding to externalities associated with joining the queues.

8.1.4 Profit

See [630] for profit-maximizing routing and pricing with parallel servers and independent breakdowns.

Lee and Lui [432] (2008) consider *Internet service providers* (ISPs) that route demand through direct links and also through profit-maximizing *higher tier ISPs*. All links are modeled as M/M/1 queues. The benefit ISP i obtains from sending flow x_{ij}, from i to j, directly or through the higher tier, is $w_{ij} \log(1+x_{ij})$. ISPs also incur delay costs and fees for transmission through the higher tier ISPs. The delay-cost rate incurred by ISP i for transmitting flow y_{ij} through the direct link with capacity c_{ij} is assumed to be $\gamma/(c_{ij}-y_{ij})$,[3] and similarly for transmitting through the higher tier.

The authors provide necessary conditions for an internal routing equilibrium to exist when the transmission costs and capacities of the links leading to higher tier ISPs are fixed. They also present an algorithm for determining the

---

[2]The authors compute and use the *extended service time* resulting from arrivals of *primary users*. A primary user preempts service of a regular customer and interrupted service must be repeated.

[3]Rather than y_{ij} times this quantity, see [674] for a similar assumption.

profit-maximizing solution when these parameters are the higher tier ISPs' decision variables.

Tran, Le, Ren, Han, and Hong [629] (2015) examine a service provider controlling M/G/1 facilities with expected service durations $\bar{\chi}_1 < \cdots < \bar{\chi}_L$.[4] There are K customer classes with potential arrival rates $\Lambda_1, \ldots, \Lambda_K$ and waiting cost rates $\theta_1 < \cdots < \theta_K$. Customer service valuations are uniformly distributed over $[0, 1]$. The provider announces class-independent prices p_l and routing probabilities s_l, $l = 1, \ldots, L$; customers decide between joining the server to which they were routed or balking.

The authors characterize equilibrium thresholds such that a k-customer with valuation α agrees to join facility l iff $\alpha \geqslant \alpha_{k,l}$. Revenue-maximizing prices and routing probabilities induce positive joining rates for servers $k = 1, \ldots K^*$ and classes $l = 1, \ldots, L^*$ for some $K^* \leqslant K$ and $L^* \leqslant L$.

The authors also consider a variation where customers are routed to one of two systems. The first system is as described above but with a single price p_1, and the second system is an M/G/∞ system with price p_2. Customers assigned to the first system choose between joining or balking before they know the M/G/1 facility to which they will be routed. Algorithms are provided for two versions of this setting, both when the systems are jointly owned and when they compete with each other.

8.1.5   Capacity allocation

Korilis, Lazar, and Orda [411] (1997) consider a system with a finite number of users, each wishing to minimize the expected delay of its own demand by dividing it among a set of parallel heterogeneous M/M/1 servers. By [506], this system has a unique equilibrium. Suppose the manager has extra capacity to allocate among the servers.

The authors prove that adding capacity to any server does not hurt any of the users (there is no Braess paradox behavior here). The minimum total wait for system users is obtained by allocating the additional capacity exclusively to the server initially having the highest capacity. Some of these results are generalized by **Altman, El Azouzi, and Pourtallier [48] (2003)** and by **Abbad, El Azouzi, and El Kamili [2] (2006)**.

Libman and Orda [443] (1999) consider *unsplittable* routing of atomic customers to parallel heterogeneous servers. Some of their main results are:

- A formula for the **SO** solution.

- Construction of a "natural" equilibrium routing and an algorithm that checks whether other equilibria exist.

---

[4]The authors start with a system having ON-OFF transitions and transform it into an equivalent system without interruptions where service time now represents the original total service time including interruptions.

- An example of a Braess-like paradox when an increase in server's capacity results in poorer system performance.

- A characterization of the optimal allocation of additional capacity when the goal is to minimize the maximum expected waiting time at a server.[5]

Chao, Liu, and Zheng [134] (2003) consider N M/M/1 service stations with dedicated demand streams at rates $\lambda_1 \leqslant \lambda_2 \leqslant \cdots \leqslant \lambda_N$. In addition, there is a stream of non-dedicated (flexible) demand that can be served at any station. A fixed amount of capacity is available for allocation among the stations. The problem is to compute **SO** capacity allocation and probabilistic routing of flexible customers to minimize average waiting time in the system. The optimal solution, which the authors denote as *one big and many small*, assigns all flexible demand to the N's facility, and waiting times under optimal capacity allocation satisfy $W_1 \geqslant W_2 \geqslant \cdots \geqslant W_N$. Consequently, routing flexible demand to the N-th server is **IC** and will result when customers are strategic.[6]

Menache and Shimkin [484] (2008) consider a manager allocating service capacities among service stations with the purpose of achieving given ratios between the respective queueing delays. The system has a finite number of users with heterogeneous waiting-cost rates, service fees, and utility functions of total demand rates. Users maximize net utilities by determining their rates of demand for each server. The authors prove that for a family of delay functions, including M/M/1 delay, there exists a unique equilibrium in this users-manager game. They also present two adaptive algorithms for capacity allocation that converge to equilibrium in the case of two servers.

Conforto, Priscoli, and Facchinei [172] (2010) consider atomic customers with linear waiting costs routing demand to a collection of M/M/1 servers. Servers are clustered, each with an initial capacity, and each cluster has an extra amount of capacity to allocate to its servers. Clusters act independently to minimize the average delay for their servers. The authors prove that an equilibrium exists and present an example where it is not unique.

8.2 Queues with different regimes

This section considers customer choice between different type of queues as reflected by their service discipline or available information. See §8.7 for

---

[5]See [412] for another model of capacity allocation associated with the Braess paradox.

[6]The authors also extend the queueing model to non M/M/1 systems such as M/G/1 and M/M/s. The flexible demand is still routed to the loaded server, but incentive compatibility is not proved.

models where customers choose between two queues, one of which provides bulk service. See [199, 202, 628] on selection between competing $M/G/1$ and $M/G/\infty$ firms.

Altman, Jiménez, Núñez-Queija, and Yechiali [50] (2004) consider a Markovian model in which customers choose between an observable queue Q_o and an unobservable queue Q_u. They extend Hassin's (1996) "gas stations" model (see [1] §7.6) by assuming heterogeneous service rates. Arriving customers estimate the expected queue length at Q_u based on the length of Q_o. The paper investigates the existence of equilibrium threshold strategies. If others were to increase thresholds then, for any given length at Q_o, one would intuitively expect a shorter queue length at Q_u. Therefore this is an ATC model and we would expect a unique mixed equilibrium. A proof of this intuition turns out to be difficult and the authors obtain related results:

- The joint probability distribution of the states of congestion in both queues for a given threshold strategy.

- A threshold minimizing a weighted sum of the mean queue lengths in both queues.

- Conditions for the best response to a *pure* threshold strategy to be again a (possibly different) threshold strategy.

The question of whether the model is indeed ATC remains open.[7,8]

Hassin [306] (2009) considers a Poisson stream of customers choosing between joining one of two servers with identical exponential service but with different queue disciplines. One server conducts an FCFS regime while the other randomly selects the next customer to be served (SIRO regime). The two queues are observable and jockeying is not allowed. Since the expected waiting time at the SIRO server depends on the joining strategy adopted by other customers (in particular, future arrivals), our interest is in the resulting equilibrium and the associated queue performance variables, such as the average joining rate to each of the queues. A (mixed) equilibrium switching curve strategy defines a threshold $x_f = n_f + p_f$ at the SIRO queue where n_f is a nonnegative integer and $p_f \in [0,1)$, for every possible queue length f at the FCFS server. The resulting behavior is that an arrival who observes f

---

[7]Footnote 10 in §7.6.1 of [1] is based on an earlier version of [50] and is incorrect.

[8]Situations where customers choose between joining a queue now or in the next period can also involve a choice between observable and unobservable queues, see §6.2.7 and §7.4. **Pazgal and Radas [524] (2008)** conducted a computerized laboratory experiment in which players choose between joining an observable $M/M/1$ queue or balking to return the next day. In the second day there is no balking option and players must wait until they are served. Thus, also in their experiment, a player sees one observable queue and chooses between joining it or joining another unobservable queue. However, the length of the latter queue is assumed to be independent of the former one and as a consequence the cost associated with balking is independent of the strategies of the other players.

customers at the FCFS server and r customers at the SIRO server joins the SIRO server with probability 1 if $r < n_f$, and with probability p_f if $r = n_f$. Otherwise the customer joins the FCFS queue. The strategy is thus defined by an infinite-length vector of thresholds. It is assumed that *arrivals to an empty system select each queue with equal probability.*[9] A nice feature of the model is that the input data consists of a single parameter, namely the system utilization $\rho = \lambda/(2\mu)$.

The paper opens with a *static version* of the model in which a finite number of customers choose their server according to a predetermined order and no new arrivals are expected. In equilibrium, the SIRO server obtains two thirds of the demand! This raises the question of whether the SIRO server would obtain a higher share of the demand in the dynamic case as well. The dynamic model is solved numerically giving for any given ρ the vector of thresholds and the resulting average joining rate for the SIRO server. The main results are:

- The market share of the SIRO server is always less than that of the FCFS server.

- If identical servers compete by choosing is between FCFS and SIRO, then the only equilibrium in this game is when both employ FCFS.

The difference between the results of the static and dynamic models can be explained by observing that the last customer to join the random queue in the static models is guaranteed an expected wait for the service of half of the customers in that queue. In the dynamic model this is not true because of the possibility of future customers joining the queue. However, for the FCFS queue there is no difference between the static and dynamic cases.

Kardeş [389] (2012) simplifies the model of [306] by assuming the choice is between an FCFS server and an egalitarian processor sharing (EPS) system where capacity is equally shared by present customers. In particular, there is no need in this version of the model to define an arbitrary decision of a customer arriving when both servers are free. The author also shows in this case that when the servers have identical capacities a higher market share is obtained by the FCFS server. The author also considers servers with different capacities and concludes that the EPS firm must be at least 15% faster than its competitor to guarantee a minimum market share of 50%.

Hayel, Quadri, Jiménez, and Brotcorne [332] (2015) assume customers choose between an unobservable $M/M/s/s$ queue and an unobservable $M/D/1$ queue. Time spent in the $M/M/s/s$ system is not costly, but a customer selecting it must pay an entry fee and incur a blocking cost if all servers are busy. Joining the $M/D/1$ queue does not require a payment, but there is a constant unit-time cost while waiting. Balking is not allowed. The authors

---

[9]It is demonstrated that giving preference to one of the servers when both are free significantly changes the outcome.

compute the equilibrium probability $p_e \in (0, 1]$ of selecting the M/M/s/s queue. Since this is an ATC situation, p_e is unique.

The authors also suggest a variation in which customers select the queue according to a logit function. Specifically, let $c_1(p)$ and $c_2(p)$ be the expected costs associated with the two options. Then, the equilibrium probability for joining the first queue is the unique solution to $p = \dfrac{e^{-\gamma c_1(p)}}{e^{-\gamma c_1(p)} + e^{-\gamma c_2(p)}}$, where $\gamma > 0$ is a parameter such that $1/\gamma$ represents the *degree of irrationality*. The authors present an example where customer welfare can increase or decrease with γ.

Bodas, Ali, and Manjunath [95] (2014) consider Poisson arrivals of customers with heterogeneous waiting-cost rates and general service time distribution. Arriving customers choose between two unobservable queues with heterogeneous service rates: a cost-free FCFS queue and a nonpreemptive highest-bidder-first (HBF) queue. Balking is not allowed. The strategy profile consists of a pair $(p(\theta), X(\theta))$ consisting of a probability $p(\theta)$ that a customer with waiting cost rate θ joins the FCFS queue with the bid $X(\theta)$ if the customer joins the HBF queue. The main results are:

- In equilibrium, customers choosing the HBF option are those with θ value above a threshold.

- Suppose both servers are owned by a monopoly having a fixed amount of capacity. Revenue can be higher if some of the capacity is allocated to the FCFS server.[10]

- Adding a cost-free FCFS server to an existing HBF server does not increase revenue (but this property does not always hold when balking is allowed).

8.3   Complementary services

When service value is conditioned upon completing a series of services we say these services are *complementary*. When the order of receiving service is predetermined we have tandem queues. As a network, tandem queues are represented by a directed path.

See [290] for tandem queues consisting of a first MTS stage and a second MTO stage, [716] for the Down-Thomson paradox in a system of tandem queues, and [28] for a model where delay information is provided to customers

---

[10]This behavior conforms with Myrdal's claim, see §6.6.4, that slowing down service at an HBF system can increase revenue.

after completing service at the first queue and before starting service at the second.

8.3.1 Equilibrium

Parlaktürk and Kumar [520] (2004) consider a system with two servers, two-stage deterministic service, and a random arrival process. The first stage of service takes time m_f, and the second takes m_s, regardless of the server involved. Both servers can perform both stages, however the two service parts for any given customer must be carried out by different servers. Servers have separate queues and the service discipline can discriminate between customers seeking to obtain their first stage of service and those arriving for their second stage of service.[11] A customer's decision of which queue to join first depends on observable queue lengths and residual service times for both servers. The queue manager decides on the priority rule at the queues, distinguishing between first-stage and second-stage customers.

Since the priority rule allows for overtaking, an arriving customer's decision also depends on the strategies of future arrivals, and hence the authors investigate (pure) equilibrium customer strategies and priority rules that yield good (social) performance in equilibrium. Two natural priority rules are shown to be unsatisfying in this respect: giving priority to second-stage customers, and first-in-system first-out. However, the authors develop a *workload-regulating rule* that induces arrivals to split evenly and avoids server idleness. The rule generally gives priority to second-stage customers. But when the number of stage-two customers in one of the queues is small, priority is given to first-stage customers *in the other queue*. It is shown that expected queue lengths under this rule are not much longer than a theoretically derived lower bound.

D'Auria and Kanta [182] (2015) consider threshold joining strategies in a Markovian model of two tandem queues with homogeneous customers. The first server serves at rate μ_1 and the second at rate μ_2. Service value is R and waiting at queue i costs C_i per unit time. The decision of whether to join or balk is taken upon arrival and reneging is not possible.

The authors compute the equilibrium joining threshold assuming customers know the total number of customers in the system but not in each of the queues. They find that the probability of the first queue having a given length conditioned on the known total length of the queues is independent of the customers' threshold. This property leads to a proof for the existence of a dominant threshold strategy.

Burnetas [114] (2013) investigates a network of N multiserver tandem queues. Staying in the nth queue costs C_n per unit time and obtaining service there is associated with reward R_n. Customers decide, *before joining*, which

---

[11]The model is simplified by allowing a server with an empty queue to give second-stage service while the first stage is still processed by the other server.

set, 1 to n, of consecutive queues they will attend. A strategy is therefore characterized by a vector $(x_1, \ldots, x_N)$ where x_n is the probability the customer joins queues 1 to *at least* n. Clearly $1 \geqslant x_1 \geqslant \cdots \geqslant x_N \geqslant 0$. The author proves that a unique symmetric equilibrium $(x_1^e, \ldots, x_N^e)$ exists. Compared with the **SO** strategy $(x_1^*, \ldots, x_N^*)$ it is shown that $x_n^* \leqslant x_n^e$ for $n = 1, \ldots, N$.

Arlotto, Frazelle, and Wei [64] (2015) consider variations of the following scenario. Customers need service from two stations, A and B, but are free to choose the order in which to visit the stations. Arrivals are Poisson, service is deterministic, and the arrival and service rates satisfy $\mu_B > \mu_A$ and $\lambda \geqslant \mu_A$. The latter assumption means the queue is overloaded, but the analysis applies to any given state of the system. Arriving customers observe the workload at the two stations before deciding whether first to join queue A or B aiming to minimize waiting time. The authors prove that the strategy *visit station A first* independently of the system state is an equilibrium.

8.3.2 Monopoly

Başar and Srikant [80] (2002) examine a network of N tandem M/M/1 queues supplying different kinds of services to users of $N + 1$ types. There are n_k k-customers $k = 1, \ldots, N$, that only need the service of server k, and there are n_0 0-customers who need all the N services. The network is owned by a profit-maximizing firm that sets one price p for a unit demand served by any of the servers. The utility of a k-customer, $k > 0$, submitting demand at rate x is $a_k \log(1 + x) - px - W_k$, where W_k is the expected delay at server k.[12] Similarly, for a 0-customer the utility is $a_0 \log(1 + x) - Npx - \sum W_k$. Customers react to price p by selecting utility-maximizing demand rates.

The authors prove that for any fixed p there is a unique customer equilibrium. They derive the optimal price for a restricted case as well as profit and customer net utilities when the number of customers is very large.

Ching, Choi, Li, and Leung [164] (2009) describe a queueing system with n customer classes and two service stages. The first stage consists of an M/M/s queue; the second consists of n independent M/M/1 queues, one for each class. The demand function for class i is linearly decreasing in the price P_i imposed on this class. Note that P_i is a nominal price and not the full price, so customers are not time sensitive. Instead, there is a uniform exogenous upper bound on the expected waiting time for each class. The authors provide an explicit solution for the profit-maximizing price when $n = 1$.

Caro and Simchi-Levi [130] (2012) consider a multiclass loss system where the service of a k-customer requires a specialized server from a pool of N_k dedicated servers and another server from a pool of N general servers. An

---

[12] As in [81], W_k is not multiplied by x.

arriving customer enters service if there are free servers of both types or is otherwise lost.[13,14] The arrival of k-customers is Poisson with a price-dependent rate (sojourn time consists only of service time which is not controlled and hence demand is not delay-dependent). The authors characterize the revenue-maximizing class-dependent prices in this system.

8.3.3   Competition

Veltman and Hassin [635] (2005) study a profit maximization model of two servers: an unobservable $M/M/1$ *service provider* charging a flat price for service and a *parking provider* providing a complementary service for the time the customer spends at the service provider. A fixed reward is received after obtaining both services, but no benefit is received from obtaining just one service. Waiting costs are linear. In contrast to competition between servers providing substitutable services, here an increase in the price charged by one of the servers reduces the profit of the other server since it reduces the common customer-equilibrium arrival rate. Some of the main observations are:

- Suppose the parking provider charges a price proportional to usage time. If potential demand exceeds a given critical value then the price equilibrium is unique and the service provider's profits will be higher than the parking provider's. Otherwise, there is a continuum of price equilibria.

- When potential demand increases, both service provider and parking provider respond by lowering prices (to compensate for the increase in expected waiting time).

- When the two servers cooperate and act as a monopoly, they charge **SO** prices (the monopoly gains all customer welfare, for the same reason as in E&H). However, competitive equilibrium prices are higher than socially desired, thus resulting in *lower* congestion.

- The parking provider can increase profits by charging a fixed price independent of usage time (in which case it can be considered as a seller of a *complementary product*).

- It is sufficient for the social planner to control one of the servers to achieve social optimality. This is done by setting a zero price at the controlled facility and letting the other server act as a monopoly.

- **Sun, Li, Tian and Zhang [609] (2009)** extend [635] by considering the arrival of batches that share the parking cost. They also add the (somewhat uncommon) possibility for both servers to charge a fee proportional to usage time. The extension to batch arrivals does not change

---

[13]Note that admission control could be used to further increase profits.

[14]Services supplied by the dedicated and general servers are complementary services but, unlike other models of this type, these services are given *simultaneously*.

the qualitative results, and the new mode of payment is undesired both socially and by the service provider.

Badrabadi and Tarokh [78] (2010) consider a Stackelberg game with three M/G/1 service providers: Two identical substitutable competing providers and a monopolistic server providing a complementary service.[15] Customers differ in service valuations and waiting-cost rates, which are uniformly distributed and perfectly correlated. Each server can choose one of two exogenous service levels (expected lead times) d_L or d_H. The monopolist acts as leader, selecting service level and price. The competing servers follow by simultaneously selecting service levels and prices. The composite lead time incurred by a customer is the maximum expected lead time offered by the two servers serving this customer.[16] There are six possible cases: the leader chooses one of d_L and d_H, and the followers either both choose d_L, both choose d_H, or choose different levels. In comparing the six solutions the authors conclude that in low-traffic intensity competitors will differentiate themselves by choosing different service levels while in high traffic they may choose the same levels.

Afèche [11] (2013) considers a tandem network of two M/M/1 servers, each operated by a different profit-maximizing service provider. The network serves three customer types: *Cross-traffic* customers requiring service from both servers, and *local-traffic* customers requiring service from only one of the two servers. For each customer type the value generated from serving demand rate λ is proportional to $\lambda^{1-\alpha}/(1-\alpha)$ for some $\alpha \in [0,1)$ and waiting-cost rates are perfectly correlated (affinely) with service valuations. The main results are:

- Optimal prices also maximize social welfare. In the absence of local traffic, decentralized price competition leads to a higher total price and a smaller demand rate (undercongestion).[17] However, the presence of time-sensitive local traffic improves performance of the decentralized operation.

- Suppose service capacities can be optimized at a linear cost and there is no local traffic. In the resulting price and capacity competition game, firms invest in capacity only if $\alpha < 0.5$. The total cross-traffic price is greater than that of a centralized monopoly which is greater than the **SO** total price. The capacities and demand rates satisfy the reversed inequalities.

---

[15]The general model assumes there are also customers interested in just one of these services.

[16]A possible interpretation is that services are provided simultaneously and the expected maximum of the two random waiting times is approximated by the maximum of their expected waiting times.

[17]As in [635].

- Peering: Suppose each server has a dedicated arrival process. Suppose also the servers and arrival processes are symmetric and each server agrees to serve the other's demand without charge. This arrangement results in a lower total price and a higher arrival rate than the monopolistic solution (overcongestion), which is contrary to the outcome under decentralized price competition.

- In the same setting as the peering model, the monopolistic solution can be achieved by an adequate price transfer agreement between servers.

Melnik [482] (2015) considers a Markovian model of profit-maximizing servers $i = 1, \ldots, m$, each providing service at n points on a linear route. Customers of type j arrive at rate λ_j and require service at points $1, \ldots, j$. Servers compete by setting prices $c_j^{(i)}$. If the rate of j-customers selecting provider i is $\lambda_j^{(i)}$, then j-customers' full cost is $c_j^{(i)} + \sum_{k=1}^{j} a_k^{(i)}$ where $a_k^{(i)} = \left(\mu - \sum_{s=k}^{n} \lambda_s^{(i)} \right)^{-1}$ is the expected waiting time at server i's queue at point k. The author computes a symmetric price-equilibrium assuming all servers provide identical full prices to each customer type and all customers evenly split their demands among all servers.

8.4 Partial control

An intermediate case between the equilibrium and **SO** solutions is reached when the social planner controls a fraction of the customer population. The planner first dictates routing behavior for the controlled fraction, thus affecting the individually optimal decisions of the uncontrolled fraction and the resulting equilibrium. It is common to refer to such games as *Stackelberg routing games*.

See [253] for partial control in a single-server queue.

Korilis, Lazar, and Orda [410] (1997) consider the model of [411] but assume part of the demand is controlled by a manager wishing to reduce the overall average system delay. The manager leads by announcing the routing of the controlled demand, presumably to some of the slower servers, and uncontrolled customers then follow. Assuming the M/M/1 latency function, the authors prove that:

- In the case of a single selfish user, the manager can always enforce the system optimum.

- There exists a threshold α_0 for the controlled portion of demand and, if exceeded, the manager can induce an equilibrium maximizing social welfare (minimizing average delay).

- In a heavily loaded system α_0 is small and it is easier to enforce the optimum.

- **Kaporis, Kirousis, Politopoulou, and Spirakis [387] (2005)** consider a simpler algorithm for the problem. The algorithm assigns the controlled demand giving priority to servers that would receive at least some demand optimally but receive no demand in equilibrium. The increased welfare obtained by the central planner's allocation is demonstrated in numerical experiments. Unexpectedly, for M/M/1 latency functions, α_0 is smaller when the algorithm is applied to instances with higher PoA, i.e., those where the role of the manager is important.

Kaporis and Spirakis [388] (2009) present algorithms for computing the fraction of demand necessary to be controlled by the system manager to induce the **SO** network routing when all other demand belongs to self-optimizing non-atomic customers. We describe the case of m parallel queues with general latency functions. Let $(o_1, \ldots, o_m)$ and $(n_1, \ldots, n_m)$ be the total assignment of demand to servers $1, \ldots, m$ in the optimal and equilibrium solutions, respectively. The algorithm initially assigns controlled demand of size o_i to servers having $n_i < o_i$, i.e., the (slower) servers which are less attractive to the users. It then discards these servers and continues recursively until encountering a system where the equilibrium solution is also optimal.

8.5   Routing with transportation costs

This section includes queueing models with customers traveling to receive service and thus incurring *transportation costs*. Similar models may include *communication costs*. Customers wish to minimize the sum of travel, waiting, and admission costs, and we extend the usual definition of full price to include this sum.

Related Hotelling-type models are reviewed in §7.5. See [225, 384] and [406] on inefficiency analyses of routing models with transportation costs. See §11.2.3 for games of location that incorporate transportation costs.

Heinhold [337] (1978) proves the existence and uniqueness of an equilibrium routing of demand to service facilities in a system where customers and multiserver facilities are spatially distributed. Customers are homogeneous except for their locations and each customer location produces Poisson demand. Balking is not allowed. In equilibrium, customers visit a facility with the minimal expected sum of traveling and waiting time. Results of an empirical study back up the theoretical predictions.

Brandeau and Chiu [106] (1994) derive an $O(n^2)$ algorithm for solving a Stackelberg location game between two servers on a tree network with n nodes. Customers are located at the nodes and incur transportation and congestion costs. Congestion cost at a facility is a nonnegative increasing convex function of its served demand. Competing servers choose locations (on nodes or edges of the tree) aiming to maximize their market share. This and earlier works on strategic facility location are surveyed by **Owen and Daskin [507] (1998)**. In particular, section §4.1.2 of their survey contains queueing models.

Brandeau and Chiu [107] (1994) consider locating service facilities on the nodes or edges of a tree network to minimize social cost. Customers are located at the nodes and can split their demand among facilities. They incur waiting and transportation costs, and act independently. For given facility locations it is shown that customer equilibrium is unique. The authors then consider the location of two facilities with given (possibly different) capacities. They prove that there exists an optimal solution such that at least one facility be located at a node and use this property to derive an $O(n^3)$ algorithm.

Grossman and Brandeau [261] (2002) consider social optimization in a variation of [337]. They show that it is possible to induce an equilibrium with the **SO** routing of demand by charging appropriate (possibly negative) tolls at the servers. Optimal toll values are not unique. For example, tolls charged by subsets of servers that do not share potential demand from a common customer location can be changed by a constant without affecting the equilibrium. This flexibility can be exploited in setting tolls with desired properties, such as revenue-neutral tolls.

Zhang, Berman, and Verter [702] (2009) develop a heuristic for the following model: There are n sites, each producing a stream of requests, and a finite set of potential locations for $M/M/1$ service facilities. Given a subset of locations of operating facilities, let T_i be the minimum i-customers' full cost (travel and congestion cost) over all operating facilities. The model requires that *all i-customers select a common facility* which achieves the full cost T_i. The demand rate λ_i is a linearly decreasing function of T_i. The goal is to find a set of locations for the service facilities that maximizes total served demand subject to a requirement that facilities cannot be operated unless they serve a given minimum demand rate.

Zhang, Berman, Marcotte, and Verter [701] (2010) extend [702] in two ways. First, they allow i-customers to split demand among several facilities that achieve the minimum full cost T_i. Second, each facility is modeled as an $M/M/s$ queue and the central planner has a given number of servers to allocate to the operating facilities. The authors develop a solution method and use it to analyze an illustrative case.

Rabieyan and Seifbarghy [539] (2010) apply heuristics to compute **SO** locations at the nodes of a network for a given number of service centers. The benefit of serving a customer from node i at location j is exogenous for every i, j. For a given choice of service-center locations customers choose centers using a logit function of distances from the centers. The selected locations are constrained such that the resulting utilization of each center is at most a given constant.

Aboolian, Berman, and Krass [4] (2012) design algorithms for computing profit-maximizing location, capacity, and customer routing to service facilities. Service price is exogenous. Each customer has upper bounds specifying maximum travel distance and waiting time. These bounds are independent random variables. Demand from a given location i must be routed to a common facility j.

Then, all i-customers who can both travel the distance d_{ij} between locations i and j and wait the expected waiting time W_j at facility j will be served. Decision variables are the routing rule and either the number of servers in an M/M/s queue with a fixed capacity or the capacity in an M/M/1 queue at each location. In most instances of a computational study, the generated solutions assigned customers to their utility-maximizing facility, creating no conflict with individual preferences.

8.6   Braess-type paradoxes

The *Braess paradox* occurs when increasing the capacity of a link or a server degrades overall system performance. Examples of the Braess paradox in unobservable queueing networks can be found in Cohen and Jeffries [171] (1997), Kameda, Altman, Kozawa, and Hosokawa [383] (2000), and in Kameda, Hosokawa, and Pourtallier [385] (2001). A simple 3-server setting for the paradox is given in [1] §3.8.1. Other related "paradoxes" associated with equilibrium routing decisions in networks are reported by Zhang, Kameda, and Shimizu [700] (1991). Specifically, the mean sojourn time, in both equilibrium and **SO** solutions, may decrease when arrival rates or communication costs increase.

See [443] for an additional example of a Braess-type paradox.

Korilis, Lazar and Orda [412] (1999) consider a (directed) network where links are associated with M/M/1 queues and a given additional capacity is available for allocation. All users enter the network at a common source and leave at a common destination. The authors prove that the following allocations avoid the Braess paradox and hence decrease the equilibrium expected time in the system: Either augment the capacities of all servers by the same factor, or only augment links that directly connect the source to the destination.

Kameda and Hosokawa [384] (2000) (see also [382]) provide a very simple and insightful example with two identical M/M/1 servers each having a dedicated user with demand rate λ. Each user incurs initially an expected waiting time of $\frac{1}{\mu-\lambda}$. Suppose users are allowed to send any part of their demand to the alternative server and that this requires a communication (or travel) cost $t > 0$ per unit demand. The authors prove that if $t < \frac{\lambda}{(\mu-\lambda)^2}$, a unique symmetric equilibrium exists in which each user sends a positive fraction of demand to the alternative server.[18] Expected waiting time at the server does not change, but both users are now worse off because they incur communication costs. Indeed, the cost degradation (and the **PoA** in the system where communication is allowed) is unbounded.

Kameda and Pourtallier [386] (2002) derive general results and provide a queueing example (Example 1 on page 427 in [386]) generalizing the example of [384].

El-Zoghdy, Kameda, and Li [225] (2006) consider multiple M/M/1 servers with a user at each server location. Users decide how much demand to process at their location and how to route remaining demand to other servers. Routing is associated with communication time, which is either constant, depends on the total amount of routed demand, or only depends on the amount of demand sent between locations. The authors investigate the possible cost degradation when communication capacity is increased. A numerical study demonstrates that it is maximal under complete symmetry.

8.7 The Downs-Thomson paradox

The *Downs-Thomson paradox* occurs when a road is expanded, customers are drawn away from public transit, and in the new equilibrium the road becomes more congested and the fare of public transit is raised to compensate for the loss of demand, resulting in all travelers being worse off.

Calvert [125] (1997) considers a Markovian system where new customers choose between a single-server queue and a facility with an infinite number of servers serving batches of size $N > 1$. In addition to this primary arrival

---

[18]The bound on t is exactly the externality, i.e., the waiting time saved when a customer balks (see §1.8.1). If the communication cost is lower, it is worth diverting some demand to the other server.

process of customers, there is also a secondary arrival process dedicated to the bulk-service facility. This assumption assures that the batch fills up with probability 1 regardless of the behavior of the primary customers.[19],[20] The Downs-Thomson effect in this context means that an increase in service rate from the single server may cause an increase in the equilibrium expected time customers spend in the system. The author demonstrates that this effect is possible in both observable and unobservable versions of the model.

Full solutions of the models considered in [125] are given by **Afimeimounga, Solomon and Ziedens [20, 21] (2005, 2010)**.

In [20] the authors treat the unobservable version. They use the term Road (R) when referring to the single server service and Train (T) when referring to the bulk service, and an arriving customer chooses these services with probabilities p_R and p_T. A nice feature of this model is that the R-system is ATC while the T-system is FTC. Thus, an increase in customer tendency towards selecting the R-server will increase the expected wait at the R-server but also, by decreasing the arrival rate to the T-server, will increase the expected wait there.

The authors show that the difference between expected waiting times at the two systems is a quadratic function of p_R. The roots in (0,1) define mixed equilibria. Additional candidates are $p_R = 0$ and $p_R = 1$. However, four equilibria are not possible because when the quadratic function has two roots it has the same sign at both $p_R = 0$ and $p_R = 1$ so only one is an equilibrium. The authors give details of all possible sets of equilibria and specify which of these are stable. An interesting result is that this richness of equilibria is actually possible only when $N = 2$. When $N > 2$, the equilibrium is unique and p_T is smaller than the **SO** value.

In [21] the authors solve the model for observable queues. The equilibrium strategy gives rise to a switching curve which defines for every queue length at one of the facilities a threshold value on the queue length of the other. This threshold separates the states that dictate joining the former from those that dictate joining the latter. The authors prove a unique equilibrium exists and that it defines a monotone switching curve. The authors provide examples where state-dependent routing mitigates the Downs-Thomson effect of the unobservable case. They find that for most parameter values the delay is less, and sometimes considerably less, for the observable model than for the unobservable model with the same parameter values. However, they also present examples where delays in the observable model are slightly greater than those in the unobservable version.

---

[19]See [111] for a similar assumption.

[20]The shuttle model of [308] and [1] §1.5 resembles this model, but with important differences. The cost alternative to bulk service is independent of customer strategy. Furthermore, no dedicated demand is assumed and therefore never joining the bulk server is always an equilibrium in the unobservable version.

Ziedins [716] (2007) considers a system with K identical observable tandem queues. The ith stage of each queue can hold at most C_i customers, $i = 1, 2$. An arriving customer can either join a queue with less than C_1 customers in the first stage, or balk. If first-stage service for a customer terminates and the second stage already is full with C_2 customers, the customer is rejected. Balking costs the customer d_1 and rejection from the second stage costs d_2. Note that there is no waiting cost in this model. Customers are only interested in the probability of successful service and may prefer one queue over another having fewer queueing customers.

The author compares the **IO** and **SO** joining policies in a small example (with $C_1 = C_2 = 2$) by considering the six possible strategies for various cost values. The Downs-Thomson paradox in this model is demonstrated by showing that for $d_2 > d_1$ the **IO** strategy may result in a higher total cost to the system when the service rate increases at the second stage.

Nobel and Stolwijk [504] (2011) consider customer choice between an M/G/s system representing a taxi stand and a bulk service M/G/∞ system representing public transportation. They consider the observable and unobservable versions and compute the equilibrium and **SO** solutions. The Downs-Thomson paradox can occur in both cases: an increase in service rate or in the number of servers in the M/G/s queue may cause an increase in the average system waiting time. The effect of bounded rationality on customer welfare is also discussed by computing system performance when customers employ heuristics rather than performing exact computations.

Chen, Holmes, and Ziedins [156] (2012) extend [125] assuming both queues provide bulk service. The batch size at the ith queue is N_i, service rate is μ_i, and there is a dedicated Poisson(σ_i) stream of customers, $i = 1, 2$. In addition, a Poisson(λ) stream of flexible customers select one of the queues with the minimum expected waiting time. The authors show that the results of [20, 21] on existence and uniqueness of the equilibrium are maintained under this extension as well. In particular, when the queue is observable the expected delay is in most cases smaller than when it is unobservable, but the reverse inequality is also possible.

Chapter 9

Supply chains, outsourcing, and contracting

This chapter surveys models where a service or a product is supplied by agents having conflicting goals while at the same time sharing common interests in the efficiency of the system. These models combine elements from models on complementary services and on cooperation, which are separately discussed elsewhere in this survey.

This book is about *queueing games*, but it seems that the models in the concluding section of this chapter especially merit this title. These models describe games between two agents with opposing objectives, each controlling a different component of a queue. These models resemble supply chains but the emphasis is on the equilibrium rather than on system control.

- In most supply-chain models, the main game is between buyers and suppliers. Customers neither obtain an explicit value nor incur costs. They either behave in a non-strategic way, or their behavior is summarized through a demand function.

- Outsourcing models consider firms that profit by letting another firm (supplier) serve their demand. Often the originator firm can also serve its demand in-house but less efficiently. In such a case, even a single supplier has incomplete monopolistic power.

- In most cases, the purpose of the research is to induce cooperation between the supply-chain's agents through contracts designed to increase system profit and distribute added revenue. Of particular interest are contracts which *coordinate the system*, i.e., they induce decisions that maximize the efficiency of the supply chain.

- It is common to (implicitly) assume a contract is for a fixed term and performance is evaluated at the end of the term to settle the contract. It is also common to use the steady-state distribution of waiting time for design and analysis of the contract. This is a reasonable approximation when the contract term is for a sufficiently long period, but a long-term contract also allows for good estimations of variables that cannot be directly observed, such as the service rate, and assumptions concerning asymmetric information may be hard to justify.

- Ren and Zhou [548] (2008) distinguish between *inventory supply chains* and *service supply chains*. In inventory supply chains there is a flow of physical storable goods and the seller obtains the revenue *directly from the customer*. In service supply chains, such as call centers, the server usually does not earn direct revenue but is compensated for investment and effort *by the user company*. Moreover, the servers are invisible to the customer who does not distinguish between them and the user company, and any costs incurred during service (waiting, unsatisfying service) will be imposed on the user company rather than on the server.

- The comprehensive surveys by Gans, Koole and Mandelbaum [245] (2003), and Akşin, Armony, and Mehrotra [25] (2007) indicate that almost all literature on call centers deals with non-strategic customers. Strategic models are briefly mentioned in §7 of [245] along with other directions for future research. Literature on queueing models motivated by staffing problems in call centers is surveyed in [422]. These models typically assume a system with many servers. Such systems can operate at high utilization, while still providing good service. This, and the fact that the stochastic system is in most cases not tractable, naturally leads to asymptotic approximations based on heavy-traffic theories.

- Agents in a supply chain provide services that are complementary, in the sense that benefit is generated only when all stages are completed. However, this differs from complementary service models described in §8.3, where customers wishing to enjoy the outcome of a service must visit a set of servers providing the different component parts of the service.

- Supply chains are also discussed in §5.5 with the focus on decentralized decision making. See also [518] for a study of a supply chain facing customer bounded rationality.

9.1   Inventory supply chains

Cachon [116] (1999) considers an $M/M/1$ *supplier* with a fixed unit-production cost that sells to a *retailer* at an exogenous wholesale unit price.

The retailer in turn sells to customers at an exogenous retail unit price. Both supplier and retailer hold inventory and incur the same fixed unit holding-cost rate. Demand cannot be backordered and there is a system penalty for lost customers which the two agents share in predetermined parts. The supplier implements base-stock level S_s and the retailer implements base-stock level S_r. *These base-stock variables are the only decision variables in the model.* The inventory of the system can be described by a single variable: if it is at most S_r then all the inventory is held by the retailer. The supplier can have positive inventory only when the retailer's inventory is S_r. Naturally, the supplier benefits from this arrangement.

The author investigates both theoretically and numerically the properties of the best-response functions when S_s and S_r are approximated as continuous functions, paying particular interest to the prevalence of multiple equilibria (which appear to be rare) and loss to the system due to competition (which can be large). The author considers several coordination schemes including subsidies and sharing of costs. The most effective method combines inventory sharing with a transfer payment for expected lost sales.

Plambeck and Zenios [533] (2003) consider a supply chain where the service rate is dynamically controlled by a *supplier* and cannot be observed by the *retailer*. The retailer's objective is to determine a payment scheme that minimizes discounted costs of inventory, backordering, and transfer payments to the supplier. The supplier incurs convex capacity costs and plans consumption over time to maximize a discounted value of an additive exponential utility function. The optimal incentive scheme combines time-dependent payments per completed job and inventory penalties, both decreasing as inventory is depleted and increasing when inventory approaches a desired base-stock level. This scheme coordinates the systems (achieves the production policy the retailer would choose in a centralized setting) if the supplier is risk neutral, but not when risk averse. The authors investigate the performance degradation caused by factors such as discounting rate, risk aversion, and convexity of the capacity costs.

Caldentey and Wein [122] (2003) consider a supply chain with exogenous price and demand. A *supplier* sets capacity μ and pays cost $c\mu$; a *retailer* sets a base-stock level s and incurs holding costs. The system also incurs backorder costs which the retailer and the supplier share in exogenous proportions α and $1 - \alpha$, respectively. The authors simplify the analysis by applying exponential approximation to the geometric distribution of the supplier's M/M/1 queue length. They prove that a unique equilibrium exists, for which they obtain a closed-form expression.

Naturally, the equilibrium does not minimize aggregate system costs. The authors derive the **PoA** which depends on the ratio of backorder to holding unit costs and the fraction α, but is independent of the unit capacity cost c. If backorder unit cost exceeds unit holding cost, **PoA** is minimized when the backorder cost is shared approximately evenly, that is, near $\alpha = 0.5$.

The authors also propose a contract that coordinates the chain by linear transfer payments such that the objectives of both agents become scaled versions of the centralized one. They conclude by solving two Stackelberg games, each with a different leader.

Cachon and Zhang [119] (2006) add contract design and asymmetric information to the model of [122]. Customers arrive to purchase a product from a *buyer*. The buyer acquires the product from an MTO *supplier* chosen from a pool of $n \geqslant 1$ suppliers. The chosen supplier will have to build up capacity μ at a cost $b\mu$, where b is a realization of a random variable with a known log-concave cdf F on an interval $[b_l, b_h]$. The realization b is unknown to the buyer. Once a unit production is completed it is delivered to the buyer. The buyer incurs inventory holding costs and backorder penalties, and uses a base-stock policy.

A *procuring mechanism* is a process where in response to the capacity cost announced by the supplier, the buyer announces the price he is willing to pay and the capacity the supplier is required to provide. If the buyer can choose from multiple suppliers, the mechanism also includes the probability of each supplier being chosen as a function of their announced capacity costs. *It is required that the announced price is such that even a supplier having the highest possible cost b_h accepts the offer.* The authors identify the optimal mechanism for the buyer and also propose simpler alternatives which are effective with regard to both the buyer's total cost and the supply chain's total cost. Specifically:

- **Late-fee mechanism:** The buyer pays the supplier a fixed price per unit and charges the supplier a fixed price per unit time for an outstanding order.

- **Lead-time mechanism:** The buyer pays a fixed price per unit and informs the supplier the lead time that must be satisfied (equivalent to the capacity the supplier must provide).

- **Reverse auction:** If the buyer can choose from a pool of suppliers, the unit price is set through an auction in which suppliers bid for the right to sell to the buyer.

An interesting property of these mechanisms is that the supplier retains most of the benefits of having a low production cost.

Gupta and Weerawat [290] (2006) consider a Markovian tandem system consisting of an MTS *supplier* **S** that produces raw components according to a base-stock level b and an MTO *manufacturer* **M** that uses them to satisfy demand. When no component is in stock, demand requests queue at **S** before proceeding to **M**'s queue. Because the arrival of raw material to **M** is not a renewal process, the delay is approximated by assuming both parts of the system are M/M/1 queues.

In the base model, the manufacturer's revenue from the sale of a unit is $\pi(b) = p_0 - \beta\bar{L}(b)$, where p_0 and β are parameters of the model, and $\bar{L}$ is the expected total delay for a customer. Each of **S** and **M** incur unit production and inventory holding costs.

Let b_0 be the value of b that maximizes combined profit of **M** and **S**. The authors investigate three incentive mechanisms for motivating **S** to adopt b_0:

- **Fixed markup contract:** **S** sells each unit to **M** at a price $(1 + \gamma)$ times its average per-unit production and inventory cost. For any $\gamma > 0$, this mechanism results in a base-stock level smaller than b_0.

- **Simple revenue sharing scheme:** This is a Stackelberg game in which **M** commits to sharing a fraction α of its revenues with **S** and **S** reacts by setting base-stock level b. Again, for any $\alpha < 1$, this mechanism results in a base-stock level smaller than b_0.

- **Two-part revenue-sharing:** If **S** chooses $b < b_0$ then **M** shares the same fraction of revenues as in the simple revenue-sharing scheme, or a certain higher fraction α otherwise. There exists an interval of α values that induce b_0; the added profit is eventually allocated according to the relative bargaining strength of the two parties.

Liu, Parlar, and Zhu [451] (2007) assume demand to be a linear function of the *retail price* p_r and PDT l. These variables are determined through a Stackelberg game where an MTO supplier sets l and *supply price* p_s, and a retailer responds with p_r. The supplier incurs linear holding and tardiness costs. The main results are:

- The authors obtain a closed-form solution for an M/M/1 supplier. The solution is also extended to the case where the service rate is determined by the supplier.

- In more general cases, the authors consider a *combination model* where the delay is composed of $g_1(\lambda)X + g_2(\lambda)$, where λ is the effective arrival rate and X is service time.[1] Under this structure they characterize the unique Stackelberg equilibrium. A good rule for quoting a lead time is to define it as a linear function of expected delay.

- The system-wide expected profit rate under decentralized and centralized settings are compared. The centralized solution has a greater PDT, a smaller retail price p_r, greater demand rate, and **PoA**$\leqslant 2$.

Gui and Ma [264] (2007) follow the economic model of [545] in a supply chain where a *retailer* sets lead time L and capacity level μ, and incurs the capacity cost while production is carried out by a *supplier* setting the supply

---

[1]The authors mention that this model includes (approximately) the case of an M/G/1 supplier.

price T. The authors show the supplier should encourage the retailer to invest more in capacity by offering a share in the resulting increased profit.

Jemaï and Karaesmen [363] (2007) treat a version of [116] with back-orders (rather than lost sales). The *supplier* and *retailer* choose individual base-stock levels to minimize their (heterogeneous) linear holding and back-order costs. These levels are always less than desired to optimize total system costs (though, a numerical study reveals that **PoA** is quite small in common scenarios). Sharing some inventory costs can encourage higher inventory levels, and the authors derive a variety of sharing contracts that coordinate the system.

Hennet and Arda [340] (2008) propose a supply-chain model with linear adjustment contracts between a *customer* and an M/M/1 MTS *producer*, and between this producer and a *supplier*. The utilities (per time unit) of the agents are:

$$\pi_c = (a - p)\lambda + \eta L - \phi(L) \text{ for the customer}$$
$$\pi_p = (p - c_p)\lambda - hI - (b + \eta)L - T \text{ for the producer}$$
$$\pi_s = T - c_v\lambda - c_\mu\mu \text{ for the supplier,}$$

where:

a is unit value of the product for the customer

p is unit price paid by the customer

h is unit holding cost

b is delay cost directly incurred by the producer

c_p and c_v are unit production costs

c_μ is unit capacity cost

η is compensation paid by the producer to the customer

$\phi(L)$ is the customer convex delay-cost function

$L = L(\lambda, \mu, S)$ is expected amount of backorders

$I = I(\lambda, \mu, S)$ is average inventory level held by the producer

S is the producer's base-stock level

$T = T(\lambda, \mu)$ is the price paid by the producer to the supplier.

The producer-customer contract is a pair (p, η) set through a negotiation between the two sides. The producer-supplier contract is a pair (r, k) determined in a Stackelberg game where the producer acts as leader and $T = \left(r - \frac{k}{\mu - \lambda}\right)\lambda$. Decision variables in the producer-customer game are λ for the customer and p, η, and S for the producer. Decision variables in the producer-supplier game are r and k for the producer and μ for the supplier.

Arda and Hennet [63] (2008) consider a tandem system, similar to [363], with two MTS M/M/1 stages and symmetric information. A *supplier* produces components that are processed in the second stage by a *manufacturer* and sold at a given market price. The supplier's decision variable is its base-stock level (treated as a continuous variable). The manufacturer determines its own base-stock level and designs a contract (p, b), where p is the price it pays the supplier per component unit and b is a per unit time and unit backlog penalty imposed on the supplier. Both agents incur fixed unit production and inventory costs, and the manufacturer also incurs backorder costs. The authors solve the Stackelberg game in which the manufacturer leads, and conclude that the solution globally optimizes system performance.

Xiao, Yang and Shen [668] (2011) consider a supply chain where an M/M/1 *manufacturer* quotes wholesale price and lead time and a *retailer* sets retail price. The manufacturer incurs a fixed unit manufacturing cost. Customers are delay sensitive, and those obtaining a defective product return the defective product to the manufacturer for *remanufacturing* at no additional charge. This affects the customers' utility and the manufacturer's total cost. Customer valuations are uniformly distributed and only those with values above a threshold join the queue. These assumptions lead to a linear demand function of retail price and delay. The authors obtain the following results:

- A closed-form solution in the centralized setting.

- A solution of the manufacturer-retailer game.

- A revenue-sharing mechanism for coordinating the system, and a necessary and sufficient condition for both agents to be better off under this mechanism.

Xiao and Shi [667] (2012) consider a *manufacturer* and a *retailer* that differentiate the market by operating two facilities producing substitutable products with different retail prices and lead times. Demand is linear in price and delay with substitution effects. The manufacturer chooses service rates μ_i costing $A_i\mu_i$, and also sets wholesale prices w_i, $i = 1, 2$. The retailer sets retail prices.

The authors consider Stackelberg games where each of the two agents can be the leader. Lead times are initially assumed to be determined according to exogenous standards, but the effect of letting the manufacturer set the lead time in the faster facility is also investigated. In all cases, the authors characterize the solution and conduct sensitivity analyses.

Xiao and Qi [666] (2012) consider a *supplier* incurring a fixed unit production cost and setting a wholesale price for an M/M/1 *manufacturer*. The manufacturer reacts by setting retail price and lead-time. Demand is linear in terms of these variables.[2] The manufacturer incurs a fixed unit cost in addition to the wholesale price. The authors suggest a mechanism for coordinating the system by splitting profit according to a predetermined ratio. The mechanism sets an appropriate wholesale price if the demand rate is at least the

---

[2]The demand function contains a term γs where s denotes reliability level, but s is not a decision variable in the base model.

centralized optimal rate, or sets a sufficiently high wholesale price otherwise, i.e., an all-unit quantity discount mechanism is adopted.

Li, Huang, Cheng, Zheng, and Ji [435] (2014) examine *after-sales service* provided by a *retailer* to customers. Retail price is exogenous and includes the after-sales service. The model assumes the following sequence of events: (i) a *manufacturer* determines a wholesale price while the retailer quotes a lead-time guarantee; (ii) given these values, the retailer sets order size q and capacity μ_0; (iii) demand D is realized according to a demand function composed of a random part and a deterministic part that linearly decreases in the quoted lead time; (iv) actual sales are $\min(q, D)$, and the demand rate for after-sales service is a given fraction ψ of sales; (v) the queue for after-sales service is modeled as an M/M/1 system with demand rate $\phi \min(q, D)$, and if μ_0 is not sufficiently large to serve the demand and satisfy the delay commitment, the retailer outsources additional capacity.[3]

The authors solve centralized and decentralized versions of this model with and without the outsourcing option. They find, for example, that the outsourcing option leads to lower lead times, lower wholesale prices, and higher sales. Sharing the capacity costs can be used as a way to increase profits of both parties. The authors also investigate a variation where the retailer is risk averse and the outsourcing price is a random variable.

Zhu [714] (2015) considers a Stackelberg game in a supply chain where an M/M/1 MTO *supplier* leads by setting capacity μ and wholesale price p_s, and a *retailer* follows setting retail price p_r and PDT l. The resulting demand rate is $\lambda_0 - \alpha p_r - \beta l$, where $\alpha, \beta > 0$. The retailer incurs linear holding costs on early completion and linear tardiness costs, the supplier incurs capacity costs $c_0 \mu$. The author finds that:

- Compared to the solution obtained when capacity is exogenous, the supplier always benefits from the ability to set the capacity level while the retailer benefits only when the supplier decides to increase capacity (i.e., the exogenous value is smaller than the equilibrium value).

- A revenue-sharing contract can be used to coordinate the system.

9.1.1   Co-production

Co-production can be considered as partially outsourcing production to customers. We describe below models of co-production where the level of customer contribution is set by the firm. See §4.7 for similar models but with the level of customer contribution set by the customer. See [618] where the firm sets the level of contribution of an MTS stage producing semi-finished items.

---

[3]This is similar to delivery expedition, see §1.8.9.

Xue and Harker [675] (2003) study a Markovian model of co-production where the firm determines the fraction ρ of self-service and performs the remaining fraction $(1-\rho)$. Fraction ρ of self-service takes an expected amount of $s = \rho e$ time units, and the expected system time for service completion is $g = \left(\frac{\mu}{1-\rho} - \lambda\right)^{-1}$, where $\mu > e$ (the firm produces at a faster rate than the customer does). This gives the relation $\mu = (g^{-1} + \lambda)(1 - \rho)$.

Demand $\lambda(f)$ depends on full price f, and capacity μ costs the server $n \cdot \mu$. When price is p, profit $\pi = p \cdot \lambda - n(g^{-1} + \lambda)(1 - \rho)$ is maximized at $g = \sqrt{n(1 - \rho)/\lambda}$. Profit is convex in ρ and therefore the optimal fraction is $\rho \in \{0, 1\}$, meaning either full service or self-service only.

Consider now two identical firms competing with nominal prices p_i and service fractions ρ_i, $i = 1, 2$. Firm i's demand $\lambda_i = a - b(f_i - f_{3-i})$, $i = 1, 2$, is a linear function of the full prices f_1 and f_2. The authors show there are three cases to be considered: both firms set $\rho = 0$; both set $\rho = 1$; or one sets $\rho = 0$ while the other sets $\rho = 1$. In addition, they give conditions for each of these possibilities to be maintained in equilibrium. In particular, the interesting solution where one firm applies self-service only and the other applies full service only is possible when customer efficiency, as reflected by the parameter e, is within a certain intermediate range. If customers act highly efficiently then both firms will choose in equilibrium the self-service only option and, contrarily, if customers act highly inefficiently both will choose full service.

9.2 Service supply chains

Ren and Zhou [548] (2008) consider contracts coordinating a service supply chain with a *user* company and a *call center* under asymmetric information. Demand and service rates are fixed. The call center's decision variables are the number of servers and *effort level*. Given an effort level e, a portion $p(e)$ of calls are satisfactorily resolved. The user gains from resolved calls, and the call center incurs costs associated with staffing and effort. The main problem is to incentivize the call center to exert effort, which is not contractible.

The authors consider a fluid model and suggest coordinating the system through a partnership contract.

Hasija, Pinker, and Shumsky [301] (2008) apply diffusion approximation in a model of a service supply chain consisting of a *client* that hires a *vendor* to provide call-center support. The client obtains a fixed revenue per completed service and incurs a penalty per unit time of customer waiting. The vendor incurs a fixed cost per server and selects the number of servers N. The client maximizes profit by designing a contract that coordinates the system, i.e., induces the vendor to employ the supply chain profit-maximizing capacity N^*. The client allocates profits so the vendor earns its reservation value.

Customers are not strategic; their arrival and reneging rates are exogenously fixed.

Two basic contract types are considered: *payment per served customer* (PPC) and *payment per unit of service time* (PPT). Contracts include a *service level agreement* that commits the vendor to a given delay standard with a penalty charged if the commitment is not satisfied. It is shown that both contract types can be used to maximize client profits under symmetric information conditions and fixed service rates.

The authors consider a model with asymmetric information in which the service rate of all servers employed by the vendor is either μ_H or μ_L, with $\mu_H > \mu_L$. The client offers a pair of H and L contracts designed so that the H-contract (L-contract) will be selected if $\mu = \mu_H$ ($\mu = \mu_L$). Optimality can be achieved in a variety of ways such as offering an H-contract with PPC and an L-contract with PPT. Such combinations enable "screening" the types of servers. The results are also extended to arbitrary probability distributions of service rate, but in this case contracts that maximize supply-chain profits do not maximize client profits, as the vendor captures some *information rents*.

Ren and Zhang [547] (2009) consider a supply chain with an *outsourcer* and a *supplier* (service provider). The supplier's type is a random variable and private information. The supplier's decision variables are the service rate μ and the *satisfaction probability* or *quality* q. The cost function associated with these variables is separable, type-dependent, and correlated. The outsourcer's cost is $c_w L(\mu) + c_g \lambda(1 - q)$, where L is average queue length.

The authors solve the outsourcer's optimal contract and discuss the way positive and negative correlation between the two parts of the supplier's cost function affect contract performance.

Akan, Ata, and Lariviere [22] (2011) consider contracting between *originator* and *service provider* under asymmetric information. The service provider operates an $M/M/s$ queue in the Halfin-Whitt **QED** regime, and customers renege at a constant rate.

Two types of originators are distinguished by demand rate, which is private information and can only be observed after contracting. The originator earns a fixed amount from each customer served and incurs loss of revenue from customers who renege. The originator also has the option, although less efficiently, of handling the service *in-house*.

The provider decides on the level of initial capacity investment and incurs additional costs if there is a need to increase capacity to satisfy demand. The provider also incurs operating costs.

Providing service costs $C(\alpha, \lambda) = c_1 \lambda + c_2 \sqrt{\lambda} \beta(\alpha, \lambda)$ where λ is realized arrival rate, α denotes service level (for example, the fraction of lost calls), and β is a *standardized excess capacity* assumed to be increasing and convex. The provider proposes contracts (α, p) where p is the fee paid by the originator. The cost of a contract aimed for a given originator's type should not exceed in-house cost (**IR**) or the cost of the alternative contract (**IC**).

The authors obtain the following main results:

- Standardized excess capacity is independent of originator type and is greater than the value that would be chosen by the originator for in-house operation.

- The service provider can obtain the expected profits it would obtain under full information by offering one contract with a per-service fee intended for low-demand originators, and one with a flat rate independent of the realized λ, intended for high-demand originators. The intuition lies in the economies of scale, which enable separating the two types of originators. For a low-demand originator it would cost less to pay per service, while the opposite would apply to a high-demand originator.

9.2.1 Marketing effort

See §6.8 where monopolistic firms search for customers.

Jiang and Seidmann [366] (2011) consider an M/M/1 system owned by an *owner* and operated by a *manager*. Service price p is exogenous and the owner incurs a delay cost of τ per customer per unit time. Customer arrival rate is $\lambda = \theta + k\alpha$, where α is the manager's marketing effort level, θ is the expected *base demand*, and $k > 0$ is a constant. The manager decides on the effort level α, but the capacity μ is set by the owner at a cost of $F(\mu)$, which is increasing and convex. The effort α is the manager's private information and the contract rewards the manager based on *realized demand* during the contract period.

The owner's objective is to maximize expected profit consisting of revenue λp minus delay costs $\lambda \tau W$ (W denotes expected waiting time), compensation s to the manager, and investment in capacity. The manager's net utility is $U(s) - V(\alpha)$, where U is increasing concave and V is increasing concave. To obtain explicit formulas for the optimal contract the authors further assume $U(s) = 2\sqrt{s}$ and $V(\alpha) = m\alpha^t$ with $t \geqslant 1.5$.

The owner's problem is to set capacity and design a compensation contract that will induce the desired effort from the manager. The contract must satisfy **IR** and **IC** constraints to assure the manager agrees to exert the desired effort level.

The authors derive comparative statics analytically and numerically. For example, they prove that an increase in the base demand θ results in a lower optimal level of effort.

Jiang and Seidmann [367] (2014) extend [366] assuming an exogenous requirement that expected waiting time does not exceed a given service standard. The *base demand* θ is a random variable obtaining the value θ_h with probability q, and $\theta_l < \theta_h$ otherwise. Both α and θ are the manager's private information. It is further assumed that $t = 2$, $F(\mu) = c\mu$, and $U(s) = s$. The sequence of events is as follows:

1. The firm announces a set of contracts and capacity levels.

2. The manager observes θ and selects a contract.

3. The firm sets capacity according to the contract.

4. The manager spends on marketing effort.

5. Demand is realized and the firm compensates the manager accordingly. If the service standard is not met the firm incurs a penalty.

By the revelation principle, it is sufficient to consider a menu of two contracts intended for the two possible realizations of the base demand. The authors propose contracts in which compensation is a linear function of realized demand. They compare the optimal menu with the first-best symmetric-information solution and show that information asymmetry reduces the firm's profit and increases the manager's expected net benefit.

The authors also suggest an alternative *charge-back* approach in which the contract delegates all the decision rights, including the capacity investment decision, to the manager. They provide an explicit formula for the optimal contract and prove that it induces the first-best effort levels. Of course the manager is able to exploit the information asymmetry in this case as well. A comparison of the two approaches shows that the firm can realize a higher expected profit by offering the alternative contract iff $q < 0.5$.

9.2.2   Value creation

See [364] for another model where the firm combines service and advertising profits.

Güneş and Akşin [267] (2004) define a *value creation* model with asymmetric information. A *firm* employs a *server* to serve its demand. Customers can be given *regular* or *extended* service. A customer is associated with probability $\hat{p}$ that an extended service will yield revenue. The probability $\hat{p}$ is a realization of a random variable with cdf $F(p)$. The server observes $\hat{p}$ but the firm only knows $F(p)$. For a critical value θ, the market is segmented and only customers with $\hat{p} > \theta$ are given extended service.

The server sets the critical value θ which maximizes payments w obtained from the firm minus a fixed cost associated with providing extended service. The firm's goal is to design a payment scheme that maximizes net profit (revenues minus payments to the server and holding costs).

Extended service takes longer on average, but the firm doesn't know whether a particular lack of generated revenue followed regular or extended service. The authors propose a payment scheme in which the firm pays the server a fee per customer that depends linearly on service duration x_1 and revenue generated x_2: $w = \alpha_1 x_1 + \alpha_2 x_2$.

The proposed incentive mechanism is quite restricted. For example, the server is not delay sensitive and consequently implements a strategy independent of queue length. Of course the first-best strategy for the firm would be a dynamic one.

The authors focus on the case where $\hat{p} \in \{p_L, p_H\}$, $p_H > p_L$. In this case, even under the restriction of a queue-length independent strategy, the optimal segmentation for the firm will, in general, correspond to a mixed strategy including either a fraction of the H-customers, or all of them plus a fraction of the L-customers. However, the server's strategy is clearly pure: providing extended service to all customers, only to H-customers, or not at all.

Gurvich, Armony, and Maglaras [294] (2009) consider a call center serving a multiclass population.[4] Customer classes differ in the probability distribution of their service value and in their willingness to listen to *cross-selling* offers. Once regular service of a customer is completed, the customer either leaves the system or enters a cross-selling phase handled by the same server. The probability the customer agrees to enter this phase is an exogenous decreasing function of queueing time. In the cross-selling phase, customers are offered a purchase depending on their type, which is accepted if the associated value is greater than the offered price. The firm decides on the number of servers, whether to attempt cross-selling to a customer, and the offered price. The analysis is asymptotic, leading to a simple heuristic solution which the authors numerically compare with the non-asymptotic Markovian case. The solution suggests cross-selling should only be offered to the most profitable customer classes and only when the system is not overly congested.

9.2.3   Outsourcing two-level services

This subject has some similarity to that of expert systems (see §6.3), but the focus there is on asymmetric information regarding customer type.

Lee, Pinker, and Shumsky [431] (2012) examine a two-level Markovian service process where the first level assesses (diagnoses) the *complexity* of the request. There are two types of servers, gatekeepers and experts. *Gatekeepers* can perform the diagnosis and either refer the job to an expert or treat it themselves, but the probability of success strictly decreases with job complexity. *Experts* are more expensive to employ, but perform the service with guaranteed success. Costs are associated with employing gatekeepers and experts, failed treatment, and customer system time. Customers are not strategic, do not balk, and incur no costs. The firm, denoted as the *client*, faces four outsourcing options to an external service provider called the *vendor*: (i) outsourcing the first level; (ii) outsourcing the second level; (iii) outsourcing both levels; (iv) eliminating the first level and outsourcing the second. The decision

---

[4]This value-creation model is placed here even though it is not about a supply chain.

variables consist of the number of servers of each type, and a threshold for diagnosed complexity above which customers are referred to an expert. While outsourcing, the firm has no control over the number of servers employed or the threshold.

The authors make simplifying and approximation assumptions which turn the two stages into independent $M/M/s$ systems, and apply the Halfin-Whitt **QED** regime of the asymptotic Halfin-Whitt model. The suggested contracts are based on variables that can be observed by the client, including the number of customers handled by the vendor, customer waiting time, and whether a customer has been satisfactorily served by the vendor. It is shown that the first-best solution can be achieved except when the two levels are outsourced. A numerical study demonstrates that the solution obtained under the approximation is close to optimal (an average error of 0.5%) for systems with load above 50.

9.3   Allocation of demand to suppliers

This section deals with demand allocation in supply chains. A different type of demand allocation is considered in Chapter 8, where strategic customers allocate demand among providers of substitutable services. The distinguishing assumption of this section is that there is just a single firm (or customer) allocating demand and trying to induce the desired suppliers' behavior.

Some of the models described here assume asymmetric information. The buyer cannot observe relevant variables, usually the service rates, and therefore these variables are not contractible. The buyer designs allocation mechanisms encouraging potential suppliers to provide fast service. Unlike when customers possess private information, here the buyer and supplier engage for a long period, and the buyer can accurately estimate service rates by observing the system. Yet, the contract may have to be signed at the beginning of the period and, in any case, achieving desired supplier behavior without the need for statistical monitoring is advantageous.

Other models assume symmetric information and allocations depend on the values set by the suppliers. Knowing the service rates, the buyer can easily design contracts inducing suppliers to select the desired rates subject to **IR** constraints. The authors focus, however, on allocations that are "smooth" in the way they penalize deviations from the desired capacity.

Gilbert and Weng [252] (1998) consider a Markovian model with exogenously fixed demand where a *coordinating agency* (the firm) hires two identical servers to serve this demand and pays them for service. The firm's goal is to achieve a given delay standard at minimum expense. The servers choose capacities to maximize profits while incurring convex capacity costs $c(\mu)$. The firm determines the amount it pays per customer served and the rule to allocate users to servers can be one of the following:

1. **KKR allocation:** This allocation, introduced by Kalai, Kamien and Rubinovitch (1992) (see [1] §8.7) maintains a single FCFS queue. Server heterogeneity is ignored and an arriving customer randomly selects a free server, never waiting for a faster server to become free.[5]

2. **Balanced allocation:** A separate unobservable queue is maintained for each server. Customers are assumed to be strategic, and in equilibrium demand is divided between servers so the expected waiting times for each server are identical.[6]

The servers choose identical rates in equilibrium under both allocation options, and these rates are larger in the case of balanced allocation. The authors characterize the cases in which the firm incurs lower costs under balanced allocation. In particular, if $c(\mu) = a\mu^2$, with $a > 0$, then despite the well-known inefficiency associated with separate queues, the firm incurs lower costs with separate queues for any given delay standard.

Grosu and Chronopoulos [262] (2004) consider demand allocation to n M/M/1 servers with heterogeneous service rates which are private information. Server i maximizes $P_i - \rho_i$, where P_i is compensation obtained from the system manager and $\rho_i = \lambda_i/\mu_i$. The authors propose a price and demand-allocation mechanism where servers report their expected service times. Given reports $\boldsymbol{b} = (b_1, \ldots, b_n)$, server i receives a compensation $P_i(\boldsymbol{b}) = b_i\lambda_i(\boldsymbol{b}) + \int_{b_i}^{\infty} \lambda_i(b_{-i}, x)dx.$ [7] Demand is allocated according to the **Bell-Stidham allocation** [85] (see [1] §3.7.1.2, [594] §1.5) assuming $\mu_i = 1/b_i$ for $i = 1, \ldots, n$. The authors show that the proposed mechanism induces truthful reports and **SO** demand allocation.

Cachon and Zhang [120] (2007) consider a Markovian model with a *buyer* and two identical *servers*. The buyer's demand λ and price per service R are exogenous. The cost to a server for establishing capacity μ is a convex function $c(\mu)$, and the capacities are observed by the buyer. The buyer wishes to reduce the system average lead time by announcing an allocation scheme.

The servers set capacities in a noncooperative way, maximizing profits $\pi_i(\mu_1, \mu_2) = R\lambda_i(\mu_1, \mu_2) - c(\mu_i)$, $i = 1, 2$. The allocation scheme affects expected lead time in two ways. First, by encouraging the servers to increase capacities towards the maximal profitable rate $\bar{\mu}$, where $c(\bar{\mu}) = R\lambda/2$.[8] Second,

---

[5]This bounded rationality assumption reduces the servers' incentive to increase capacity.

[6]The model may have no server equilibrium, see [120] p. 412.

[7]The vector (b_{-i}, x) is obtained from $\boldsymbol{b}$ by substituting $b_i = x$.

[8]The authors mention it is either optimal for the system to have one server allocated all jobs or two servers allocated half of the jobs. They consider the latter case.

by achieving efficiency for any given pair of capacities. These goals are shown to be contradictory. (i) **Bell-Stidham allocation** [85, 262], which minimizes the buyer's lead time for any choice of capacities does not guarantee maximal capacities and an equilibrium may even fail to exist. (ii) **Balanced allocation** [252], where the active servers' lead times are equal (this is the outcome when the queues are unobservable and non-atomic customers act independently). This allocation does not guarantee maximal capacities and an equilibrium may fail to exist. (iii) **Linear allocation**, which is a two-parameter allocation, and more than one choice of parameter values can be used to induce an equilibrium with the maximal capacities. (iv) **Proportional allocation**, where there is a single parameter allocation with a unique choice that induces an equilibrium with optimal capacities if R is sufficiently large and $c(\mu)$ is quadratic.

The authors show examples where linear allocation results in the smallest lead time among these options. They also consider state-dependent allocations. Here, in addition to achieving desired capacity levels, allocation can take advantage of queue-length information so a server is not kept idle when the queue is not empty.

Choi, Huang, and Ching [166] (2012) extend the model of [120] to multiple servers and prove that the main results still hold.

Benjaafar, Elahi, and Donohue [86] (2007) consider a general model with a *buyer* wishing to outsource a fixed demand rate to N identical *suppliers*. Supplier i commits to a *service level* s_i which is a contractible variable and can be observed by the buyer. The buyer's goal is to maximize the expected service level. Revenue realized by a supplier per served unit is exogenous and the supplier's goal is to offer a service level maximizing expected profits after deducting variable and fixed costs associated with the offered service level.

A **supplier allocation scheme** (SA) announces a criterion for allocating demand among suppliers, allocating greater portions α_i to those offering higher service levels s_i. A **supplier selection scheme** (SS) allocates the entire demand to supplier i with probability α_i which increases in s_i.

The authors focus on proportional allocation functions of the form $\alpha_i(s_1, \ldots, s_N) = s_i / \sum s_j$, and prove the existence of a unique equilibrium. We describe here the two special cases of their model where each supplier behaves as an M/M/1 queue.

- **Make-to-order suppliers:** The service level is defined as the probability of fulfilling a service request within a given lead time τ, $s_i = \Pr(W_i \leqslant \tau) = 1 - e^{-(\mu_i - \alpha_i \lambda)\tau}$, which can be translated to a commitment on μ_i. For example, under the SS scheme, supplier i commits to $\mu_i = \lambda + \ln[1/(1 - s_i)]/\tau$. There exists a unique and symmetric equilibrium for both SA and SS, and the service level will be higher under the SS scheme.

- **Make-to-stock suppliers.** In this model, suppliers maintain a fixed *target utilization level* ρ_i, so the supplier sets capacity to $\mu_i = \alpha_i \lambda / \rho_i$. The service level is defined as the *fill rate* which is the probability of fulfilling an order from on-hand inventory. The supplier incurs variable production costs, capacity costs, and inventory costs, and decides on a base-stock level b_i, which is determined by service level $1 - \rho^{b_i}$, or equivalently, $b_i(s_i) = \ln(1 - s_i)/\ln(\rho)$. The analysis and insights are similar to those in the MTO example.

Ching, Choi, and Huang [163] (2011) enhance the model of [86] by adding the option of imposing penalties on suppliers when they fail to satisfy promised levels of service. Such penalties induce suppliers to establish higher capacity levels.

Ozdaglar [509] (2008) considers a single user submitting service requests at rates x_i to parallel servers with heterogeneous delay functions $l_i(x_i)$. Servers maximize profits by setting prices p_i; the user reacts by setting net-utility maximizing demand rates x_i. The user's utility from service $u(\sum x_i)$ has negative second and third derivatives.[9] The author shows that **PoA** $\leqslant 1.5$. **Dube and Jain [208] (2014)** provide conditions for the existence of an equilibrium in this model with two users assuming that in addition to price, servers decide on splitting their capacity between the two users. As in the single-user case, also here **PoA** $\leqslant 1.5$.

Ching, Choi, and Huang, [162] (2010) provide a multiserver extension to the KKR model (see [252]), and **Ching, Huang, Choi, and Huang [167] (2012)** apply this extension to the separate queues model of [252]. An interesting result is that an increase in the number of servers induces higher service capacities for both common and separate queues. This results from more intense competition, though the same demand is split among more servers. The authors also find that separate queues induce higher service capacities than a single queue.

Gong, Wang, Deng, Murthy, and Cai [258] (2010) consider a model similar to [120].

They suggest a **residual proportional allocation** (RPA) in which suppliers' market shares are proportional to $(\mu_i - \lambda \alpha_i)^\beta$, where $(\mu_i - \lambda \alpha_i)$ is the *residual capacity* of supplier i. When $\beta = 1$ a proportional allocation results, and when $\beta \to \infty$, it converges to the balanced allocation. The main result of this study is that RPA stimulates high equilibrium capacity when balanced allocation does not have a Nash equilibrium.

Wee and Iyer [657] (2011) consider allocation of holding costs to two competing servers with convex capacity costs, a fixed revenue R per completed service, and a constant holding-cost rate per unit of demand. The goal of the

---

[9]We refer to §5 in [509] that allows $l_i(0) > 0$.

allocation is to motivate the servers to select the highest service rate that gives them nonnegative profits.

Two variations of demand allocation are examined. **Split allocations**, where arriving jobs are immediately allocated to servers, and **pooled allocations**, where there is a single queue and jobs are assigned to the first server that becomes free. The authors suggest a holding-cost allocation policy that achieves the maximum possible service rate for both split and pooled systems. Interestingly, depending on the model's parameters. the holding costs charged to a server may increase or decrease with selected capacity

Toktaş-Palut and Ülengin [624] (2011) consider a supply chain with a single MTO *manufacturer* and n heterogeneous MTS M/M/1 *suppliers* each supplying a different component. Demand is Poisson. When demand occurs the manufacturer can start the production of a unit only after receiving all components. The authors model the manufacturer as a GI/M/1 queue and derive an approximation for its coefficient of variation. Suppliers incur holding and backorder costs. The manufacturer incurs backorder costs on finished products and holding costs on components while waiting for missing components to arrive.

The authors consider three types of contracts: (i) a backorder (holding) cost subsidies contract to adjust the base-stock levels when they are smaller (greater) than the optimal ones; (ii) a transfer payment contract based on Pareto improvement (i.e., after payment the suppliers are at least as well-off as they would be under the decentralized solution); (iii) a cost sharing contract. All these contracts coordinate the chain and improve supplier welfare relative to the decentralized solution where each supplier independently minimizes costs, but only the third guarantees that all members of the supply chain participate in the contract.

Jin and Ryan [369] (2012) consider a *buyer* allocating demand to MTS *suppliers*, similar to [86], but with the important difference that the assumed-fixed price in [86] is now a decision variable. The authors use an exponential score function $a(s_i, p_i) = e^{s_i - \alpha p_i}$, where p_i and s_i are price and service level (fill rate) at supplier i. The buyer applies proportional allocation and outsources each unit of demand to supplier i with probability $\beta_i = a(s_i, p_i)/\sum a(s_j, p_j)$. The buyer's goal is to minimize a weighted sum of backorder costs and transfer payments to the suppliers. An advantage of this strategy is that the buyer has only one single control parameter, α, which measures the relative weight of price to service level. Some of the qualitative conclusions are:

- The equilibrium price and service level both increase as the number of suppliers increases.

- The buyer's cost increases in the number of suppliers. Thus, the buyer prefers fewer suppliers!

- Single sourcing, in which the buyer is the Stackelberg leader, outperforms multisourcing.

- If the buyer owns the suppliers, the optimal service level would be higher than in the decentralized model.

Elahi [222] (2013) assumes a single *buyer* allocating demand to a group of MTS M/M/1 *suppliers*. The buyer's goal is to maximize the average *fill rate* (the probability of fulfilling demand from on-hand inventory). Price is exogenous and the buyer's decision is the allocation mechanism. Suppliers incur linear holding, production, and capacity costs. Their only decision variable is the base-stock level. They participate in the competition only when they can earn a nonnegative expected profit.

Let s_i denote fill rate and z_i base-stock level at supplier i. In *service competition*, supplier i is allocated a proportion $s_i / \sum s_j$ of the buyer's demand, while in *inventory competition* the proportion is $z_i / \sum z_j$. The author shows that:

- If the buyer controls the suppliers' strategy it allocates the entire demand to a single supplier. This is the buyer's first-best solution.

- In both types of competition there exists a unique equilibrium.

- Inventory competition always results in a higher average service level.

- When suppliers are homogeneous,

  - the equilibrium allocation is explicitly computed and
  - the buyer prefers fewer suppliers.

- If demand is allocated proportionally to a measure combining service and base-stock levels, which the author names *optimal competition*, then the competition results in the buyer's first-best average service level.

Gopalakrishnan, Doroudi, Ward, and Wierman [260] (2014) investigate dispatching rules and server workload incentives in a large multiserver system where strategic servers value idle time. Servers are homogeneous, with utility composed of the fraction of time idle minus a convex increasing capacity cost. Service capacities μ_i chosen by the servers are observable by the dispatcher, whose objective is to minimize the system's linear staffing and expected delay costs. Customers are not strategic and, in particular, are not delay sensitive.

An *r-routing policy* assigns a new arrival to server $i \in I$ with probability $(\mu_i)^r / \sum_{j \in I} (\mu_j)^r$, where I denotes the set of idle servers. Some special cases are noteworthy: When $r = 0$ we obtain a *random* strategy; with $r \to \infty$ we obtain the *fastest-server-first* (FSF) strategy; and with $r \to -\infty$ we obtain the *slowest-server-first* (SSF) strategy.

Another interesting class consists of *idle-time-based policies*: *Longest idle server first* (LISF), *shortest idle server first* (SISF), and the random policy are examples of this class. The authors prove the remarkable property that all idle-time-based policies result in the same unique symmetric equilibrium. Since the random policy belongs to both classes, idle-time-based policies do not give better results than the best r-routing policy. For the latter, assuming an M/M/2 system, the authors provide conditions for the existence of a symmetric equilibrium.

Zhan and Ward [693] (2015) consider a firm assigning customers to strategic servers in an M/M/N queue with customer abandonments. Servers choose service rates from a given finite interval. The probability of successful service is a decreasing function $p(\mu)$, where $\mu p(\mu)$ is strictly concave. The firm pays P_S for every completed service and penalizes the servers P_F for each failed service. When a customer arrives to more than one idle server, the customer is sent to the server that has been idle the longest (longest-idle-server-first rule).

The authors characterize the symmetric equilibrium service rate and provide an approximate equilibrium solution when N and λ are large. These results are used to solve the firm's cost-minimization problem, assuming fixed costs per customer abandonment and per failed service in addition to payments to the servers. The decision variables are N, P_S and P_F, and must satisfy **IR** constraints. The authors leave for future research the interesting extension of the model where the firm can also control the routing policy.

9.3.1   Quality inspection and rework

Lu, Van Mieghem, and Savaskan [458] (2009) consider a *principal* hiring two *servers* and setting a *quality inspection precision* $p \in [0,1]$, which is the probability of identifying a bad output. Each server selects a first-pass mean service time $t \geqslant \underline{t}$. The probability of producing good quality is an increasing concave function $F(t)$, with $F(\underline{t}) = 0$ and $F(\infty) = 1$. Output diagnosed as bad is sent to rework, which always generates a good output and takes less time than the minimum first-pass effort, $\underline{t}$. The probability of a good output at the end of the process is $Q = F(t) + [1 - F(t)]p$. Rework is routed according to one of the following schemes: (i) *Self-routing*; rework by the agent who generated the bad output. (ii) *Dedicated routing*; one server does all first-pass work and the other does the rework. (iii) *Cross-routing*; each server reworks the bad output from the other.

The principal earns a fixed value from a good output and incurs costs related to failed jobs. Each server incurs a fixed cost per unit time of work (first-pass and rework) and earns a wage rate w with an additional fee b when completing a good output. The server's decision variable is t and the principal's decision variables are w, b and p. Note that the amount of first-pass demand routed to the servers is assumed to be independent of their choice of t.

The model is formulated in terms of queueing network routing, but with no delay or holding costs. The variable t is interpreted as mean service time, but delay is actually not part of the model and queues are explicitly treated only in the cross-routing scheme. The authors note that waiting costs reduce the principal's incentives for quality effort because it leads to longer delays.

Let V^{FB} denote first-best profit under self-routing when the principal also controls t. Let V^S, V^D and V^C denote the profits under self routing, dedicated routing, or cross routing, respectively.

The authors show that:

- In the case of *limited demand*, where all demand can be served, $V^{FB} = V^D = V^C > V^S$.

- In the case of *unlimited demand*, the throughput of the system depends on the agents' decisions and the servers are continuously busy (except for the rework server in the dedicated scheme). First-best profits cannot be achieved by any of the three routing schemes. The first-pass service times and resulting quality performance are greatest under cross-routing, but this doesn't necessarily imply that maximum profit is also achieved under this scheme, and the ranking depends on the parameters.

9.4 Competition

Bernstein and de Véricourt [89] (2008) consider competition between two M/M/1 MTS *suppliers* with heterogeneous service rates in a market with n *buyers*. Unit price p_k and rate of demand λ_k for buyer k are exogenous. Each supplier quotes each buyer a contract specifying a unit backorder penalty per unit time of delay. Buyers interest is in product availability and therefore each buyer selects the supplier with the highest quoted penalty (or randomly selects a supplier if the backorder penalty quotes are equal).[10] The supplier also decides on a supply strategy. For instance, backordering current demand permits the supplier to reserve current stock for future demands associated with higher backorder penalties.

The state of the system consists of on-hand inventory and backorders. The suppliers adopt a multi-threshold policy which dictates when to produce and when to supply an order submitted by a given buyer. It is shown that selecting the supplier with the higher backorder penalty quote does not mean selecting the supplier with the shortest average lead time, rather *it is equivalent to choosing the one with the highest percentage of on-time deliveries*, i.e., the maximum fill rate..

The authors dedicate special attention to the case of two buyers. Because

---

[10]The buyer does not enjoy the backorder penalty directly but it serves as an indication (or signal) of supplier incentive for on-time delivery.

a slight change in the penalty offered by a supplier may fundamentally change the choice of the buyers and prevent the existence of an equilibrium, the authors define a game with a lower bound on the smallest possible change in the quote. The solution is obtained at the limit when this bound goes to zero. Suppose that $\mu_1 > \mu_2$ and $\lambda_1 \geqslant \lambda_2$. Then there exists μ^* such that:

- If $\mu_2 < \mu^*$, then in the unique equilibrium supplier 1 establishes contracts with both buyers.

- If $\mu_2 > \mu^*$, then in the unique equilibrium supplier i establishes a contract with buyer i, $i = 1, 2$.

Tang and Chen [617] (2014) consider a supply chain with heterogeneous service providers hiring homogeneous servers from a firm that sets a unit price p and incurs a fixed unit cost. Provider i decides on the number n_i of servers and the demand rate λ_i to serve, establishing n_i M/M/1 service facilities each serving demand λ_i/n_i, such that the delay in each facility is at most an exogenous standard d_i. The provider's profit consists of utility $v_i \log(1 + \lambda_i)$ minus linear holding and capacity costs.

The authors solve the equilibrium in this game assuming all providers are active. They also consider an extension of the model where several homogeneous firms compete, and characterize conditions for the existence of a unique equilibrium.

Hong, Xu, and Zhang [344] (2015) solve the following Stackelberg game in a two-stage network. Two *service providers* with dedicated demand λ_i compete for capacity allocated by a *facility provider* owning K units of capacity. The authors consider delays at both stages as M/M/1 delays, such that if capacity K_i is allocated to service provider i then the i-customers' expected delay is $t_i = (K_i - \mu_i)^{-1} + (\mu_i - \lambda_i)^{-1}$. Thus, an increase of μ_i has the opposite effect on t_i, increasing at the first stage and decreasing at the second.

The demand of i-customers is a function $\lambda_i = A_i(1 - \theta_i t_i)$. Note that λ_i also appears in the right-hand side of the equation as part of t_i. Given K_i, service provider i sets μ_i to maximize λ_i. The facility provider allocates capacities K_i, $K_1 + K_2 = K$, to maximize aggregate demand $\lambda_1 + \lambda_2$. The outcome is compared to that obtained in a pooling version where $K_i = K - \mu_{3-i}$ for $i = 1, 2$ and also to the centralized solution.

9.4.1   Outsourcing under competition

Cachon and Harker [118] (2002) suggest that outsourcing can mitigate competition resulting from the economies of scale inherent in their model (see §7.1). To make this point they consider two firms having the option of outsourcing to a single supplier. The firms and supplier possess the same technology, and the supplier establishes dedicated capacity for each outsourcing firm. Thus, the supplier cannot pool demand to gain efficiency. The game consists of two stages. In the first stage, each firm can sign a contract with the supplier given by a fee the firm pays for each customer processed and the average waiting time the supplier guarantees for the firm's clients. A contract must be profitable for both the firm (that has the option of insourcing) and the supplier. The second stage is price competition, where demand depends on the full prices of the two firms.

The authors do not analyze the negotiation process between the firms and supplier but focus on showing that although the supplier is not more efficient than the firm, there exists a set of outsourcing contracts that generates more profits to both firms and positive profit to the supplier. Thus, the firms may profit when they both outsource as compared to the case where they both insource. However, as in the prisoner's dilemma game, this does not mean that in equilibrium both firms outsource.

Allon and Federgruen [35] (2011) assume each of N competing retailers quotes price p_i and expected delivery time w_i. Serving a customer *in-house* costs c_i and unit capacity costs γ_i per unit time. Alternatively, the retailers have the option of outsourcing to a common outside supplier incurring costs c_0 and γ_0. The demand for retailer i is according to the separable demand model (see [32]). Two factors give advantage to outsourcing. First, pooled service at the outside supplier reflects *economies of scope*. That is, the capacity required to satisfy the demand and waiting time of each of the outsourcing retailers is smaller than the aggregate service rate required to perform these tasks by dedicated servers. Second, outside suppliers can be more efficient and operate at lower rates.

The authors consider a two-stage game. In the first stage, each retailer decides whether to outsource or process in-house, and in the second stage the aggregate profits earned by the service chain are shared according to the Nash bargaining solution. The disagreement point is exogenous for the supplier, and for each retailer it is the profit value when all serve in-house. Some of the main results are:

- Different from [118], firms do not necessarily benefit from outsourcing. This difference arises because the number of firms outsourcing rather than providing in-house service is endogenized, and the contract is determined using a Nash bargaining solution.

- The supply chain may be unstable, i.e., it may not be immune to defection, even when all firms benefit from collective outsourcing.

- The authors demonstrate how the benefits and stability of the supply chain depend on the number of firms and the demand function.

9.4.2   Competition between supply chains

Pekgün, Griffin, and Keskinocak [526] (2015) examine competition between two M/M/1 firms facing linear demand functions

$$\lambda_i = [a_i - b_i p_i - c_i L_i + \beta_{ij} p_j + \gamma_{ij} L_j]^+ \quad j \neq i,$$

where p is price and L is lead-time. The price-competition intensity is measured by the ratio of cross-effect to own-effect coefficients β_{ij}/b_j and, similarly, the intensity of lead-time competition is γ_{ij}/c_j. Firms choose whether to operate in a centralized or decentralized way. In the decentralized case, marketing first sets the price to maximize income while ignoring the unit production costs, and then production quotes the best lead time that can be satisfied.

The main conclusions are:

- When price competition is less intense than lead-time competition, decentralization can be an equilibrium strategy leading to higher profits for both firms.

- When firms have identical parameters except for capacity, decentralization can greatly reduce profits under high capacity or against a competitor with higher capacity.

- When firms are identical, if production also sets capacity at a constant unit cost, not only can decentralization lead to a significantly more profitable equilibrium than a centralized one but in addition competition may no longer imply lower prices or longer lead-times for a decentralized firm. This applies even when price competition is more intense than lead-time competition

- An example demonstrates that in equilibrium identical firms may choose the centralized option, though both could profit from decentralization, analogous to the prisoner's dilemma.

Narenji, Fathian, Teimoury, and Jalali [498] (2013) consider two competing supply chains, each consisting of an M/M/1 *manufacturer* and a *distributor*. Each firm first decides whether to operate in a centralized or decentralized way. In the decentralized case, the manufacturer quotes a delay guarantee and sets a unit price to the distributor, who in turn determines the unit price to the customers. The manufacturer incurs unit manufacturing and capacity costs and the distributor incurs unit distribution costs. In the centralized case, the decisions on delay and price are made by the chain's manager. The demand rate for each firm is determined by a linear demand function of prices delay at both firms. The authors compute the best response functions when each firm conducts a centralized or decentralized regime and an example is numerically solved under each scenario.[11]

---

[11]An interesting question, which the authors do not treat, concerns what scenario should be expected in the game of choosing the regime played between the chains, as in [526].

Fathian, Narenji, Teimoury, and Jalali [232] (2013) obtain closed-form solutions to a Stackelberg version of [498] where the manufacturers additionally incur holding and tardiness penalties.

9.5 Internet service provision

Demirkan, Cheng, and Bandyopadhyay [193] (2010) consider a supply chain consisting of an *infrastructure provider* (IP) supplying capacity to an M/M/1 *Internet service provider* (ISP) at unit price w. The ISP sells service to consumers at price p. The model involves two exogenous functions. The IP's cost of supplying capacity μ is $c\mu + e\mu^2$, and the marginal service value of a customer when demand is λ is $k/\sqrt{\lambda}$, where c, e and k are constants. In equilibrium, this marginal value is equal to the full price $p + \nu/(\mu - \lambda)$, where ν is the waiting cost rate. The decision variable of the IP is the capacity price w whereas the ISP determines the service fee p and capacity μ.

Four scenarios are considered and compared:

- **Overall coordination:** The IP and ISP cooperate to maximize joint profits.

- **The IP leads** by setting w, then the ISP sets p and μ to maximize profit.

- **The ISP leads** by quoting a price-capacity schedule (i.e., w as a function of μ) from which the IP selects its profit-maximizing combination, and the ISP then optimally sets its profit-maximizing p.

- **Aligned coordination:** For a given w, the IP chooses its profit-maximizing μ and the ISP chooses its profit-maximizing μ and p. The authors the existence of a unique w such that these μ values are equal, and this w is mutually agreeable.

The authors report on computational experiments with several interesting conclusions. In particular, aligned coordination generates the same total surplus for the IP and ISP as does overall coordination. In the other two scenarios, the leader receives a higher profit than if the other party were leading.

Cheng, Bandyopadhyay, and Guo [159] (2011) introduce a queueing model of net neutrality, while **Choi and Kim [165] (2010)** use the same framework but make some changes that result in less clear-cut conclusions.

A single monopolistic *Internet Service Provider* (ISP) and two *content providers* (CPs) located at the ends of the unit interval provide *complementary services.*[12] The ISP is modeled as an M/M/1 queue. Customers have a common service value and, as in the Hotelling model, are uniformly distributed over the unit interval and incur linear travel and waiting costs. CPs do not charge a fee for service but gain a fixed revenue (r_k for CPk, $k = 1, 2$, and w.l.o.g., $r_1 \geqslant r_2$) per customer from external sources (advertising). The ISP can charge customers an *access fee* F, subject to the constraint that the market is covered, i.e., every customer can obtain nonnegative net utility by selecting one of the CPs.

The central topic of this research deals with the effects of *net neutrality* regulation on investment incentives of a monopolistic ISP and the CPs:

- **Neutrality:** Suppose the regime must be FCFS. By symmetry, customers located at $x < 0.5$ prefer CP1, while those with $x > 0.5$ select CP2. The constraint on F is $v - C/(\mu - \lambda) - t/2 - F \geqslant 0$, where t is the unit distance transportation cost.

- **Discriminatory solution:** The ISP charges an optimal access fee and price for priority, subject to **IR** and **IC** constraints. Two outcomes are possible. If r_1 is significantly greater than r_2, only CP1 buys priority. Otherwise, both will purchase.[13]

- In the short run, allowing non-neutrality implies that social welfare and consumer surplus would either increase or remain unchanged; CPs are usually worse off, whereas the ISP would be better off.

- In the long run, the ISP may invest less in capacity in a non-neutral setting because expanding capacity reduces the CPs' willingness to pay for prioritized service.[14] The ISP invests at the **SO** level under net neutrality, but may overinvest or underinvest when discrimination is allowed.

- If capacity choices for the ISP are discrete (as opposed to continuous), the ISP will always underinvest in the long run when discrimination is allowed.[15]

Guo, Cheng, Bandyopadhyay, and Yang [271] (2010) consider the possibility of integration between the ISP and one of the CPs. As a result of such an integration, social welfare may increase or decrease. In some cases the ISP may prefer to prioritize the competing CP when maximizing its own revenues.

---

[12]The model resembles that of [635]. The ISP acts similar to the parking provider and delay is generated at the CPs queues. However here the CPs' profits come from an external source and not from customer payments.

[13]This is a case of *rent dissipation*. See [1] §4.2 for another strategic priority-queue model that leads to a similar result. This option is not allowed in [165].

[14]Cf. Myrdal's claim§6.6.4.

[15]Private communication with H. Cheng.

Krämer and Wiewiorra [416] (2012) study market-expansion effects caused by a discriminatory regime. Their model is without competition and every CP is visited by every customer.

Potential CPs are characterized by a parameter $\theta \sim \mathrm{U}[0,1]$. Revenue per customer received by an active θ-CP with expected waiting time w is $\lambda r(1 - \theta w)$, where λ is the demand rate produced by each customer for all CPs, and r is the unit revenue (from advertisements).

The CP's decisions are whether to be active and also whether to purchase priority from the ISP in the discriminatory case. Buying priority reduces the waiting time offered by the CP and increases revenue.

The active CPs are those with θ smaller than a threshold $\bar{\theta}$ and therefore total demand handled by the ISP is $\Lambda = \lambda \bar{\theta}$. In the discriminatory case there is an additional threshold $\tilde{\theta} \leqslant \bar{\theta}$ that distinguishes between CPs that buy priority and those that do not.

A monopolistic M/M/1 ISP processes the demand of the whole system. Revenue consists of fees collected from customers and also payments from CPs that buy priority in the discriminatory case. The ISP's decision variables are price, and, in the long run, also capacity, for which it incurs a convex cost $c(\mu)$.

Customer utility equals $U = b + v\bar{\theta} - Cw - a$, where $\bar{\theta}$ is the mass of active CPs, w is average waiting time, and a is the price charged by the ISP. Clearly, the profit-maximizing price satisfies $U = 0$. This provides the ISP with an incentive to invest in capacity and reduce w. The authors find that:

- In the short run, while discrimination has no effect on the number of active CPs (called here *content variety*), it increases ISP's profits and social welfare.

- In the long run, discrimination increases investment in capacity and content variety. This result holds when θ is uniformly distributed, and also for general distributions of θ unless there are "too many" congestion-sensitive CPs waiting to enter the market when priority becomes available.

Reggiani and Valletti [546] (2012) assume CPs pay the ISP a connection fee independent of traffic volumes. A continuum of non-atomic CPs are distributed along a line. The ISP is located at 0. A CP located at $x > 0$ pays transportation costs proportional to x. Additionally, one large CP supplying several *applications* pays the ISP a fee per application. The arrival rate to the ISP's M/M/1 queue consists of the mass of small CPs entering the market and the mass of applications by the large CP.

A novel assumption of this model is that differences in congestion and priority affect the profitability of advertising. With this assumption, the authors give conditions under which prioritization increases capacity investment as well as user welfare.

Bourreau, Kourandi, and Valletti [103] (2015) extend [416] by assuming two competing M/M/1 ISPs. There is a continuum of CPs distinguished by waiting costs $h \in [0, \infty)$. A CP of type h that connects to ISP i gains advertising revenue of $a\lambda x_i(1 - hw_i)$, where x_i is the number of end users subscribing to ISP i, λ is the rate of service requests per user, and w_i is the congestion at ISP i. The profit of an ISP charging price p_i, serving demand x_i, and building capacity μ_i is $p_i x_i - C(\mu_i)$ where C is increasing and convex. Users are uniformly distributed over [0,1] and the two ISPs are located at the ends of the interval. The utility of a user subscribing to an ISP at distance x from its location is $R + v\bar{h} + \frac{d}{w} - p - tx$, where $\bar{h}$ is the mass of CPs connected to this ISP, w is waiting time (so $1/w$ represents speed of service), and p is price. A user subscribes to a single ISP. CPs can connect to two, one, or no ISPs. ISPs set price and capacity, and when discrimination of CPs is allowed (no net neutrality) also set priority fees. The results include:

- Allowing discrimination increases investment in capacity, decreases congestion, and increases social welfare (the sum of firms' profits and users' utility).

- ISP's profits can be higher under net neutrality, but a prisoner's dilemma situation may occur where both prefer net neutrality but each has a unilateral incentive to switch to the discriminatory regime.

Guo, Cheng, and Bandyopadhyay [269] (2012) utilize the Hotelling demand model. The ISP selects from among four options: selling priority to one of two CPs, to neither of them, or to both (at the same price). In each case the ISP selects the optimal fixed price for customers and priority fee for CPs. The main results are: in some cases the ISP will subsidize the customer fixed fee to expand the market and increase income obtained from CPs. The option of priority selling to CPs increases the ISP's profits, but usually hurts the CPs and never decreases customer surplus or social welfare. However, it may drive the less effective CP out of the market and thus reduce competition.

Guo, Cheng, and Bandyopadhyay [270] (2013) expand [269] by adding the option of priority selling to *customers*. There is an M/M/1 ISP with fixed capacity and two CPs, Y and G, with revenue $r_G > r_Y$ per unit demand. There are two customer types: H and L, with different service values. H-customers constitute a fraction $\alpha < 0.5$ of the customer population. H- and L-customers generate demand at rates $\lambda_H > \lambda_L$, where $\alpha\lambda_H > (1-\alpha)\lambda_L$. The server charges customers a fixed fee per unit demand. Customer preferences of both types are uniformly distributed on the unit interval where Y and G are located at the ends. Customers incur "transportation costs" to their CP in addition to nominal and delay costs. Each selects the CP that maximizes his utility.

The authors solve eight variations of the model depending on CP discrimination (neither buy priority, both, G, or Y) and customer discrimination (both types pay the same fixed fee and obtain equal priority, or H-customers pay a higher fee and obtain priority). CP priority dominates customer priority. Depending on (r_Y, r_G), profit-maximizing is achieved in the four options where G buys priority. Social welfare also requires one of these options (in different (r_Y, r_G) regions), but is indifferent when both CPs or neither buy priority and whether H-customers obtain priority.

Krämer, Wiewiorra, and Weinhardt [417] (2013) survey literature concerning the net-neutrality debate.

9.6 Queueing games

We describe here games with two agents controlling different components of the queue.

Yao [682] (1995) investigates the convergence of an algorithm for computing equilibria in S-modular games that exhibit both ATC and FTC behaviors. The paper contains interesting queueing applications including the following two:[16]

- Consider an M/M/1 queue jointly managed by two profit-maximizing agents, **A1** and **A2**. **A1** sets the arrival rate $\lambda \leqslant u$ and **A2** sets the service capacity $\mu \leqslant u$ where the upper bound u is exogenous. When $\lambda < \mu$, **A1**'s profit is $p_1 \lambda - gL$ where $L = \lambda/(\mu - \lambda)$ is the expected number

---

[16]These are Examples 2.4 and 2.7 in [682]. The terminology used by the author in Example 2.4 differs from ours. We denote (λ, μ) what the author denotes (μ_1, μ_2).

of customers in the system, **A2**'s profit is $p_2\lambda - c_2\mu - gL$. Otherwise, profits are 0. Thus, the agents split profits and share waiting costs, but capacity cost is incurred only by **A2**. The best-response functions are monotone increasing and therefore this is an FTC case. $\lambda = \mu = 0$ is always an equilibrium, and if u is large enough there are also two other solutions: $(\lambda, \mu) = \left[\frac{gc_2}{(p_1-c_2)^2}, \frac{gp_1}{(p_1-c_2)^2}\right]$, and $(\lambda, \mu) = \left[u - \sqrt{\frac{gu}{p_1}}, u\right]$.

- Suppose two users send jobs at rates λ_1 and λ_2 to an M/M/s/s loss system. Let $B(\lambda)$ be the blocking probability, where $\lambda = \lambda_1 + \lambda_2$. User i maximizes a payoff function $f_i = r_i\lambda_i - c_i\lambda B(\lambda)$. Thus, the two users share the cost of lost traffic of both job types. In this case we have ATC behavior with a unique equilibrium.

Altman [41] (1996) considers two queueing games.

The first game has N parallel M/M/1/L_i queues, $i = 1, \ldots, N$, with homogeneous service rates but heterogeneous buffer sizes. The system is controlled by two cost-minimizing players. One player, *the controller*, dynamically distributes a given amount of *extra service rate* among the queues. The second player, *the router*, directs every new arrival to one of the queues. The author provides sufficient conditions on the state-dependent cost functions of the two players such that the following pair of strategies define an equilibrium:

- Route a new arrival to the shortest non-full queue.

- Fully allocate the extra rate to the shortest non-empty queue.

The second game assumes a discrete-time model of an observable queueing system operated by two players. Once service ends and the queue is empty, the server starts a vacation and a *service controller* **S** decides when to resume serving. During the active server phase an *admission controller* **A** decides whether to accept new arrivals. All arrivals are admitted during the vacation phase. Waiting costs incurred during active phases are paid by **A** whereas those incurred during vacations are paid by **S**. **A** obtains a reward for each unit time of the active phases (equivalently, for every served customer) and **S** pays a setup cost whenever service restarts.

The service controller **S** wishes to reduce the number of setups, but also has an incentive to terminate the vacation when the queue becomes long. However, **S** ignores the externalities that long queues have on **A** when service restarts. These considerations make the resulting game interesting. The author proves that an equilibrium exists with both agents applying (mixed) threshold strategies.

Guo and Hernández-Lerma [288] (2005) consider two-person zero-sum games defined on continuous-time Markov chains with discounted payoffs. They illustrate their findings using the following single-server queueing example.

The basic arrival and service rates with i customers in the system are $i\lambda$ and $i\mu$. Player 1, representing the demand side, selects an action $a \in A(i)$ and player 2, representing the server's side, selects an action $b \in B(i)$. Then, the service rate is changed by $u(a)$, the arrival rate changes by $v(b)$, and player 1 pays an amount $c_1(a) - c_2(b)$ to player 2 (all of these quantities can be negative). In addition, player 2 pays a usage fee of $i \cdot p$ to player 1. The authors derive conditions for the existence of a pair of optimal stationary strategies.

Kardeş, Ordóñez, and Hall [390] (2011) consider a general model of discounted stochastic games where players assign values to certain parameters but different values may be realized. The authors consider the equilibrium obtained when each player's strategy is a min-max best response to the other players' strategies. In other words, each player responds assuming the worst possible realized parameter values. The authors give conditions for the existence of such an equilibrium and demonstrate the applicability of their findings using the following zero-sum M/M/1 queueing game.

A *service provider* dynamically controls the service rate and a *router* dynamically controls the arrival rate. The payments transferred between router and provider are determined by exogenous functions of queue length and the realized arrival and service rates. The authors compute and examine the equilibrium through a numerical example.

Chapter 10

Vacations

This chapter is about queueing systems with strategic customers and server vacations. We start by defining the various types of vacations.

- **Planned vs. forced vacations**: Server vacations may be strategically planned, like for scheduled maintenance. *Strategic (planned) vacations* typically start when the server becomes idle. In other cases vacations are *forced* as a result of random events, for example breakdowns caused by mechanical failure, or when the server is reassigned higher priority customers, called *primary users* (PUs), whereas customers in the model are considered *secondary users* (SUs). These terms, primary and secondary users, are especially popular in the literature on *cognitive radio networks*.

- **Single vs. multiple vacations:** A *vacation* model assumes a vacation lasts a predetermined (stochastic) duration. In the case of *multiple vacations*, if there are no customers in the queue when the vacation ends a new vacation starts.

- **Active vs. independent breakdowns:** It is useful to distinguish between two types of forced vacations. *Active breakdowns* (or simply, breakdowns) can occur only when the server is busy. *Independent* or *accidental* breakdowns can occur at any time and correspond to the system oscillating between ON and OFF (or up and down) states. In particular, systems where vacations are forced by the appearance of primary customers having preemptive priority over regular (secondary) customers, experience independent breakdowns.

- **Setup periods:** Restarting service after a time of idleness may require a setup period. If the duration of the vacation is exponentially distributed, multiple vacations would be equivalent to a setup time that starts upon arrival of a customer to an empty system.

- **N-policy:** More generally, the system may follow an N-policy, meaning work is resumed when the queue length reaches a threshold, N.

- **Working vacations and breakdowns:** During *working vacations* or *working breakdowns* service is not completely shut down but rather is still provided at a lower rate.

- **Information:** The system state is often a two-dimensional vector (n, I) where $n = 0, 1, \ldots$ denotes the number of customers in the system, $I = 0$ if the server is on vacation and $I = 1$ otherwise. We use the terminology of [115]: The system is (fully) *observable* if the state is observable, *almost observable* when n but not I is observable, *almost unobservable* when only I is observable, and (fully) *unobservable* when neither n and I are known to customers.

- **Base stock:** Vacation models generalize common idleness of the server when the queue empties in an MTO system or reaches the base-stock level in an MTS model. See [356] for an MTS generalization of the N-policy that also requires a setup period before production can be resumed.

- **Reneging:** Observable and almost-unobservable models assume customers know the state of the server. Assuming this information is also available after joining, a rational customer who joined when the server was active may wish to renege should a breakdown occur. The models with breakdowns we discuss assume customers cannot renege when breakdowns occur. Allowing for reneging would make the model more realistic.[1]

- **Capacity control:** Working vacations can be regarded as a special case of *dynamic capacity control*. For example, the models of [196] and [440], where service slows down when queue length is below a threshold bear similarities to working vacations.

- **Probabilistic joining:** Several papers on queueing vacation models consider customers with exogenous probabilistic joining functions. As mentioned in §1.2, such models that are otherwise non-strategic are not

---

[1] See [19] for an observable queue where low priority can renege upon arrival of a high-priority customer.

included in this survey. See [447] for a recent representative of this literature, including a detailed bibliography.

- See [24, 401] for observable models of vacation queues, and [602, 608] for multiple-vacation models with uncertain service rates.

10.1 Strategic vacations

10.1.1 N-policy

See [41] for an N-policy queue with two controllers, one controls admissions when the server is active and the other sets N. See [552] for a system applying N-policy under bounded rationality.

Dellaert [192] (1991) describes a Markovian queue employing N-policy with the server's decision variables being N and a PDT d. Customer behavior is given by the following rule: As long as d is below a threshold $d_{\max}$, the customer joining probability is $1 - d/d_{\max}$ and therefore the arrival rate is $\lambda(d) = \lambda(1 - d/d_{\max})$. The server earns a fixed amount per served customer and incurs a fixed setup cost. If service ends before the PDT, the server incurs a holding cost, and if service concludes after the PDT the server incurs a tardiness cost.

The author provides an example showing considerable difference in maximum profits when the PDT must be constant or allowed to be state-dependent.

Guo and Hassin [272] (2011) consider an M/M/1 queue operating according to N-policy. Customers have heterogeneous service valuations and linear waiting costs. When the server is idle, in contrast to a regular queue, a customer's decision to join influences both those who will arrive later and previous arrivals. Hence, a new customer should consider future arrivals and also the current state of the queue. In fact, joining the queue is also associated with positive externalities by helping to postpone and shorten vacations.

- **The unobservable case:** A customer may be more inclined to enter the system when seeing more customers do the same, leading to FTC behavior and multiple equilibrium arrival rates. Obviously, "all balk" is always an equilibrium strategy, and there exist at most two positive equilibrium arrival rates. As opposed to a regular queue, the **SO** arrival rate may be larger than some of the equilibrium arrival rates.

- **The observable case:** If conditions are satisfied for the server to be active a positive fraction of the time, there exists a unique equilibrium threshold joining strategy. A uniform fee may not be sufficient to co-ordinate the system. Furthermore, in the **SO** solution, some customers should join the queue even though incurring negative utility.

- N **is a decision variable:** Suppose the system incurs a fixed operating-cost rate when the server is busy, but there are no start-up or shut-down costs. In the unobservable case with a arrival rate the **SO** decision is to have the server active at all times ($N = 1$), while with heavy traffic, N should be as large as possible. In the case of an observable queue, $N = 1$ is always **SO**.

- **Tian, Yue, and Yue [622] (2015)** generalize the results of [272] for the unobservable case to general service distributions. They also solve the almost-unobservable model obtaining the **SO** and equilibrium joining probabilities when the server is busy. When the server is on vacation, the **SO** strategy dictates always joining, and there are at most two positive equilibrium joining threshold strategies.

- **Sun, Li, and E [606] (2015)** consider a variation where an idle server reviews the queue at intervals of $\exp(\theta)$ length and reactivates when observing at least N customers. The authors demonstrate that decreasing θ (less frequent reviews) has similar effects on equilibrium and social welfare as does increasing N.

- **Chen, Zhou, and Zhou [152] (2015)** generalize both [115] and [272] by assuming a setup period starts when the system has N customers and the server is reactivated only when it ends. The qualitative results of this model resemble those of [272].

Guo and Hassin [273] (2012) introduce customer delay-cost heterogeneity into the model of [272]. When the queue is unobservable and there are two customer types the number of equilibrium solutions can be at most five, but when the delay sensitivity distribution is continuous the number of solutions may be unbounded and even uncountable. The authors give explicit solutions when the distribution is uniform on $[0,1]$. Analogous results are derived when the queue is observable. In contrast to results in the same models but without vacations, the equilibrium joining rates may be smaller than the **SO** rates.

Guo and Zhang [283] (2013) study strategic customer behavior in a Markovian model where customers choose between a *public facility* with c servers and free service, or obtaining the service elsewhere at a *private facility* for a full price τ. For some constants $n < c, N$, when the number of customers in the public system falls to n, it inactivates $c - n$ servers. The servers are reactivated when the number of customers reaches N.

- **The observable case:** The authors solve an example where ATC behavior with a unique equilibrium arrival rate exists in light traffic, FTC behavior with three equilibria in moderate traffic, and again ATC behavior in heavy traffic.

- **The almost-unobservable case:** Equilibrium is characterized by two arrival rates, one when all servers are active, and the other when some are inactive. Again, FTC behavior and multiplicity of equilibria exist. If $N \leqslant c$ it is possible that the number of equilibria is not bounded or even countable. If $N > c$ there are at most three solutions.

- **Comparison:** Information can significantly degrade system performance because of customers ignoring both positive and negative externalities associated with their actions.

- **System cost:** The authors define system cost as the sum of customer waiting costs, activation costs, stuffing costs, and full price costs at the private server, and investigate its sensitivity to changes in τ and N. An interesting finding is that a slight drop in τ can drastically increase the use of the private facility and significantly increase system cost.

Guo and Li [278] (2013) complement [272] by considering two partial-information variations of the N-policy model:

- **Almost unobservable case:** When the server is busy, the situation is ATC and a unique equilibrium arrival rate exists. When the server is idle, the situation is FTC and multiple equilibria are possible. As in the unobservable case [272], "all customers balk when observing an idle server" is always an equilibrium and there can be 0,1, or 2 positive equilibrium arrival rates. When the server is busy and N is small, the equilibrium arrival rate is larger than the **SO** rate and a tax can coordinate the system. In all other cases, the arrival rate is smaller than the **SO** rate and a subsidy is needed to coordinate the system.

- **Almost observable case:** "All customers balk" is always an equilibrium, and the authors give necessary and sufficient conditions for the existence of other threshold equilibria.

- **Value of information:** A comparison of maximal social welfare under observable and almost-observable cases finds that the value of information decreases with $R\mu/C$ (customers are less patient).

Ioannidis, Jouini, Economopoulos, and Kouikoglou [356] (2013) consider two Markovian models, but we only describe the one with more strategic behavior, which they identify as Model B.[2]

---

[2]Model A is similar but has arbitrary probabilistic joining, reneging and is without setup. A similar model is also considered by **Economopoulos, Kouikoglou, and Grigoroudis [214] (2011)**.

Model B considers an observable M/M/1 production system where customers have random due times and a customer places an order iff the probability of missing the due time, given the current queue length, is less than some common value. A customer who places an order cannot renege even when the order is delayed beyond the due time.

The system operates according to a joint *base-stock / base-backlog* (BS-BB) policy defined by three thresholds (c, s, σ):

- When the backorder level reaches c new orders are rejected.

- When the inventory level reaches s production is stopped.

- When the inventory level drops to $\sigma < s$ the system is reactivated and commences a costly setup process of an exponentially distributed duration.

Thus, the state of the system is defined by the state of the machine (working, idle, or setup) and the inventory/backlog size. A final feature of the model is that items kept in stock have exponential lifetimes.

The authors derive expressions for the system steady-state performance measures and use them to compute cost-minimizing threshold levels.

10.1.2   Multiple vacations

Burnetas and Economou [115] (2007) characterize customer equilibrium join-or-balk strategies in an M/M/1 system with multiple vacations under various levels of information. Denote by $N(t)$ the number of customers in the system at time t.

- **The observable case:** There exists a unique (pure) equilibrium switching curve strategy (one threshold for each state of the server) which is also dominant.

- **The almost-observable case:** This is an FTC case because for any given $N(t)$ the probability the server is in an active state is higher when the threshold used by others is higher. Therefore, there exists an interval $\{n_L, \ldots, n_U\}$ of (pure) equilibrium threshold strategies and a mixed equilibrium between each pair of consecutive integers.

- **The unobservable and almost unobservable cases:** In contrast to the E&H benchmark model, it is not a priori clear here if the unobservable model is ATC type. A small joining probability may lead to a shorter queue when the server is active, but may also increase the probability the server is on vacation. However, the authors show that in both cases there exists a unique mixed equilibrium strategy.

- **Numerical analysis** shows:

– If $N(t)$ is observable, customers are more likely to join the system when the arrival rate is higher.

– If $N(t)$ is unobservable, a higher arrival rate implies the system is more loaded and customers are less likely to join.

- **Huang, Wang, and Fu [350] (2012)** prove analogous results assuming a two-phase Erlang service distribution.

- **Ma, Liu, and Li [461] (2013)** solve a discrete-time version of this model. For the almost-unobservable case these authors only compute the pure equilibria. **Liu, Ma, and Zhang [455] (2015)** complete the analysis by computing the mixed solutions.

- **Zhang, Wang, and Liu [695] (2013)** generalize [115] by considering working vacations. For each of the four information scenarios, the authors either derive an explicit solution for the balking threshold or provide an algorithm for its computation.

- **Sun and Li [605] (2014)** consider the same working-vacations model as in [695] for the observable, unobservable, and almost-unobservable cases. The authors present examples where equilibrium solutions lead to overcongestion relative to the **SO** solutions.[3]

- **Yang, Hou, Wu, and Liu [677] (2014)** consider the observable, unobservable, and almost unobservable versions of the Geo/Geo/1 queue with multiple working vacations. For each case the authors present an example with **SO** and equilibrium strategies.

- See [153] for an extension of [115] combining breakdowns and multiple vacations.

Sun, Guo, and Tian [601] (2010) consider a Markovian system where if the queue is empty when service ends the system enters a *closedown period*. If a customer arrives during this period service starts immediately. Otherwise, a *setup period* begins upon arrival of a new customer prior to starting service.

Customer behavior in the *observable case* is straightforward and is defined by two thresholds, one for active periods and one for setup periods, such that an arrival joins if queue length does not exceed the relevant threshold.

In the *almost-observable case* there is single threshold. The authors derive the probabilities for the system being in a setup period given queue length and use them to derive an interval of pure threshold equilibrium strategies. The model has an FTC spirit as the threshold increases in the arrival rate – the higher the arrival rate the more likely the server is active for any given queue length. However, there is no proof that the best response to the threshold adopted by the others is monotone increasing. The authors numerically demonstrate several interesting properties:

---

[3]As in [272], the model has a mixture of positive and negative externalities which could lead to the opposite outcome.

- The equilibrium threshold lies between the two thresholds of the observable case.

- Social welfare is a unimodal function of the arrival rate with a single maximum.

- Information about the state of the server need not increase social welfare in equilibrium.

The authors also solve a variation of the observable model where the close-down period is not interrupted with a new arrival and must be carried on to conclusion once started.

Economou, Gómez-Corral, and Kanta [215] (2011) consider a single-server queue with multiple vacations, Poisson arrivals, and general service and vacation time distributions.[4] Customers receive a constant reward when completing service and incur linear waiting costs.

- **The unobservable case:** The equilibrium strategy is the probability q_e of joining. As in [115], it is not a priori clear this is an ATC model. The authors prove that the expected benefit from joining decreases when the joining probability of the others increases, and hence the case is indeed ATC. They derive an explicit formula for this probability, depending on the expectations $E[R_B]$ and $E[R_V]$ of the stationary versions of the residual service and vacation time, respectively. They also derive the **SO** joining probability q_s and show that $q_s \leqslant q_e$. This inequality is not as obvious as in [221], since an arriving customer generates both positive externalities (activating the server) and negative externalities (increasing the queue length), with the negative externalities stronger.

- **The almost-unobservable case:** The equilibrium strategy is a pair of joining probabilities; $q_e(0)$ when the server is on vacation and $q_e(1)$ when the server is active.

  - The authors prove the uniqueness of the equilibrium strategy recursively.[5] The probability $q_e(0)$ can be determined first because the joining decision during a vacation period is independent of the strategy customers adopt when the server is active, which clearly is an ATC case. In contrast, when the server is active the decision to join depends on the probability of customers joining an idle server, because the queue formed during the vacation remains when the vacation ends. However, for a given $q_e(0)$ we obtain another ATC situation with respect to joining an active server. Hence, there is a unique solution, for which the authors derive an explicit solution.

---

[4]If vacation time is exponential it is equivalent to assuming it starts with the first arrival. Hence, this model generalizes [115].

[5]Similar to [324].

The authors also derive the **SO** strategy, which turns out to be more complicated.

- In general, the intuitive expectation that $q_e(0) \leqslant q_e(1)$ fails to hold. The reason is that with long vacations (or short service time) a customer may prefer joining a busy server to joining an idle one. An interesting property, shown numerically, is that the value of q_e in the unobservable case is always between $q_e(0)$ and $q_e(1)$.[6]

- $q_e(1)$ exhibits interesting behavior when $E[R_V]$ increases: first decreasing to zero, staying there for a while, and then increasing. This is explained as follows: increasing $E[R_V]$ yields a more congested system only if $q_e(0)$ is large enough. Then, when $E[R_V]$ further increases, $q_e(0)$ becomes smaller and almost all customers arriving to a server on vacation balk, service will start with a short queue, and customers who find an active server are more willing to join.

Sun, Wang, and Tian [612] (2012) compute equilibrium, **SO**, and profit-maximizing arrival rates in an unobservable Markovian model with *closedown and setup* periods (similar to [601]). The authors consider three variations:[7]

1. *Interruptible setup/closedown policy*: If a customer arrives during closedown, service immediately starts. Otherwise, the setup phase starts at the instant of the first arrival.

2. *Skippable setup/closedown policy*: An arrival during a closedown period cannot be served before the end of the period, but in this case setup is not necessary.

3. *Insusceptible setup/closedown policy*: Once closedown starts, the server's vacation cannot be interrupted until both closedown and setup are completed.

Zhang, Wang, and Liu [696] (2013) extend [115] and present recursive algorithms for computing the equilibrium joining strategies in an M/G/1 system having generally distributed setup times. The authors consider both the observable and almost-observable cases. In general, the equilibrium strategy is not of the threshold type. The joining strategy consists of a joining probability for each queue length in the almost-observable case, and two such vectors in the observable case, one for each possible state of the server. For both cases, the authors provide algorithms that identify the equilibrium joining strategies.

---

[6]This is similar to the result of [601] for the observable models.

[7]It is interesting to note that in all three cases the authors prove that the equilibrium arrival rate in their version is unique, so ATC behavior prevails even though the model exhibits both ATC and FTC behavior (see [115]).

10.1.3 Single vacation

Tian and Yue [620, 621] (2011, 2012) solve the single-vacation variation of [115] for the (fully and almost) observable and unobservable cases, respectively. The qualitative results derived are similar to [115]. Note that in this model the server can be in one of three possible states: busy, on vacation, or idle. However, the decision a customer makes while observing an idle server is straightforward and the solution consists of only two thresholds.

Liu, Ma, and Li [453] (2012) solve the single-vacation variation of [115] for the observable and almost-observable cases with similar qualitative results. The authors also consider a discrete-time version of the model. For the almost-observable case these authors consider only pure strategies, while **Liu, Ma, and Zhang [455] (2015)** complete the analysis for the discrete-time version by computing the mixed solutions.

Yue, Tian, Yue, and Qin [690] (2013) solve the observable single-vacation model with closedown and setup periods. The solution consists of three thresholds corresponding to the possible states of the server. The authors also consider the unobservable version, but the possibility of multiple equilibria in this case remains unsettled.

Wang, Wang, and Zhang [641] (2014) derive the equilibrium joining probability (threshold strategy) in the unobservable (observable) model assuming a Geo/Geo/1 model with a single *working vacation*.

Sun, Li, and Li [607] (2014) compute equilibrium solutions in an M/M/1 system in which a single two-stage *working vacation* starts when a server becomes idle. Each stage is exponential with the same parameter. Service rates in the two stages can be different, but both are lower than the regular service rate. The authors numerically find a uniqueness of the equilibrium in the *observable, unobservable* and *almost-unobservable* cases. In the almost-unobservable case the equilibrium joining probabilities during vacations are not necessarily smaller than in the regular busy state.

10.2 Forced vacations, breakdowns and catastrophes

10.2.1 Independent breakdowns

See [644] for a system with ON-OFF states and retrials, and [199] for routing decisions when firms oscillate between ON-OFF states.

Cheng [157] (1997) considers independent breakdowns in an unobservable M/M/1 system with heterogeneous service valuations given by the aggregate value function $V(\lambda)$ (see §1.5). The durations of ON and OFF periods are exponentially distributed with rates η and γ, respectively, such that $\eta, \gamma \ll \lambda, \mu$. This assumption enables a simple and accurate approximation formula for expected sojourn time. Given a price p, the equilibrium λ satisfies $V'(\lambda) = p + vT(\lambda, \mu)$, where T is expected waiting time and v is unit-time cost. The author derives formulas for the **SO** price both in the short-run problem where μ is fixed and in the long-run problem where μ can be changed at a linear cost. An extension of the long-run model allows for *backup capacity* at a higher than regular unit cost. Backup capacity can be used while the main server is down, resulting in a system with *working breakdowns*. The main results are:

- The firm should equate backup and main capacities and their optimal size is determined by the sum of their marginal capacity costs.

- The firm should charge users the total marginal capacity costs of main and backup capacities.

Cheng [158] (1999) compares two systems with the same breakdown and repair parameters. The *twin system* maintains an M/M/2 queue with identical independent servers, each operating at rate μ. In the event of failure in one server, the other server continues serving. The *consolidated* (pooled) system has a single server with capacity 2μ. Capacity is a decision variable having a constant marginal unit cost. The author numerically computes the **SO** service price in the two systems assuming demand is *isoelastic*, i.e., $V(\lambda) = A\lambda^{1-\alpha}/(1 - \alpha)$ with $0 < \alpha < 1$, or $V(\lambda) = A \ln \lambda$ in the case of unit elasticity ($\alpha = 1$). The conclusion is that:

- In general, the twin system induces higher welfare, larger arrival rates, larger capacity, and lower pricing than in the pooled system.

- The difference between the systems becomes more meaningful as delay costs increase.

Capar and Jondral [128] (2004) suggest an observable M/M/1 queue with independent breakdowns. The queue is managed by a profit-maximizing firm that dynamically sets entry prices; the demand rate is a negative exponential function of the product of price and delay.

Economou and Kanta [219] (2008) solve a Markovian extension of Naor's model where the queue alternates between ON and OFF periods. They consider two information settings:

- **The observable case:** As in Naor's model, customer welfare, given the state of the system, is independent of decisions made by later arrivals and there exists a dominant pure threshold strategy for which the authors derive an explicit solution.[8]

- **The almost-observable case:** Queue length also serves as a signal about the state of the server. The authors prove this is an FTC situation. Consequently, there can be multiple equilibrium solutions and the pure equilibrium thresholds constitute an interval of integers with a mixed equilibrium between every two consecutive pure solutions.[9]

- **Li and Han [436] (2011)** consider a discrete-time almost-observable version of [219]. They derive the solution for the equilibrium joining threshold and show how to compute the **SO** threshold. As expected, given the negative externalities associated with joining, also here self-optimization leads to excessive joining.

- **Do, Tran, Nguyen, Hong, and Lee [201] (2012)** consider the unobservable version of [219], and **Li, Wang, and Zhang [441] (2014)** derive a closed-form solution to the equilibrium joining probability.

- **Do, Hong, and Hong [198] (2012)**, and **Do, Tran, Hong, and Lee [200] (2012)** extend previous analyses of the unobservable and observable cases, respectively, by allowing for working breakdowns.

- **Yang, Wang, and Zhang [681] (2014)** consider a discrete-time version of [219]. The joining threshold in the observable case is as in the continuous-time version solved in [219] but with a modified service parameter. The authors also solve the equilibrium joining probability in the unobservable case and conduct sensitivity analysis.

Shiang and van der Schaar [572] (2008) examine a model where N users with pre-assigned priorities route demand to heterogeneous M/G/1 servers with independent breakdowns. User i's request is lost if delayed more than d_i time units, and the user's goal is to minimize his loss probability. The authors suggest a *dynamic strategy learning* algorithm for routing demand, while allowing probabilistic jockeying from a longer queue to a shorter one.

Wang, Zhang, and Tong [649] (2010) consider N servers each associated with a primary user (PU). PUs experience independent transitions of exponentially distributed duration between ON and OFF states. A PU generates deterministic demand at a given rate when the state is ON and no demand

---

[8]This solution is reproduced in [645].

[9]The cause for FTC behavior is different from that observed in [272] for the N-policy model. In [272], when the server is on vacation arriving customers have a higher tendency to join if they expect a higher future arrival rate that would bring the server quickly back to operating mode. In contrast, here the vacation's duration is independent of the arrival process. The cause for FTC behavior is that the probability the server is down at a given queue length can be higher if it is at maximum possible length. Thus, a higher threshold encourages joining at higher queue lengths.

otherwise. There are M secondary users (SUs), each maintaining a queue with Poisson arrivals of service requests. At each period, each SU whose queue is not empty randomly chooses a server; if the server is idle the SU sends a unit demand to that server, i.e., it *contends*, with a state-independent probability p. The unit is served if it is the only request sent to the server in that period. If more than a single SU sends a request to the server there is a *collision* and no service is granted. SUs cannot observe queue lengths, *including their own*. The research question is to compute a *contention probability p* that minimizes the mean queue length.[10]

Wang and Zhang [646] (2011) generalize [219] by assuming that repair time consists of two exponential phases. They investigate two cases:

- **The observable case:** The state of the server is w (working), d (delay – first stage of repair), or r (repair – second stage). The authors provide explicit formulas for the respective equilibrium thresholds, $(n(w) > n(r) > n(d))$.

- **The almost-observable case:** The authors solve the steady-state probabilities in terms of the roots of a polynomial equation, and characterize an interval of integer equilibrium thresholds. They observe that these thresholds are between $n(d)$ and $n(w)$.

Jagannathan, Menache, Modiano, and Zussman [357] (2012) consider an almost-unobservable version of the model of [219].[11] Customers choose between joining the queue or obtaining service by a different service provider at a fixed full price. The strategy is a pair (p, q) where p denotes the probability of joining when the server is ON and q denotes the probability of joining when the server is OFF. The authors obtain a closed-form solution for the unique equilibrium strategy.

An interesting addition to the model emerges when it is assumed the alternative service provider sets the service fee to maximize profits. The authors provide an explicit solution for this variation.

The same model is also independently solved, but with different techniques, by **Li, Wang, and Zhang [441] (2014)**.

Li, Wang, and Zhang [437] (2013) consider a Markovian single-server model with independent working breakdowns. In the observable case customers follow a threshold strategy for each of the two possible server states. The authors provide a closed-form solution but do not discuss uniqueness. In the unobservable case, the strategy consists of a joining probability which the authors prove to be unique, as expected, because the situation is ATC.

---

[10]An interesting variation not considered in this paper is the equilibrium outcome derived when SUs act strategically knowing their own queue lengths.

[11]The same model is considered in [692], however their equation (21) giving the equilibrium joining probability deviates from the solution in equation (17) in [357].

Tran, Tran, Le, Han, and Hong [630] (2014) examine multiple $M/G/1$ servers with independent breakdowns. Customers of class k have waiting-cost rates θ_k and uniformly distributed valuations α per unit quality of service. Servers are also heterogeneous, having different service quality levels and different service distributions. Thus, the utility of a (θ_k, α)-customer who waits W units of time to obtain service from server l charging price p_l and having quality r is $\alpha r - \theta_k W - p_l$. Decision variables are prices p_l and the load balancing vector s that specifies the probability s_l that a new arrival is routed to server l. Customers are informed about these values and decide whether to arrive based on their type. The authors propose an algorithm for computing the profit-maximizing solution.

Zhao, Jin, and Yue [708] (2015) consider a Markovian discrete-time unobservable queueing model with ON-OFF transitions, dynamic service-rate control, and a finite buffer of size K.[12] Service rate is proportional to the number of customers in the system. In each period there can be at most one event of any given type (e.g., a state transition and an arrival of a customer). Customers attempting to join the system incur a cost T, are accepted if the buffer is not full, and gain a fixed amount R upon service completion. Therefore, the expected utility for a customer attempting to join is $\epsilon R - T$ where ϵ is the probability that an attempting customer is admitted and successfully completes service. As in the E&H model, the behavior is ATC and there is a unique equilibrium attempting probability. The authors numerically solve examples and compute the admission fee that coordinates the system.

Economou and Manou [217] (2016) consider a semi-deterministic model with independent working breakdowns. While transitions between fast and slow service modes are governed by a Markovian continuous-time process, the input-output process is modeled as a continuous fluid queue. The authors compute the unique subgame-perfect equilibrium in the observable game, consisting of a threshold for each of the two server states. A similar result is derived for the almost-observable case, where the threshold's dependence on the arrival rate is restricted to whether it is smaller than the slower service rate, higher than the fast service rate, or between the two. The authors also study the problem of social optimization and quantify the discrepancy between the equilibrium and socially optimal strategies.

10.2.2   Active breakdowns

See [683] for a system with breakdowns that involve setup costs and instantaneous repairs (i.e., there are no vacations).

---

[12]Transition to the OFF state occurs in this model due to an arrival of a prioritized *primary user*, so when an ON-to-OFF transition occurs while the buffer is full the customer queueing at the tail of the buffer is expelled.

Zhang, Wang, and Liu [694] (2012) extend earlier results on M/G/1/1 retrial queues ([1] §6.4) by incorporating breakdowns and *delayed vacations*. Each time the server completes a service or vacation, a random amount of time is reserved before another vacation can be taken if there is no new arrival. Random breakdowns occur during server's activity. Customers finding the server busy or down retry after a random time. These customers are not informed of the system state. The duration of vacations and repair times follow general distributions, while reserved times, retrial times and time to a breakdown are exponentially distributed. Customers are homogeneous and incur waiting and retrial costs.

The authors obtain closed-form solutions for the equilibrium and **SO** retrial rates and show the former to be greater than the latter. The system can be coordinated by an appropriate retrial toll.[13]

Zhang, Wang, and Zhang [703] (2014) consider a Markovian system with active breakdowns under the following assumptions. (i) When a breakdown occurs the customer whose service is interrupted stays at the service area until service is restored. (ii) Customers arriving to an idle server always join. Customers arriving to a busy server either balk or join an FCFS orbit queue. (iii) The customer at the head of the orbit queue, and only this customer, keeps retrying at exponentially distributed intervals until finding the server idle. (iv) *Customers arriving when the server is broken always balk.* The only decision is whether a new customer reaching a busy server balks or joins the orbit queue. The main results are:

- As in Naor's model, when the orbit queue is observable the **SO** threshold is at most that of the **IO** threshold and at least as large as the profit-maximizing threshold.

- The same relations hold for joining probabilities when the orbit queue is unobservable.

- **Wang, Wang, and Zhang [642] (2015)** obtain similar results assuming the served customer leaves the system when a breakdown occurs.

Wang and Zhang [643] (2015) consider a Markovian variation of [694], where customers arriving when the server is not available decide whether to pay an entry fee and retry at a later time. All customers have the same service value and linear waiting costs.

The authors derive a closed-form solution for the joining probability in the unique symmetric equilibrium for a given fee as well as the **SO** joining probability. They also present a procedure for computing the profit-maximizing fee and conduct sensitivity analyses. For example, they find that an increase

---

[13]Excessive equilibrium retrial rate was already observed by Hassin and Haviv (1996) and explained in [1] §6.4.3. These authors also suggested partial compensation for waiting as an alternative to retrial tolls.

in μ (or a decrease in λ) has opposite effects on the profit-maximizing and **SO** prices. The profit-maximizing price increases while the the the **SO** price is non-increasing. This property is similar to that observed in [141].

Chen and Zhou [153] (2015) consider a Markovian model combining breakdowns and multiple vacations. Breakdowns occur when the server is busy and vacations start when the system becomes empty. *Customers cannot join the system during repair times* although they can join when the server is on vacation. The authors provide solutions for the equilibrium thresholds and joining probabilities in the observable, almost observable, and unobservable cases.

10.2.3   Duopoly competition under independent breakdowns

Tran, Hong, Han, and Lee [628] (2013) consider price competition between two facilities. The first has a single server that becomes unavailable during intervals of random lengths while providing service to higher-priority "primary customers." The second facility has an infinite number of servers and offers each customer an exclusive dedicated server. The service distribution of all servers is the same (general) distribution. Customers have homogeneous service valuations and heterogeneous uniformly distributed delay-cost rates. A customer can join one of the facilities or balk. In general, customers with low delay sensitivity join the shared facility, those with intermediate delay sensitivity choose a dedicated server, and highly sensitive customers balk.

The authors prove that for fixed prices, there is a unique customer equilibrium. Moreover, there exists a unique price equilibrium for which the authors provide an explicit formula. They also derive a function $u(c)$ such that coexistence of the two firms in the market is possible only if the price c_2 set by the infinite-server firm and the price c_1 set by the single-server firm satisfy $c_1 < c_2 < u(c_1)$.

Do, Tran, Han, Le, Lee, and Hong [199] (2014) consider an unobservable system with independent breakdowns, which they model as an $M/G/1$ facility. They compute the first two moments of the service distribution and derive the customer joining probability in equilibrium. The authors then solve a duopoly in this market, both when the firms compete and when they cooperate. Equilibrium arrival rates are determined so customers are indifferent between joining either of the firms or balking. The authors also consider a variation where one of the firms offers a dedicated server to each customer (similar to [628]). In both cases, the Nash bargaining solution is derived (with an exogenous agreement point) and compared to the best non-cooperative price equilibrium.

10.2.4 Catastrophes

Boudali and Economou [100] (2012) consider an M/M/1 queue subject to *catastrophes* which occur according to a Poisson process with rate ξ.[14] When a catastrophe occurs, all customers abandon the system without being served. Recovery time is exponential and arrivals are not accepted during recovery. Customers obtain reward R_s upon service completion or obtain compensation R_f if forced to abandon the system due to a catastrophe, and incur waiting cost C per unit time.

The observable version generalizes Naor's model. The main qualitative difference is that if $R_f \geqslant \frac{C}{\xi}$ then the unique dominant strategy is always to enter. When the queue is very long the motivation to enter is based on the expectation of getting the compensation R_f. Similarly, the unobservable version generalizes E&H with a new possibility, that if R_f is sufficiently large all customers join regardless of potential demand. As in the above-mentioned special cases, the **SO** threshold in the observable case and the joining probability in the unobservable case are smaller than their equilibrium counterparts due to negative externalities.

The authors question whether informing customers as to the state of the queue is socially desired. They find that in most such models, social welfare is greater when customers are informed as opposed to when they are uninformed, even when behaving selfishly when they are informed and according to the **SO** way when they are uninformed.

Boudali and Economou [101] (2013) solve a variation of [100] allowing for arrivals when the server is unavailable. The main results are:

- **The observable case:** The joining strategy for customers finding a functioning server is always a threshold strategy. However, depending on the relative value of service reward and mean repair waiting cost, the equilibrium strategy of those finding the server down may be either a threshold or a reverse-threshold strategy. The latter means joining if the number of customers in the system *exceeds* a certain threshold.

- **The unobservable case:** The main difference with respect to [100] is that when compensation to customers experiencing a catastrophe is high, there is FTC behavior which results in three equilibrium solutions (pure balking, pure joining, or a mixed strategy equilibrium). The authors indicate some counterintuitive situations. For example, the equilibrium joining probability is not necessarily increasing with the service reward and not necessarily decreasing with the waiting cost.

---

[14]Some of these results are reproduced in [133].

10.3   Clearing systems

Economou and Manou [216] (2013) consider M/M/1 *clearing systems* in an *alternating environment*. The model, motivated by transportation applications, assumes *all* present customers are removed at the completion of service. In other words, this is bulk service with infinite batch size. The system alternates between two environments according to a continuous-time Markov chain. Environment i is characterized by arrival rate λ_i and service rate μ_i. Customers receive a reward R from service and incur waiting costs C per unit time. Their decision is whether to join or balk. Let $E(t)$ and $N(t)$ denote the environment and queue length at customer arrival time t. There are four information cases in which $E(t)$ and $N(t)$ can be observable or unobservable.

While $E(t)$ is directly relevant to customer decisions, $N(t)$ only serves as a *signal* providing information about $E(t)$ so its importance is indirect. When $E(t)$ is observable, $N(t)$ is not relevant and consequently analysis is relatively straightforward except for in the *almost-observable* case where $N(t)$ is known and $E(t)$ is not:

- Suppose $(\mu_1 - \mu_2)(\rho_1 - \rho_2) < 0$, where $\rho_i = \lambda_i/\mu_i$. This means the slow service environment is more congested and a short queue is a sign of fast service. In such a case customers prefer to join short queues and it is natural to consider threshold *joining* strategies. Consider a tagged customer, and suppose that all others increase their thresholds. As a result, the signal associated with a short queue becomes even stronger and the tagged customer's best response threshold tends to increase (cf. [218]). Hence this is an FTC situation. The authors provide a complete characterization of the interval of integer equilibrium thresholds and the associated mixed thresholds.

- Suppose $(\mu_1 - \mu_2)(\rho_1 - \rho_2) > 0$. The above arguments are reversed and customers prefer to join long queues. The authors in this case analyze threshold *balking* strategies. This case leads to an ATC situation with a unique equilibrium.

Manou, Economou, and Karaesmen [469] (2014) consider a *stochastic clearing system* model allowing for generally distributed inter-service times and varying capacity. Customers are homogeneous with a common service value and linear waiting costs and arrive according to a Poisson process. At any instant of service, the number of served customers is limited by the (random) capacity at that instant and *all unserved customers balk*. Arriving customers have no information on the time elapsed since the last service instant and *cannot renege*.

In the observable case, arriving customers base join-or-balk decisions on the number of waiting customers which signals how much time has elapsed since the last service instant (similar to [49, 324, 397]). The authors conduct a probabilistic analysis which completely characterizes the equilibrium solutions and demonstrates the possibility of both ATC and FTC behavior depending on the nature of the underlying inter-visit time distribution. In particular, multiple solutions are possible. The unobservable case is ATC and the authors compute the unique equilibrium. Specific solutions are given for the case of exponential inter-service times, where the observable case leads to a unique threshold strategy. An interesting insight is that the equilibrium in the observable case threshold is **SO** in contrast to the unobservable case where, in general, it exceeds the **SO** threshold.

Chapter 11

Bounded rationality

Strategic queueing models often include elements of bounded rationality. Several authors explicitly mention this fact, for example when the outcomes of lab studies differ from theoretical rational equilibrium. Others leave this fact implicit, for example the KKR model [252].

This chapter describes three groups of models exhibiting bounded rationality behavior. The first pertains to *heuristic strategies* used when the purely rational strategy turns out to be complex. The second group includes *attraction models* where customers select inferior options with a positive probability. This behavior can also be attributed to factors not included in the model, such as heterogeneity of customers and service options. Models in the third group assume firms manipulate customers by promises that customers accept at face value rather than deducing correct information from them. This behavior can be attributed to an inability to make the required deduction, psychological reasons, or customer naivety. Some of these models exhibit another type of bounded rationality, assuming promises are state dependent but this information is ignored by customers.

11.1   Heuristic strategies

Heuristic strategies simplify the computational effort and reduce the amount of required information. Decision makers may use simplifying assumptions on the others' strategies or ignore past information that could be used to better estimate relevant variables.

Sanders [557] (1988) assumes user i, $i = 1, \ldots, N$, sends demand at rate λ_i along a given path of a network.[1] The entire demand flowing through link l is served by an M/M/1 server with service rate μ_l. User i has utility function $h_i(\lambda_i, W_i)$ when submitting demand at rate λ_i and the total of the expected waiting times along the links of its path is W_i. The author suggests an iterative process where at each iteration every user i is asked to declare the values of $\frac{\partial h_i}{\partial \lambda_i}$ and $\frac{\partial h_i}{\partial W_i}$, where (λ_i, W_i) are the values associated with the current demand vector and the demand rate allocation $(\lambda_1, \ldots, \lambda_N)$ is modified according to a gradient hill-climbing scheme aiming to maximize aggregate utility.

Supposing the utility functions are private information, the author suggests a mechanism with side payments to induce truthful reporting by the customers under the following *myopic* bounded-rationality assumption: *At each moment a user attempts to maximize the change in his utility while acting as if the current iteration is the last one.* Under the proposed payment scheme every user will state the true gradient with respect to the rate vector regardless of whether other users will or will not report truthfully, that is, reporting the truth is a dominant strategy.

Chakravorti [132] (1994) considers the single link model of [557] where the utility $u_i(\lambda_i, W)$ of user i is a function of arrival rate λ_i and expected delay W. This function is private information. The manager controls arrival rates with the objective of generating a Pareto optimal solution subject to a balanced-budget constraint. The proposed solution requires each user i to continuously report the ratio $\alpha_i = \frac{\partial u_i/\partial W}{\partial u_i/\partial \lambda_i}$ where λ_i and W are the current arrival rate and system delay. This information is used by the queue manager to control arrival rates, and it is shown that under the myopic assumption (or *local incentive compatibility*) the proposed mechanism converges to a Pareto optimal solution and induces truthful reporting.

Ruelas-Gonzalez, Limon-Robles, and Smith-Cornejo [552] (2010) consider a system with one server continuously active and another server activated according to an N-policy. An arrival observing n customers in the system considers the average number of customers per active server n_p (that

---

[1]The author considered the single-link special case in an earlier paper, see [1] §4.6.

can be either n or $n/2$) and joins with probability $\left[\left(\frac{n_p}{N}\right)^\alpha + 1\right]^{-1}$, for constants ν and α. [2] The authors solve a case study, estimate the parameters ν and α, and use them to compute the value of N that minimizes lost sales and operations costs.

Li, Jiang, and Liu [438] (2012) consider price and capacity competition between two observable M/M/1 queues with capacity costs $c_i(\mu_i)$, $i = 1, 2$. Customers cannot balk, but there is an upper bound $\bar{P}$ on price. Servers know each other's service rate but customers do not know these values. When they observe queue length n_i and price p_i they estimate full price at $p_i + \beta n_i$. Customers apply this approximation when choosing a server upon arrival and when jockeying. When server i completes a service, the last customer in the other queue can costlessly jockey to the end of this queue, thus limiting the difference in queue lengths to $|p_1 - p_2|/\beta$.

- The authors give an explicit expression for the unique symmetric price and capacity equilibrium when $c_i(\mu_i) = c\mu_i^2$ and $2\lambda c < \bar{P}$. In particular, equilibrium prices are $p_1 = p_2 = \bar{P}$.

- Results for the asymmetric case, $c_i(\mu_i) = c_i\mu_i^2$, are obtained numerically.

These results are compared with the two following models:

- The model of Lee and Li (1994) (see [1] §7.3) is similar, but with customers knowing the service rates. It turns out that the lack of information and bounded rationality of the customers is exploited by the servers to obtain higher profits.

- In the symmetric capacity cost case the servers build less capacity but earn higher profits as compared to So [584] (2000), where it is assumed the queue is unobservable but service rates are known and jockeying is not allowed. Thus, servers prefer that customers be irrational and informed rather than rational and uninformed.

Manjrekar, Ramaswamy, and Shakkottai [468] (2014) consider the following discrete-time queueing game with N servers and $N \cdot M$ customers:

- At each period all customers are randomly assigned to servers.

- Customers incur convex holding costs.

- A random workload is added to the customer's workload at the beginning of the period.

---

[2] While customers rationally consider the number of active servers (unlike in [459]), those arriving to a single active server ignore the possibility that another server will be activated at a later time.

- Auctions are held by every server with winners obtaining one unit of service. Customers bid to minimize the expected sum of holding and service costs.

The authors use the idea of mean field approximation assuming customers ignore how their actions affect the behavior of others. Under this simplified model, a customer's best response turns out to be monotone increasing in his workload, and therefore the customer who obtains service has the maximal workload in the queue. For a large number of customers, the mean field equilibrium is an accurate approximation of the perfectly rational equilibrium.

Huang and Chen [352] (2015) consider a variation of the E&H model where customers do not know the service rate and base decisions to join on *anecdotal reasoning*. Specifically, time is divided into *generations* which are long enough allowing for the system to approximately attain steady state. An arriving customer acquires information about realized system time W from one former customer chosen from a past generation, but with recent generations having higher probabilities of being sampled. The customer joins if the net utility of the sampled customer was positive.

Unlike in the rational E&H model, the profit-maximizing price is not **SO**. The authors also prove, contrary to the Chen and Frank observation (§6.2.1), that:

- In the short run, with μ fixed and λ sufficiently close to μ, the profit-maximizing price increases in λ.

- If service rate is a decision variable associated with a linear cost to the firm, and if λ is sufficiently small, an increase in λ results in a lower price.

11.2   Quantal response and attraction demand functions

See [342, 567] where delay-standard sensitivity is combined with logit choice.

11.2.1   A single server

Liu, Methapatara, and Wynter [452] (2010) consider a profit-maximizing M/M/1 firm operating during N periods with demand rates λ_t $t = 1, \ldots, N$. The firm offers multiple service classes and allocates a fraction $\phi(s, k)$ of capacity μ for serving class k customers in period s. The resulting delay for these customers is $z(s,k)$. The price charged to customers arriving in period t, joining service class k, and wishing to be served in period $s \geqslant t$ is $r(t,s,k)$. The resulting utility for these customers is $U(t,s,k) = v_t - r(t,s,k) - \eta_t z(s,k) - \zeta_t \cdot (s - t)$. Given these utility values and the utility associated with balking, t-customers randomly choose s and k, or balk, according to a logit function.

The authors provide numerical evidence the firm can increase profits by providing customers sufficient incentive to shift demand to off-peak periods and maintaining more than one service class.

Huang, Allon, and Bassamboo [351] (2013) investigate the effect of irrationality in an M/M/1 queue, assuming that when the expected benefit of joining is b and the benefit of balking is 0, the probability a customer joins is $\frac{\exp(b/\beta)}{1+\exp(b/\beta)}$ (1 represents e^0). The parameter $\beta > 0$ measures the *degree of irrationality*; $\beta \to 0$ refers to rational behavior, and the probability of choosing worse options increases as β grows. The authors consider Naor's observable model and the E&H unobservable model under equilibrium, social optimization, and profit maximization, and find the following fundamental differences:

- In the unobservable model, the equilibrium is mixed, in general, and therefore with adequate pricing (which may be negative when β is high) it is possible to optimally regulate the queue. This is not possible in the observable case because the optimal equilibrium is pure and any price will still lead to a mixed outcome.

- In general, high irrationality reduces social welfare, but can be used by the firm to increase revenues.

- An interesting result in the unobservable model (with $\beta > 0$) is that higher arrival rates always lead to higher optimal revenue, unlike in the E&H model.

11.2.2   Competition

So [584] (2000) considers price competition among M/M/1 service providers in a market of fixed size λ, where the demand rate λ_i of firm i is determined by an attraction model

$$\lambda_i = \lambda \left(\frac{L_i p_i^{-a} t_i^{-b}}{\sum L_j p_j^{-a} t_j^{-b}} \right),$$

where p_i is price, t_i is delay, and L_i, a, and b are constants. Firm i has capacity μ_i and operating cost γ_i, per customer. The author proves the existence of a unique equilibrium and presents an iterative procedure that leads to it. Additional qualitative features of the model are:

- With a monopolistic firm, higher capacity increases profit, as expected. However, when identical firms compete, higher capacity for all competitors leads to lower pricing and lower profits.

- When the number of competing firms increases while total capacity remains fixed, the result is a lower price and a longer delay.

- All other factors being equal, a firm with higher capacity exploits this advantage by offering a shorter delay and charging a higher price. In contrast, a firm with a lower operating cost offers a lower price and a longer time guarantee. When two firms compete, a firm having both higher capacity and lower operating costs would offer both shorter delay and lower price. However, this is not necessarily true when more than two firms compete.

Gallego, Huh, Kang, and Phillips [243] (2006) derive sufficient conditions for the existence of a unique equilibrium in a general price-competition model with an (asymmetric) attraction model of demand. In particular, the assumptions are satisfied for the logit, Cobb-Douglas, and linear attraction functions (see §1.5.2) of the price vector, and for the following two queueing examples:

- Firm i is an $M/M/1/\kappa_i$ system incurring a fixed operating cost per customer and a fixed penalty per lost (rejected) customer.

- An $M/D/1$ special case of the $G/G/1$ model of [33].

Zhang, Dey and Tan [705] (2008)[3] consider two competing servers with heterogeneous Poisson demand. Customer service valuations $v \sim U[0,1]$ are perfectly correlated with the waiting-time cost rate γv. Each server offers two priority classes, $j = 1, 2$. The decision variables of server i are delay guarantee d_i and prices p_{ij}. Demand and balking rates are determined by a logit demand function where the probability a v-customer buys priority i from provider j is

$$\Pr(i, j, v) = \frac{e^{(v-p_{ij}-\gamma v d_i)/\Delta}}{1 + \sum_{k=1}^{2} \sum_{m=1}^{2} e^{(v-p_{km}-\gamma v d_k)/\Delta}}$$

and the probability of balking is $\left[1 + \sum_{k=1}^{2} \sum_{m=1}^{2} e^{(v-p_{km}-\gamma v d_k)/\Delta}\right]^{-1}$, where Δ is a parameter.[4] The authors (numerically) solve the price equilibrium in the duopoly model and compare it to the monopolistic solution. They conclude that:

---

[3]The first part of this paper is summarized in §7.1.

[4]The authors observe that when $\Delta \to 0$ customers join a given class if both **IR** and **IC** conditions hold and the continuous demand function subsumes the traditional discrete form.

- A monopoly provider charges a higher price but provides a lower delay relative to the duopoly equilibrium.

- Customers gain from the competition only under low traffic.

- Competition *reduces* social welfare.

Allon and Federgruen [33] (2008) examine the attraction model $\lambda_i = M v_i(p_i, \theta_i) / (v_0 + \sum v_j(p_j, \theta_j))$, where p_i is price, θ_i is *service level*, $v_i(p_i, \theta_i)$ is the firm's attractiveness function, and M and v_0 are positive constants. The authors assume a uniform price increase by all firms cannot result in an increase in demand volume to any firm, which translates into

$$\frac{v_i}{v_0 + \sum v_j} < \frac{\partial v_i / \partial p_i}{\sum \partial v_j / \partial p_j}.$$

The capacity cost $C_i(\lambda, \theta)$ associated with providing service level θ when firm i faces demand at rate λ is convex in λ. The authors show that an equilibrium exists both in price competition and in service (delay) competition.

Allon and Gurvich [37] (2010) consider a many-server heavy-traffic model of competition where service providers set price and the number of servers they employ. Demand faced by each firm depends on price and service level offered by all firms in the market, with additional assumptions that are satisfied, for example, by the multinomial logit and Cobb-Douglas models.

An equilibrium does not exist, in general, in this model and therefore the authors relax the requirements and deal with ϵ-equilibria. They examine two types of approximations, a deterministic fluid model and a more refined stochastic diffusion model. The main results include:

- The firms can be fairly close to optimality by first solving the price-competition game, assuming customers are insensitive to delay, and then setting the optimal number of servers given this price vector.

- Conditions for **QED** and **ED** regimes to emerge in the competitive equilibrium.

- Bounds on quality of fluid and diffusion approximations.

Allon, Federgruen, and Pierson [36] (2011) study a model of competition among servers using data from the fast-food industry. Their model assumes a linear relation between customer utility and price, waiting time, chain identity, and demographic factors. The market share of servers and balking probability are determined through a multinomial logit model. Each server's costs are assumed to be linearly dependent on demand rate and the reciprocal of expected waiting time. (The latter dependence is also applied and justified in [33].)

The authors estimate the impact of various factors on price equilibrium and associated demand. In particular, they estimate how sales of a certain firm are affected by price and waiting time of all service providers. These estimates indicate customers attribute a very high cost to the time and therefore firms can often significantly improve their absolute and relative market shares by modestly reducing waiting times.

Li, Guo, and Lian [442] (2015) model duopoly competition among service providers where server selection (and balking) is determined by the logit choice model. The probability of selecting server i is $e^{U_i/\beta} / \left(1 + \sum e^{U_j/\beta}\right)$, where $\beta > 0$ measures the *degree of irrationality* and $U_i = V_b + \alpha(\mu_b - \mu_i) - p - \frac{C}{\mu_i - \lambda_i}$. In this expression, V_b and μ_b are benchmark service value and service rate, price p is exogenous, and service rate μ_i is a decision variable.

The authors prove that:

- When customer utility is required to be non-negative, the welfare-maximizing price is lower than the competitive price, which, in turn, is less than the monopoly revenue-maximizing price.

- When customer utility is allowed to be negative, the welfare-maximizing price can be larger than the other two prices in a large-size market. In this case firms do not benefit from a higher level of bounded rationality.

Blake and Elahi [91] (2015) experimentally investigate the theoretical model of [222], with a single buyer and two suppliers. The authors explain observed differences between laboratory experiments and theoretical results by relaxing the model's rationality assumptions. *Gamesmanship behavior* occurs when players use the profit of their competitors as a benchmark and direct their decisions according to the difference between this and their own profit.

The authors obtain the following results:

- In most cases, empirical average base-stock levels are greater than the corresponding equilibrium values but smaller than the gamesmanship equilibrium. This suggests that subjects maximize their own profit while at the same time try to beat out their competitors.

- To explain their experimental findings the authors use a logit quantal response equilibrium. For $i = 1, 2$, let $EU_i(z)$ be a weighted sum of the expected profit of supplier i and the difference (representing gamesmanship) between that profit and that of his competitor's when supplier i chooses base-stock level z and expectation is taken over the random decisions made by his competitor. Let Ω_i be the set of possible decisions for supplier i. Then, for every $z \in \Omega_i$ the probability of supplier i choosing base-stock level z is $\Pr(z) = \exp(EU_i(z)/\beta) / \sum_{\omega \in \Omega_i} \exp(EU_i(\omega)/\beta)$. The authors examine the weights on gamesmanship and identify conditions where their influence are significant.

11.2.3 Location

Marianov, Ríos, and Icaza [472] (2008) consider a firm establishing a given number of new $M/M/s/K$ facilities on a network while facing competition from existing facilities. Given a complete set M of established facilities, travel times (t_{ij}) for customer i to location $j \in M$, and equilibrium expected waiting time w_j at j, customer i's cost of selecting $j \in M$ is $c_{ij} = \alpha t_{ij} + (1-\alpha)w_j$ for a fixed $\alpha \in [0,1]$. The probability customer i chooses service from the facility at location j is assumed to be $e^{-\gamma c_{ij}} / \sum_{k \in M} e^{-\gamma c_{ik}}$.

The authors propose a heuristic algorithm for locating new facilities to maximize market share.

Saidi-Mehrabad, Teimoury, and Pahlavani [556] (2010) suggest a model of competition among $M/M/s/K$ service facilities. The full cost incurred by a customer from location i traveling to facility j for service is $c_{ij} = p_{ij} + f(t_{ij} + w_j)$, where p_{ij} is price, t_{ij} is travel time, and w_j is expected waiting time. Customers choose a facility according to a logit model. The authors consider a process where customers learn the values w_j from experience. They also consider variations where customers have the option of *veering* (similar to jockeying).

Pahlavani and Saidi-Mehrabad [512] (2011) develop a model for a firm's profit-maximizing pricing given the fixed prices of its competitors. The locations of all facilities are fixed; demand is routed according to a logit function of full price (the sum of price and transportation costs). Customers balk from the selected facility with a probability that linearly increases in queue length. Balking customers choose (only once) a different firm according to a logit function depending on price differences and transportation costs. The problem is formulated as a mathematical program and a heuristic is offered.

Zarrinpoor and Seifbarghy [691] (2011) propose heuristics for a variation of [472] including facility installation costs, customer waiting costs and transportation costs. The firm minimizes *combined costs to the firm and its customers* subject to a constraint that the firm serves a given market share.

Abouee-Mehrizi, Babri, Berman, and Shavandi [5] (2011) compute profit-maximizing locations for a given number of $M/M/1$ facilities. Fixed costs associated with establishing a facility and linear capacity costs are location-dependent. Service capacity and common price p at all facilities are decision variables. Demand for service is set in three steps. (i) Customers at location i select facility j with probability $P_{ij} = e^{-d_{ij}} / \sum_k e^{-d_{ik}}$, where d_{ij} denotes distance. (ii) Arrival rate to facility j is $\sum_i w_i P_{ij} e^{-\alpha_j p}$ where w_i is the demand rate originating at i. (iii) Queues are observable and customers join only if queue length is below an exogenous threshold.

11.3   Quotation sensitivity

This section describes models where a firm quotes an unreliable promised delivery time (PDT) and customers react as if it were reliable. In these models it is the firm that incurs penalties when the realized delivery time deviates from the one promised. These costs usually consist of tardiness (lateness) costs for late delivery and often also holding (earliness) costs for early delivery. Section 11.3.3 is devoted to another type of quotation sensitivity where the firm manipulates customers by the way it presents information, even though the information is reliable.

PDT sensitivity is commonly assumed in the context of supply chains [451, 714], decentralized systems [228], and server vacations [192]. Sensitivity to PDT is also assumed in models with strategic delays, see §2.6.2.

11.3.1   PDT sensitivity

Weng [659] (1996) considers a profit-maximizing M/M/1 server handling two customer classes. P-customers agree to pay price P_P independent of waiting time while L-customers are sensitive to PDT. When an L-customer is quoted PDT L_L and price P_L and actual delivery time is x, profit generated by an L-customer is $P_L - b_L L_L - C_w x - C_t(x - L_L)^+ - C_e(L_L - x)^+$. Profit generated by a P-customer under these conditions is simply $P_P - C_w x$. It is assumed, ignoring penalties, that L-customers are more profitable to the server (i.e., $P_L - b_L L_L > P_P$), and therefore obtain preemptive priority. Server decision variables are arrival rates λ_L and λ_P and PDT L_L.

The author presents a closed-form solution to the profit-maximizing values of the decision variables.

Hatoum and Chang [317] (1997) consider a multiclass M/M/1 queue with class demands that linearly decrease in price and PDT. The server incurs class-dependent production, holding, and tardiness costs. The authors numerically investigate the sensitivity of the profit-maximizing solution to the cost parameters and conclude that tardiness costs constitute a more influential component than holding costs.

Palaka, Erlebacher, and Kropp [514] (1998) consider an M/M/1 system with linear demand function $\lambda(p, l) = a - b_1 p - b_2 l$ of price p and PDT l. The server incurs linear earliness and tardiness penalties. PDT is measured as the s fractile of the waiting-time distribution, where s is an exogenous reliability level.

The server maximizes profits by setting l and p (and in the long-run version also μ). The server can also control demand by rejecting arrivals, but the authors prove this never to be optimal. The paper contains extensive comparative statics with some interesting conclusions. For example:

- When reliability level s increases, the firm defends itself against incurring higher penalties and acts in two ways, increasing the PDT and also decreasing the expected lead time.

- When the tardiness penalty increases from small to large values, the firm first reacts by limiting arrival rate and increasing price, and then by increasing PDT to avoid tardiness penalties and reducing price.

- It is possible that when delay cost, lateness penalty, or demand lead-time sensitivity increase, the firm achieves a shorter lead time by both decreasing arrival rate and increasing capacity.

Chatterjee, Slotnick, and Sobel [135] (2002) consider a firm that cannot observe queue length and sets PDT based on the known service duration of the current job. A customer offered PDT l (excluding service) remains for processing with probability $e^{-\xi l}$ where ξ is a constant. A job of duration s has revenue $r(s)$ and a unit time tardiness penalty $\tau(s)$. The firm maximizes profit (revenue minus tardiness costs).

The authors prove the optimal PDT function $L(s)$ in the M/M/1 model is log-linear: $L(s) = \max\{a - b\ln[r(s)/\tau(s)], 0\}$.. When revenue net of processing costs increases in proportion to processing time, there is a critical value above which all jobs should be assigned a zero PDT. If the unit tardiness penalty depends on processing time, and revenue is nonlinear with respect to processing time, then a job with a longer processing time should receive a longer PDT.

Armony and Maglaras [65, 66] (2004) consider an M/M/s firm offering two service classes, Q_1 and Q_2, to a continuum of customer types. A PDT D_2 is quoted for class Q_2 customers. A τ-customer has two utility functions: a delay-sensitive function $u_1^\tau(W_1)$ for choosing Q_1 when expected waiting time is W_1, and a PDT-sensitive function $u_2^\tau(D_2)$ for choosing Q_2.[5] The example used by the authors has $u_1^\tau = R_1^\tau - C_1 W_1$ and $u_2^\tau = R_2^\tau - C_2 D_2$.[6] The utility associated with balking is 0. Customers select a server or balk, either by employing the utility-maximizing option or according to a multinomial logit function. In the unobservable case [65] W_1 is based on the long-run information, while in the observable case [66] it is based on state-dependent information.

The firm dynamically allocates servers to the two service classes. For a given D_2, the firm's goal is to minimize W_1 while maintaining its commitment to Q_2-customers.

The authors focus on asymptotic analysis. They consider the Halfin-Whitt

---

[5] The authors recognize in [65] that this behavior reflects bounded rationality saying, "In principle, the utility may depend on the entire distributions of W_1 and W_2, however, this does not appear to be very realistic (due to bounded rationality arguments)." However, they also note in [66] that the quoted lead time turns out to be asymptotically exact.

[6] The class Q_2 is motivated by a call center's *call-back* option, and therefore the numerical examples assume $C_2 \ll C_1$.

regime and derive approximations for the expected value and variance of waiting time for both queues. This leads to insights on the trade-off between the performance measures of the two service classes and enables computing the optimal value of D_2 and the minimum number of servers N for desired performance.

Slotnick and Sobel [583] (2005) characterize the optimal policy in a model where customers know their service time upon arrival,[7] but not the service times of those already in the queue. The service provider has the complete information which it uses to quote PDTs to new arrivals. Customers do not deduce this information from the quote and balk with a probability that depends on their service time and the PDT. The firm's goal is to maximize expected revenue minus tardiness penalties. Special attention is given to the case where revenue is proportional to processing time, the tardiness penalty is proportional to lateness, and balking probability is an exponential function of PDT. Under these conditions, profit decreases as queue length increases, and increases as processing time increases; a PDT increases as backlog increases and decreases as processing time increases. The results are extended to classes of heterogeneous customers.

Hong and Lee [345] (2013) introduce unreliable PDTs into the model of [104]. The authors consider both the capacity allocation problem of dividing a given capacity μ between the two servers and the variation where total capacity becomes a decision variable.

The price for regular service is exogenous while the price for express service is a decision variable. The server incurs penalty R_i whenever the service duration of an i-customer exceeds L_i. This penalty is independent of the duration of the delay. Numerical results demonstrate the advantage of segmenting the market. Similar to [104], the sensitivity of the solution with respect to guarantees L_1 and L_2 depends on the market sensitivity to price and expected service time differences.

11.3.2 Dynamic PDT quotation

We describe here models where the PDT quoted by the firm is state-dependent. These models usually assume that customers' reaction follows an exogenous probabilistic joining function of the PDT. Rational customers can, however, use dynamic (state-dependent) PDTs to deduce relevant information on the system state. Ignoring this information may be attributed to bounded rationality. Dynamic PDT quotation is also discussed in [192] in a model with strategic server vacations.

---

[7] A similar assumption appears in [322], while in [135] it is assumed *the server* knows the customer's service time.

Duenyas [209] (1995) considers an M/M/1 multiclass system with functions $p_i(a)$ for the probability of an i-customer joining when quoted PDT a. Revenue from serving an i-customer is R_i. If an order is x units of time late, the firm incurs a class-independent penalty cx. PDTs are class dependent and dynamically adjusted to the state of the system. The author proves structural properties of profit-maximizing PDTs when the service order is FCFS and when the firm is free to sequence orders. In the latter case it is optimal to follow the earliest due date rule (EDD). **Duenyas and Hopp [210] (1995)** solve single-class variations of this model.

Savaşaneril, Griffin, and Keskinocak [562] (2010) consider profit-maximizing PDTs in an M/M/1 MTS system. The probability that a customer receiving a PDT quote d will place an order is an exogenous decreasing function $f(d)$. The firm earns a fixed revenue per customer and incurs linear tardiness and holding costs. The firm's decision variables are the state-dependent PDT and the base-stock level.

The main results are:

- The profit-maximizing base-stock level increases as customer PDT-sensitivity increases.

- When the optimal base-stock level is positive, optimal PDTs in states with zero on-hand inventory are higher as compared to the MTO system where the firm cannot hold inventory.

- Increasing the base-stock level does not necessarily decrease the expected number of customers waiting in the system.

- If the firm rejects arriving customers when necessary, quoting lead times with lower precision will only marginally affect profits. However, if all customers must be accepted the negative impact on profits can be high.

Feng, Liu, and Liu [234] (2011) examine dynamic price and lead-time quotes in a G/M/1 system with heterogeneous service valuations. When a customer arrives, the server quotes state-dependent price p and PDT l. The cost a customer associates with the PDT is a convex function $x(l)$; the customer joins the queue if his service value exceeds the full price $p + x(l)$. The server incurs a fixed cost when a customer balks and a cost per unit time of tardiness, and obtains a fixed revenue per customer served. The main finding is that the profit-maximizing solution is obtained when, *sequentially*, the server first determines the PDT by maximizing the profit from the current customer and then determines the price while taking into consideration the effect of the current customer's joining on future arrivals.

Slotnick [581] (2011) considers a multiclass service system where service requests are characterized by size a and type g. A g-customer promised lead time l accepts the offer with probability $e^{-\xi(g)l}$ and balks otherwise. Joining requests are grouped according to type. When the total group size reaches a given value the group joins a second M/M/1 queue as a single batch. Termination of processing an (a, g)-request rewards the system by $w(g) \cdot a$. If the order is tardy the firm incurs a fixed tardiness penalty.

The author applies dynamic programming to a finite-horizon version of this model, and also suggests a faster heuristic algorithm. A computational study indicates that optimal PDT is a decreasing function of the size of the group of g-requests waiting to reach the minimum group size, and also a decreasing function of the arrival rate for group g.[8]

Kaman, Savaşaneril, and Serin [380] (2013) consider an M/M/1 MTS manufacturer obtaining a fixed revenue per served customer and incurring constant penalty rates for holding inventory and late delivery. A decreasing exogenous function $f(d)$ gives the probability of a customer placing an order when quoted a PDT of d. The manufacturer has imperfect information on the state of the system (the number of orders in the queue or the number of items in stock), and this is expressed by probabilities that the manufacturer observes signals i' when the actual state is i. The manufacturer maximizes profits by setting base-stock level and dynamic PDTs.[9]

The authors conduct a numerical study and investigate the value of perfect information. They conclude, for example, that imperfect information is likely to increase the stock level in the system and that the value of information is likely to be higher under high holding cost and low traffic.

Slotnick [582] (2014) assumes customers are sensitive to the firm's *reputation* for on-time delivery, presented as a weighted average of past tardiness values. The probability a customer quoted PDT L places an order is $e^{-(\xi L + \gamma T)}$, where T is the firm's reputation at the time of decision. Jobs have heterogeneous sizes, price per unit size is exogenous and the quoted PDT depends on job size, queue length, and reputation of the firm at present. Processing time is proportional to job size and job arrivals are Poisson. The problem is to determine a PDT quotation policy maximizing the firm's discounted profits (revenue minus linear tardiness costs). Results of a computational study include:

- PDT is longer when the shop workload (including the current job) is large, but this relation does not necessarily hold with respect to job size alone.

- If the firm has a poor reputation for on-time delivery, it should quote a shorter PDT so customers will be more likely to stay (rather than

---

[8]Note the similarities in this situation, the model in [125], and the *shuttle model* in [308] and [1] §1.5.

[9]The actions prescribed by the policy depend on the observed state, but are independent of historical data which could be used to obtain improved predictions on the current state.

quoting a longer lead time to improve its reputation for on-time delivery in the future).

- For a customer who is more sensitive to reputation or length of PDT, the firm should quote a longer PDT.

Savaşaneril and Sayin [563] (2015) consider a profit-maximizing M/M/1 MTS firm facing n customer classes with heterogeneous arrival rates, revenue potential, and distribution of maximal accepted PDT. The system incurs class-independent holding and tardiness penalties, and the queue discipline is FCFS. The authors characterize the profit-maximizing dynamic admission and lead-time quote policy.

The main results are as follows:

- A lower-than-optimal base-stock level increases expected waiting time but does not necessarily increase expected tardiness. A higher-than-optimal base-stock level may or may not improve expected tardiness.

- Deviations from the optimal quotation scheme, e.g., a static quotation scheme, result in higher stock levels.

- In a two-customer class setting, a single server MTS queue is contrasted with a system of two MTS queues each dedicated to a demand class. Total server capacity is identical in both systems. The increase in the profit due to resource pooling can be considerably high when server utilization is close to 100%. Interestingly, when the utilization is very low and customer class unit revenues significantly differ pooling may harm the firm.

- The benefit of pooling is lower in the presence of an effective lead time quotation policy.

11.3.3   Sensitivity to delay guarantee and reliability level

Consider a queue with known arrival and service distributions except that the service rate parameter μ is a decision variable which cannot be directly observed by customers. The firm may commit to serving customers within lead time l with probability s. Note that l and s are sufficient information for deducing μ. Moreover, assuming a fixed μ, for every l there is a corresponding reliability level $s(l)$ and the choice of the particular pair $(l, s(l))$ to be quoted is arbitrary and should not affect the demand of rational customers. This section concerns models where the firm manipulates non-rational customers by the selection of the pair (l, s).[10]

---

[10]Customer behavior can be related to the *anchoring effect* described by A. Tversky and D. Kahneman, "Judgment under uncertainty: heuristics and biases," *Science* **185** (1974) 1124-1131.

Ho and Zheng [342] (2004) consider competition where m servers maximize market share. Each server i announces delay guarantee T_i and reliability level Q_i, and the utility of a customer joining this server is $U_i = \beta_0 - \beta_T T_i + \beta_Q Q_i$. A customer selects server i with probability $S_i = e^{U_i} / \sum e^{U_j}$.

Suppose utilities at servers $j \neq i$ are fixed, and server i announces T_i. For any arrival rate λ_i to server i there are corresponding $Q_i(T_i, \lambda_i)$ and U_i. In equilibrium the resulting selection probability satisfies $\lambda_i = S_i \Lambda$, where Λ is market size. The authors prove that for any given T_i there is a unique equilibrium $\lambda_i(T_i)$ and provide an algorithm for computing the best response T_i^* that maximizes $\lambda_i(T_i^*)$. The authors also consider duopoly competition, prove an equilibrium (T_1, T_2) exists, and present a prisoner's dilemma type example when capacity can be added at a cost.

Boyaci and Ray [105] (2006) extend the model of [104] by considering the following demand function:

$$\lambda_i = a - \beta_p p_i + \theta_p (p_j - p_i) - \beta_\alpha (1 - \alpha_i) + \theta_\alpha (\alpha_i - \alpha_j) - \beta_L L_i + \theta_L (L_j - L_i),$$

where the lead time L_i is the α_i-fractile of the waiting-time distribution at server i, α_i is the corresponding *reliability level*, and j is the other server. Since the waiting-time distribution can be represented by different L and α values, this means that customers are affected by the way in which the information is presented.

The authors show that optimal solutions qualitatively depend on whether demand is more sensitive to price or to delay, and whether demand is more sensitive to lead time or to reliability.

Shang and Liu [567] (2011) extend [342] by discussing competition in delay quotation and capacity. Servers have heterogeneous revenue from serving a customer and unit capacity costs, and maximize profits. The game has two stages. In the first stage the firms compete in terms of capacities and in the second they compete in terms of delay quotations. The results include:

- A proof of the equilibrium's uniqueness in the second phase.

- Sufficient conditions for the existence of an interior equilibrium and a characterization of boundary equilibria in the first phase.

- An example with two competing firms where, as already observed in [342], a uniform capacity cost reduction harms both firms by intensifying the competition between them.

Jouini, Akşin, and Dallery [377] (2011) analyze a Markovian multi-server model combining probabilistic joining and exponential reneging. A customer arriving when all servers are busy and there are n queueing customers is informed of the β-fractile d_n of his waiting-time distribution. The customer joins the queue iff $d_n \leqslant t$ where $t \sim \exp(\gamma)$ represents the customer's *patience*.

Customers who decide to join update their patience to a value $\theta t + (1-\theta)d_n$[11] and use the updated patience to set a reneging time if service doesn't start by then. The authors show how the queue manager can influence the trade-off between balking and reneging by controlling β.

Park and Hong [518] (2014) consider a Stackelberg game in a supply chain. The leader is an M/M/1 *supplier* setting guaranteed lead time l, reliability level s, and capacity μ. The follower is a *retailer* setting retail price p. The supplier gains $(w_s - m)\lambda - A\mu$, where $\lambda = \lambda(p, l, s) = a - bp - cl + gs$ is the demand function, w_s is an exogenous wholesale price, m is unit production cost, and A is unit capacity cost. The retailer gains $p - w_s$ per customer.

The authors prove a unique subgame-perfect equilibrium exists and present a numerical example.

11.4 Joining and reneging

The first attempt to formulate a rational decision model of reneging was by **Haight [295] (1959)**. This model does not explicitly involve cost or value parameters, and, moreover, customers do not take into account the reneging strategies of others. Customers have a maximal tolerable waiting time W, the queue is observable, but the customer does not have prior knowledge of the service process. Consider a customer who arrives at time 0, and let $n(t)$ be the number of customers queueing in front of him at time t. The reneging decision is guided by a nondecreasing function $\Phi(t)$ such that the customer reneges when $\Phi(t) = n(t)$. The initial function is a hyperbola with asymptotes $t = 0$ and $n = n(0)/W$ (the minimum attrition rate necessary to keep waiting time below W), and the customer periodically revises it according to the observed departures from the queue.

Parkan and Warren [519] (1978) and **Martin and Pankoff [473] (1982)** consider reneging decisions in an observable G/M/1 queue with an unknown service rate. A customer's prior belief about the value of μ is a random variable with a gamma distribution. Upon arrival and at any service termination customers compute their conditional expected utility and renege if it is negative. The model reflects several aspects of bounded rationality. Customers assume those ahead of them will not renege. Customers ignore the option value of inspecting more service terminations and reneging later (the customer behaves as if the reneging option exists only at the current time of decision). In their estimation of μ, customers ignore the behavior of those who arrived earlier and possess more information. Specifically, the model

---

[11]The parameter θ is not restricted to [0,1].

allows customers ahead in the queue to renege, but the customer in question ignores the signals that such behavior transmits.

Shimkin and Mandelbaum [573] (2004) assume customers of an $M/M/s$ system differ in two functions. They differ in the utility $R(t)$ obtained when starting service t units of time after arrival, and in the nonlinear cost $C(t)$ of queueing t units of time. The queue is unobservable but customers know when service starts. Customers decide on reneging time while maximizing expected value from service minus waiting cost.

The authors develop conditions on the waiting cost function guaranteeing the existence and uniqueness of an equilibrium. They also give broader conditions for a different concept of *myopic equilibrium*. This new concept assumes customers adopt a *myopic decision rule* choosing the abandonment time as the first (weak) local maximum of the utility function (rather than choosing a global maximum). The authors provide two justifications for this rule:

- It is natural when abandonment decisions are taken based on the customer's assessment of the current situation and the utility of further waiting.

- Often customers lack information regarding the waiting-time distribution for long waiting.

Turhan, Alanyali, and Starobinski [632] (2012) consider an $M/M/C/C$ loss system with two customer types, primary users (PUs) and secondary users (SUs). PUs have preemptive priority over SUs. The system state (x, y) consists of the number of users of each type in the system. SUs pay a price $u(x, y)$ upon arrival. When preempted by a PU, an SU receives a compensation $K > \max u(x, y)$ and leaves the system. PUs have a fixed demand rate of λ_1, while that of SUs is $\lambda_2(u)$, which depends on the current price but not on the probability of eventually obtaining service. This behavior reflects bounded rationality. The server's goal is to maximize profit from SUs payments by selecting an appropriate dynamic pricing policy.

The authors prove the optimal price depends solely on the total number of customers in the system, $x + y$, and that it is increasing in this variable.

Bibliography

[1] Hassin, Refael and Moshe Haviv, *To Queue or Not to Queue: Equilibrium Behavior in Queueing Systems* Kluwer Academic Publishers, 2003. Also available online: **http://www.math.tau.ac.il/~hassin/book.html**. Errata related to the printed version can be found in **http://www.math.tau.ac.il/~hassin/errata.pdf**. (Cited on page 1, 2, 5, 6, 11, 13, 14, 24, 26, 28, 29, 30, 31, 33, 38, 41, 44, 54, 56, 59, 66, 84, 88, 90, 92, 93, 94, 97, 100, 108, 110, 112, 113, 115, 119, 122, 133, 137, 140, 141, 143, 165, 169, 170, 172, 177, 178, 187, 193, 208, 209, 212, 216, 226, 228, 245, 256, 277, 284, 285, and 296.)

[2] Abbad, Mohammed, Rachid El Azouzi, and Mohamed El Kamili, "The problem of capacity addition in multi-user elastic demand communication networks," *Mathematical Methods of Operations Research* **63** (2006) 461-471. (Cited on page 214.)

[3] Abhishek, Vineet, Ian A. Kash, and Peter Key, "Fixed and market pricing for cloud services," *INFOCOM* (2012).(Cited on page 178.)

[4] Aboolian, Robert, Oded Berman, and Dmitry Krass, "Profit maximizing distributed service system design with congestion and elastic demand," *Transportation Science* **46**(2) (2012) 247-261. (Cited on page 226.)

[5] Abouee-Mehrizi, Hossein, Sahar Babri, Oded Berman, and Hassan Shavandi, "Optimizing capacity, pricing and location decisions on a congested network with balking," *Mathematical Methods of Operations Research* **74** (2011) 233-255. (Cited on page 291.)

[6] Acemoglu, Daron and Asuman Ozdaglar, "Competition and efficiency in congested markets," *Mathematics of Operations Research* **32**(1) (2007) 1-31. (Cited on page 134.)

[7] Adler, Ilan and Pinhas (Paul) Naor, "Social optimization versus self-optimization in waiting lines," (1969). (Cited on page 26.)

[8] Afanasyev, Maxim and Haim Mendelson, "Service provider competition: delay cost structure, segmentation, and cost advantage," *Management Science* **12**(2) (2010) 213-235. (Cited on page 192.)

[9] Afèche, Philipp, "Incentive-compatible revenue management in queueing systems: optimal strategic idleness and other delaying tactics," (2004). This is an early version of [12]. (Cited on page 50, 51, 173, and 174.)

[10] Afèche, Philipp, "Economics of data communications," Chapter 2 in Terrence Hendershott (ed.), *Handbooks in Information Systems* Vol 1, Elsevier B.V., 2006. (Cited on page 3.)

[11] Afèche, Philipp, "Decentralized service supply chains with multiple time-sensitive customer segments: pricing, capacity, decisions and coordination," (2013). (Cited on page 222.)

[12] Afèche, Philipp, "Incentive-compatible revenue management in queueing systems: optimal strategic delay," *Manufacturing & Service Operations Management* **15**(3) (2013), 423-433. (Cited on page 50, 51, 169, 173, 174, 175, and 302.)

[13] Afèche, Philipp, Mojtaba Araghi, and Opher Baron, "Customer acquisition, retention, and queueing-related service quality: optimal advertising, staffing, and priorities for a call center," (2015). (Cited on page 206.)

[14] Afèche, Philipp and Barış Ata, "Bayesian dynamic pricing in queueing systems with unknown delay cost characteristics," *Manufacturing & Service Operations Management* **15**(2) (2013) 292-304. (Cited on page 77.)

[15] Afèche, Philipp, Opher Baron, and Yoav Kerner, "Pricing time-sensitive services based on realized performance," *Manufacturing & Service Operations Management* **15**(3) (2013) 492-506. (Cited on page 55, 145, and 150.)

[16] Afèche, Philipp, Opher Baron, Joseph Milner, and Ricky Roet-Green, "Pricing and prioritizing time-sensitive customers with heterogeneous demand rates," (2015). (Cited on page 163.)

[17] Afèche, Philipp and Haim Mendelson "Pricing and priority auctions in queueing systems with a generalized delay cost structure," *Management Science* **50**(7) (2004) 896-882. (Cited on page 146 and 177.)

[18] Afèche, Philipp and Michael Pavlin, "Optimal price/lead-time menus for queues with customer choice: segmentation, pooling and strategic delay," *Management Science* forthcoming. (Cited on page 170 and 175.)

[19] Afèche, Philipp and Vahid Sarhangian, "Rational abandonment from priority queues: equilibrium strategy and pricing implications," (2015). (Cited on page 29 and 264.)

[20] Afimeimounga, Heti, Wiremu Solomon, and Ilze Ziedins, "The Downs-Thomson paradox: existence, uniqueness and stability of user equilibria," *Queueing Systems* **49** (2005) 321-334. (Cited on page 228 and 229.)

[21] Afimeimounga, Heti, Wiremu Solomon, and Ilze Ziedins, "User equilibria for a parallel queueing system with state dependent routing," *Queueing Systems* **66** (2010) 169-193. (Cited on page 228 and 229.)

[22] Akan, Mustafa, Bariş Ata, and Martin A. Lariviere, "Asymmetric information and economies of scale in service contracting," *Manufacturing & Service Operations Management* **13**(1) (2011) 58-72. (Cited on page 240.)

[23] Akan, Mustafa, Bariş Ata, and Tava Lennon Olsen, "Congestion-based leadtime quotation for heterogeneous customers with convex-concave delay costs: optimality of a cost-balancing policy based on convex hull functions," *Operations Research* **60**(6) (2012) 1505-1519. (Cited on page 54.)

[24] Akgun, Osman T., Douglas G. Down, and Rhonda Righter, "Energy-aware scheduling on heterogeneous processors," *IEEE Transactions on Automatic Control* **59**(3) (2014) 599-613. (Cited on page 33, 34, 137, and 265.)

[25] Akşin, O. Zeynep, Mor Armony, and Vijay Mehrotra, "The modern call center: a multi-disciplinary perspective on operations management research," *Production and Operations Management* **16**(6) (2007) 665-688. (Cited on page 232.)

[26] Aktaran-Kalayci, Tûba and Hayriye Ayhan, "Sensitivity of optimal prices to system parameters in a steady-state service facility," *European Journal of Operational Research* **193**(1) (2009) 120-128. (Cited on page 44.)

[27] Ali, Murtuza, Tejas Bodas, and D. Manjunath, "Optimal and equilibrium allocations in a discriminatory processor sharing system," *NetG-Coop* (2014). (Cited on page 105 and 117.)

[28] Allon, Gad and Achal Bassamboo, "The impact of delaying the delay announcements," *Operations Research* **59**(5) (2011) 1198-1210. (Cited on page 79, 80, and 218.)

[29] Allon, Gad, Achal Bassamboo, and Eren Başar Çil, "Large-scale service marketplaces: the role of the moderating firm," *Management Science* **58**(10) (2012) 1854-1872. (Cited on page 199.)

[30] Allon, Gad, Achal Bassamboo, and Eren Başar Çil, "Skill and capacity management in large-scale service marketplaces," (2013). (Cited on page 199.)

[31] Allon, Gad, Achal Bassamboo, and Itai Gurvich, " 'We will be right with you:' managing customer expectations with vague promises and cheap talk," *Operations Research* **59**(6) (2011) 1382-1394. (Cited on page 18, 79, and 80.)

[32] Allon, Gad and Awi Federgruen, "Competition in service industries," *Operations Research* **55**(1) (2007) 37-55. (Cited on page 99, 187, 195, 196, 197, and 253.)

[33] Allon, Gad and Awi Federgruen, "Service competition with general queueing facilities," *Operations Research* **56**(4) (2008) 827-849. (Cited on page 196, 288, and 289.)

[34] Allon, Gad and Awi Federgruen, "Competition in service industries with segmented markets," *Management Science* **55**(4) (2009) 619-634. (Cited on page 196.)

[35] Allon, Gad and Awi Federgruen, "Outsourcing service processes to a common service provider under price and time competition," (2011). (Cited on page 125 and 253.)

[36] Allon, Gad, Awi Federgruen, and Margaret Pierson, "How much is a reduction of your customers' wait worth? An empirical study of the fast-food drive-thru industry based on structural estimation methods," *Manufacturing & Service Operations Management* **13**(4) (2011) 489-507. (Cited on page 289.)

[37] Allon, Gad and Itai Gurvich, "Pricing and dimensioning competing large-scale service providers," *Manufacturing & Service Operations Management* **12**(3) (2010) 449-469. (Cited on page 289.)

[38] Allon, Gad and Eran Hanany, "Cutting in line: social norms in queues," *Management Science* (2012) **58** 493-506. (Cited on page 101.)

[39] Alperstein, Hanna, "Optimal pricing for the service facility offering a set of priority prices," *Management Science* **34**(5) (1988) 666-671. (Cited on page 140.)

[40] Alptekinoğlu, Aydin and Charles J. Corbett, "Leadtime-variety tradeoff in product differentiation," *Manufacturing & Service Operations Management* **12**(4) (2010) 569-582. (Cited on page 181.)

[41] Altman, Eitan, "Non zero-sum stochastic games in admission, service and routing control in queueing systems," *Queueing Systems* **23** (1996) 259-279. (Cited on page 41, 260, and 265.)

[42] Altman, Eitan, "Applications of dynamic games in queues," *Advances in Dynamic Games* **7** (2005) 309-342. (Cited on page 3.)

[43] Altman, Eitan, Konstantin Avrachenkov, and Urtzi Ayesta, "A survey on discriminatory processor sharing," *Queueing Systems* **53** (2006) 53-63. (Cited on page 115.)

[44] Altman, Eitan, Urtzi Ayesta, and Balakrishna J. Prabhu, "Load balancing in processor sharing systems," *Telecommunication Systems* **47** (2011) 35-48. (Cited on page 135 and 136.)

[45] Altman, Eitan, Tamer Başar, Tania Jiménez, and Nahum Shimkin, "Competitive routing in networks with polynomial costs." *IEEE Transactions on Automatic Control* **47**(1) (2002) 92-96. (Cited on page 208.)

[46] Altman, Eitan, Tamer Başar, and R. Srikant, "Nash equilibria for combined flow control and routing in networks: asymptotic behavior for a large number of users," *IEEE Transactions on Automatic Control* **47**(6) (2002) 917-930. (Cited on page 105, 107, and 208.)

[47] Altman, Eitan, Thomas Boulogne, Rachid El Azouzi, Tania Jiménez, and Laura Wynter," A survey on networking games in telecommunications," *Computers & Operations Research* **33** (2006) 286-311. (Cited on page 207 and 208.)

[48] Altman, Eitan, Rachid El Azouzi, and Odile Pourtallier, "Avoiding paradoxes in multi-agent competitive routing," *Computer Networks* **47** (2003) 133-146. (Cited on page 214.)

[49] [∗ ∗ ∗] Altman, Eitan and Refael Hassin, "Non-threshold equilibrium for customers joining an M/G/1 queue," *ISDG* (2002).

[50] Altman, Eitan, Tania Jiménez, Rudesindo Núñez-Queija, and Uri Yechiali, "Optimal routing among ·/M/1 queues with partial information," *Stochastic Models* **20**(2) (2004) 149-171. Correction in *Stochastic Models* **21**(2) (2005) 981. (Cited on page 26, 27, 70, and 281.)

[51] Altman, Eitan and Nahum Shimkin, "Worst-case and Nash routing policies in parallel queues with uncertain service allocations," unpublished manuscript (1993). (Cited on page 9 and 216.)

[52] Altmann, Jörn, Hans Daanen, Huw Oliver, and Alfonso Sánchez-Beato Suárez, "How to market-manage a QoS network," *INFOCOM* (2002). (Cited on page 42.)

[53] Alves, Vasco F., "Endogenous queue number determination in M/M/2 systems," (2015). (Cited on page 30.)

[54] Anand, Krishnan S., M. Fazil Paç, and Senthil Veeraraghavan, "Quality-speed conundrum: trade-offs in customer-intensive services," *Management Science* **57**(1) (2011) 40-56. (Cited on page 102.)

[55] Andritsos, Dimitrios A. and Sam Aflaki, "Competition and the operational performance of hospitals: the role of hospital objectives," *Production and Operations Management* **24**(11) (2014) 1812-1832. (Cited on page 155, 156, 157, and 187.)

[56] Andritsos, Dimitrios A. and Christopher S. Tang, "Introducing competition in healthcare services: the role of private care and increased patient mobility," *European Journal of Operational Research* **234**(3) (2013) 898-909. (Cited on page 123.)

[57] Anily, Shoshana and Moshe Haviv, "Cooperation in service systems," *Operations Research* **58**(3) (2010) 660-673. (Cited on page 123.)

[58] Anily, Shoshana and Moshe Haviv, "Regular games: characterization and total balancedness of regular market games," (2013). (Cited on page 126 and 127.)

[59] Anily, Shoshana and Moshe Haviv, "Subadditive and homogeneous of degree one games are totally balanced," *Operations Research* **62**(4) (2014) 788-793. (Cited on page 127.)

[60] Anselmi, Jonatha, Urtzi Ayesta, and Adam Wierman, "Competition yields efficiency in load balancing games," *Performance Evaluation* **68** (2011) 986-1001. (Cited on page 126.)

[61] Anselmi, Jonatha and Bruno Gaujal, "The price of forgetting in parallel and non-observable queues," *Performance Evaluation* **68** (2011) 1291-1311. (Cited on page 136 and 208.)

[62] Araman, Victor F. and Kristin Fridgeirsdottir, "A uniform allocation mechanism and cost-per-impression pricing for online advertising," unpublished manuscript (2011). (Cited on page 11 and 119.)

[63] Arda, Yasemin and Jean-Claude Hennet, "Inventory control in a decentralized two-stage make-to-stock queueing system," *International Journal of Systems Science* **39**(7) (2008) 741-750. (Cited on page 146.)

[64] Arlotto, Alessandro, Andrew E. Frazelle, and Yehua Wei, "Strategic open routing in queueing networks," (2015). (Cited on page 236.)

[65] Armony, Mor and Constantinos Maglaras, "On customer contact centers with a call-back option: customer decisions, routing rules, and system design," *Operations Research* **52**(2) (2004) 271-292. (Cited on page 220.)

[66] Armony, Mor and Constantinos Maglaras, "Contact centers with a call-back option: and real-time delay information," *Operations Research* **52**(4) (2004) 527-545. (Cited on page 15 and 293.)

(Cited on page 15 and 293.)

[67] Armony, Mor and Erica Plambeck, "The impact of duplicate orders on demand estimation and capacity investment," *Management Science* **51**(10) (2005) 1505-1518. (Cited on page 104.)

[68] Armony, Mor, Nahum Shimkin, and Ward Whitt, "The impact of delaying the delay announcements in many-server queues with abandonment," *Operations Research* **57**(1) (2009) 66-81. (Cited on page 38 and 72.)

[69] Ata, Bariş, Anton Skaro, and Sridhar Tayur, "OrganJet: overcoming geographical disparities in access to deceased donor kidneys in the United States," *Management Science* forthcoming. (Cited on page 104.)

[70] Ata, Bariş and Tava Lennon Olsen, "Near-optimal dynamic lead-time quotation and scheduling under convex-concave customer delay costs," *Operations Research* **57**(3) (2009) 753-768. (Cited on page 52, 54, and 168.)

[71] Ata Bariş and Tava Lennon Olsen, "Congestion-Based leadtime quotation and pricing for revenue maximization with heterogeneous customers," *Queueing Systems* **73** (2013) 35-78. (Cited on page 54.)

[72] Ata, Bariş and Shiri Shneorson, "Dynamic control of an M/M/1 service system with adjustable arrival and service rates," *Management Science* **52**(11) (2006) 1778-1791. (Cited on page 43 and 50.)

[73] Atar, Rami, Israel Cidon, and Mark Shifrin," MDP based optimal pricing for a cloud computing queueing model," *Performance Evaluation* **78** (2014) 1-6. (Cited on page 48.)

[74] Atar, Rami and Subhamay Saha, "An ϵ-Nash equilibrium with high probability for strategic customers in heavy traffic," manuscript (2015). (Cited on page 9 and 43.)

[75] Avi-Itzhak, Benjamin, Boaz Golany, and Uriel Rothblum, "Strategic equilibrium versus global optimum for a pair of competing servers," *Journal of Applied Probability* **43** (2006) 1165-1172. (Cited on page 32, 38, 74, 125, and 200.)

[76] Ayesta, Urtzi, Olivier Brun, and Balakrishna J. Prabhu, "Price of anarchy in non-cooperative load balancing games," *Performance Evaluation* **68** (2011) 1312-1332. (Cited on page 135.)

[77] Azad, Amar Prakash, Eitan Altman, and Rachid El Azouzi, "From altruism to non-cooperation in routing games," *WiOpt* 2010 528-537. (Cited on page 36 and 210.)

[78] Badrabadi, Ali Habibi and Mohammad Jafar Tarokh, "Web service providers' game on price and service level," *International Journal of Industrial Engineering & Production research* **21**(4) (2010) 181-195. (Cited on page 222.)

[79] Banker, Rajiv D. and Stephen C. Hansen, "The adequacy of full-cost-based pricing heuristics," *Journal of Management Accounting Research* **14** (2002) 33-58. (Cited on page 55 and 146.)

[80] Başar, Tamer and R. Srikant, "A Stackelberg network game with a large number of followers," *Journal of Optimization Theory and Applications* **115**(3) (2002) 479-490. (Cited on page 220.)

[81] Başar, Tamer and R. Srikant, "Revenue-maximizing pricing and capacity expansion in a many-users regime," *INFOCOM* (2002). (Cited on page 150 and 220.)

[82] Bayram, I. Safak, Muhammad Ismail, Mohamed Abdallah, Khalid Qaraqe, and Erchin Serpedin, "A pricing-based load shifting framework for EV fast charging stations," *IEEE International Conference on Smart Grid Communications* (2014). (Cited on page 55.)

[83] Bearden, J. Neil, Amnon Rapoport, and Darryl A. Seale, "Entry times in queues with endogenous arrivals: dynamics of play on the individual and aggregate levels," In A. Rapoport and R. Zwick (eds.), *Experimental Business Research* Vol. II. Springer, 2005, pp. 201-221. (Cited on page 93.)

[84] Bell, Colin E., "On competition to join a simple queueing system before facility opens," *Management Science* **31**(3) (1985) 358-368. (Errata: p. 644.) (Cited on page 84 and 85.)

[85] [∗ ∗ ∗] Bell, Colin E. and Shaler Stidham Jr., "Individual versus social optimization in allocation of customers to alternative servers," *Management Science* **29** (1983) 831-839. (Cited on page 11, 41, 119, 132, 133, 134, 135, 208, 212, 245, and 246.)

[86] Benjaafar, Saif, Ehsan Elahi, and Karen L. Donohue, "Outsourcing via service competition," *Management Science* **53**(2) (2007) 241-259. (Cited on page 246, 247, and 248.)

[87] Ben-Shahar, Israel, Ariel Orda, and Nahum Shimkin, "Dynamic service sharing with heterogeneous preferences," *Queueing Systems* **35** (2000) 83-103. (Cited on page 98.)

[88] van den Berg, Hans, Michel Mandjes, and Rudesindo Núñez-Queija, "Pricing and distributed QoS control for elastic network traffic," *Operations Research Letters* **35** (2007) 297-307. (Cited on page 112.)

[89] Bernstein, Fernando and Francis de Véricourt, "Competition for procurement contracts with service guarantees," *Operations Research* **56**(3) (2008) 562-575. (Cited on page 251.)

[90] Besbes, Omar and Costis Maglaras, "Revenue optimization for a make-to-order queue in an uncertain market environment," *Operations Research* **57**(6) (2009) 1438-1450. (Cited on page 52 and 77.)

[91] Blake, Roger and Ehsan Elahi, "Supplier competition: theory vs. experiment," *IEOM* (2015). (Cited on page 290.)

[92] Blocq, Gideon and Ariel Orda, "How good is bargained routing?" *INFOCOM* (2012). (Cited on page 125 and 211.)

[93] Blocq, Gideon and Ariel Orda, "Worst-case coalitions in routing games," manuscript (2014). (Cited on page 125 and 211.)

[94] Blume, Andreas, John Duffy, and Ted Temzelides, "Self-organized criticality in a dynamic game," *Journal of Economic Dynamics & Control* **34** (2010) 1380-1391. (Cited on page 98.)

[95] Bodas, Tejas, Murtuza Ali, and D. Manjunath, "A system with a choice of highest-bidder-first and FIFO service," *Valuetools* 2014. (Cited on page 178 and 218.)

[96] Bodas, Tejas, Ayalvadi Ganesh, and D. Manjunath, "Tolls and welfare optimization for multiclass traffic in multiqueue systems," (2014) (Cited on page 213.)

[97] Bodas, Tejas and D. Manjunath, "On load balancing equilibria in multiqueue systems with multiclass traffic," *NetGCooP* (2011). (Cited on page 212.)

[98] Borgs, Christian, Jennifer T. Chayes, Sherwin Doroudi, Mor Harchol-Balter, and Kuang Xu, "The optimal admission threshold in observable queues with state dependent pricing," *Probability in the Engineering and Informational Science* **28** (2014) 101-119. (Cited on page 28.)

[99] Borst, Sem, Avi Mandelbaum, and Martin I. Reiman, "Dimensioning large call centers," *Operations Research* **52**(1) (2004) 17-34. (Cited on page 15.)

[100] Boudali, Olga and Antonis Economou, "Optimal and equilibrium balking strategies in the single server queue with catastrophes," *European Journal of Operational Research* **218**(3) (2012) 708-715. (Cited on page 279.)

[101] Boudali, Olga and Antonis Economou, "The effect of catastrophes on the strategic customer behavior in queueing systems," *Naval Research Logistics* **60**(7) (2013) 571-587. (Cited on page 279.)

[102] Boudali, Olga and Antonis Economou, "Strategic behavior in batch service systems." manuscript (2015). (Cited on page 73.)

[103] Bourreau, Marc, Frago Kourandi, and Tommaso Valletti, "Net neutrality with competing Internet platforms," *The Journal of Industrial Economics* **LXIII**(1) (2015) 30-73. (Cited on page 258.)

[104] Boyaci, Tamer and Saibal Ray, "Product differentiation and capacity cost interaction in time and price sensitive markets," *Manufacturing & Service Operations Management* **5**(1) (2003) 18-36. (Cited on page 165, 166, 294, and 298.)

[105] Boyaci, Tamer and Saibal Ray, "The impact of capacity costs on product differentiation in delivery time, delivery reliability, and price," *Production and Operations Management* **15**(2) (2006) 179-197. (Cited on page 298.)

[106] Brandeau, Margaret L. and Samuel Chiu, "Location of competing facilities in a user-optimizing environment with market externalities," *Transportation Science* **28**(2) (1994) 125-140. (Cited on page 225.)

[107] Brandeau, Margaret L. and Samuel Chiu, "Facility location in a user-optimizing environment with market externalities: analysis of customer equilibria and optimal public facility locations," *Location Sciences* **2**(3) (1994) 129-147. (Cited on page 225.)

[108] Breinbjerg, Jesper, Alexander Sebald, and Lars Peter Østerdal, "Strategic behavior and social outcomes in a bottleneck queue: experimental evidence," (2014). (Cited on page 87.)

[109] Brooks, James D., "Equilibrium behavior in bipartite queueing networks," *ISERC* (2014). (Cited on page 99.)

[110] Brooms, Anthony C., "Individual equilibrium dynamic routing in a multiple server retrial queue," *Probability in the Engineering and Informational Sciences* **14** (2000) 9-26. (Cited on page 90.)

[111] Brooms, Anthony C., "On the Nash equilibria for the FCFS queueing system with load-increasing service rate," *Advances in Applied Probability* **37** (2005) 461-481. (Cited on page 24 and 228.)

[112] Brooms, Anthony and E.J. (Sean) Collins, "Stochastic order results and equilibrium joining rules for the Bernoulli feedback queue," (2013). (Cited on page 29.)

[113] Buche, Robert and Harold J. Kushner, "Stochastic approximation and user adaptation in a competitive resource sharing system," *IEEE Transactions on Automatic Control* **45**(5) (2000) 844-853. (Cited on page 97 and 98.)

[114] Burnetas, Apostolos, "Customer equilibrium and optimal strategies in Markovian queues in series," *Annals of Operations Research* **208** (2013) 515-529. (Cited on page 219.)

[115] Burnetas, Apostolos and Antonis Economou, "Equilibrium customer strategies in a single server Markovian queue with setup times," *Queueing Systems* **56** (2007) 213-228. (Cited on page 9, 264, 266, 268, 269, 270, 271, and 272.)

[116] Cachon Gérard P., "Competitive and cooperative inventory management in a two-echelon supply chain with lost sales," unpublished manuscript (1999). (Cited on page 232 and 236.)

[117] Cachon, Gérard P. and Pnina Feldman, "Pricing services subject to congestion: charge per-use fee or sell subscription?," *Manufacturing & Service Operations Management* **13**(2) (2011) 244-260. (Cited on page 43 and 162.)

[118] Cachon, Gérard P. and Patrick T. Harker, "Competition and outsourcing with scale economics," *Management Science* **48**(10) (2002) 1314-1333. (Cited on page 193, 194, 252, and 253.)

[119] Cachon, Gérard P. and Fuqiang Zhang, "Procuring fast delivery: sole sourcing with information asymmetry," *Management Science* **52**(6) (2006) 881-896. (Cited on page 43 and 234.)

[120] Cachon, Gérard P. and Fuqiang Zhang, "Obtaining fast service in a queueing system via performance-based allocation of demand," *Management Science* **53**(3) (2007) 408-420. (Cited on page 14, 245, 246, and 247.)

[121] Calbert, Greg, "Adaptive weighting of time costs in queueing systems," *IDC* (1999) 125-129. (Cited on page 104.)

[122] Caldentey, René and Lawrence M. Wein, "Analysis of a decentralized production-inventory system," *Manufacturing & Service Operations Management* **5**(1) (2003) 1-17. (Cited on page 233 and 234.)

[123] Caldentey, René and Lawrence M. Wein, "Revenue management of a make-to-stock queue," *Operations Research* **54**(5) (2006) 859-875. (Cited on page 160.)

[124] Callander, Steven and Johannes Hörner, "The wisdom of the minority," *Journal of Economic Theory* **144** (2009) 1421-1439.e2. (Cited on page 62.)

[125] Calvert, Bruce, "The Downs-Thomson effect in a Markov process," *Probability in the Engineering and Informational Science* **11** (1997) 327-340. (Cited on page 227, 228, 229, and 296.)

[126] Campos-Nánez, Enrique, Natalia Fabra, and Alfredo Garcia, "Dynamic auctions for on-demand services," *IEEE Transactions on Systems, Man, and Cybernetics - Part A: Systems and Humans* **37**(6) (2007) 878-886. (Cited on page 57.)

[127] Cao, Xi-Ren, Hong-Xia Shen, Rodolfo Milito, and Patrica Wirth, "Internet pricing with a game theoretical approach: concepts and examples," *IEEE/ACM Transactions on Networking* **10**(2) (2002) 208-216. (Cited on page 131 and 188.)

[128] Capar, Fatih and Friedrich Jondral, "Spectrum pricing for excess bandwidth in radio networks," *PIMRC* (2004) 2458-2462. (Cited on page 273.)

[129] Cardellini, Valeria, Vittoria de Nitto Personé, Valerio Di Valerio, Francisco Facchinei, Vincenzo Grassi, Francesco Lo Presti, and Veronica Piccialli, "A game-theoretic approach to computation offloading in mobile cloud computing," *Mathematical Programming, Series A*, forthcoming. (Cited on page 97 and 210.)

[130] Caro, Felipe and David Simchi-Levi, "Optimal static pricing for a tree network," *Annals of Operations Research* **196** (2012) 137-152. (Cited on page 220.)

[131] Çelik, Sabri and Costis Maglaras, "Dynamic pricing and leadtime quotation for a multi-class make-to-order queue," *Management Science* **54**(6) (2008) 1132-1146. (Cited on page 51 and 52.)

[132] Chakravorti, Bhaskar, "Optimal flow control of an M/M/1 queue with a balanced budget," *IEEE Transactions on Automatic Control* **39**(9) (1994) 1918-1921. (Cited on page 284.)

[133] Chang, Zheng, Tapani Ristaniemi, and Zhu Han, "Queueing game for spectrum access in cognitive radio networks," manuscript (2015). (Cited on page 279.)

[134] Chao, Xiuli, Liming Liu, and Shaohui Zheng, "Resource allocation in multisite service systems with intersite customer flows," *Management Science* **49**(12) (2003) 1739-1752. (Cited on page 215.)

[135] Chatterjee, Subimal, Susan A. Slotnick, and Matthew J. Sobel, "Delivery guarantees and the interdependence of marketing and operations," *Production and Operations Management* **11**(3) (2002) 393-410. (Cited on page 293 and 294.)

[136] Chau, Chi-Kin, Qian Wang, and Dah-Ming Chiu, "On the viability of Paris Metro Pricing for communication and service networks," *INFOCOM* (2010). (Cited on page 168.)

[137] Chayet, Sergio and Wallace J. Hopp, "Sequential entry with capacity, price, and leadtime competition," unpublished manuscript (2007). (Cited on page 201.)

[138] Chayet, Sergio, Panos Kouvelis, and Dennis Z. Yu, "Product variety and capacity investments in congested production systems," *Manufacturing & Service Operations Management* **13**(3) (2011) 390-403. (Cited on page 144.)

[139] Chen, Feng and Vidyadhar G. Kulkarni, "Admission control in the presence of priorities: a sample path approach," in *Stochastic Processes, Optimization, and Control Theory: Applications in Financial Engineering, Queueing Networks, and Manufacturing Systems*, H. Yan, G. Yin, and Q. Zhang (eds.), Springer US 2006. (Cited on page 24.)

[140] Chen, Feng and Vidyadhar G. Kulkarni, "Individual, class-based, and social optimal admission policies in two-priority queues," *Stochastic Models* **23** (2007) 97-127. (Cited on page 24.)

[141] [∗ ∗ ∗] Chen, Hong and Murray Frank, "Monopoly pricing when customers queue," *IIE Transactions* **36** (2004) 569–581. (Cited on page 59, 141, and 278.)

[142] Chen, Hong and Yat-wah Wan, "Price competition of make-to-order firms," *IIE Transactions* **35** (2003) 817-832. (Cited on page 188, 189, and 190.)

[143] Chen, Hong and Yat-wah Wan, "Capacity competition of make-to-order firms," *Operations Research Letters* **33** (2005) 187-194. (Cited on page 189 and 190.)

[144] Chen, Jian, Shuo Huang, Refael Hassin, and Nan Zhang, "Two back-order compensation mechanisms in inventory systems with impatient customers," *Production and Operations Management* **24**(10) (2015) 1640-1656. (Cited on page 44, 146, and 177.)

[145] Chen, Jian and Nan Zhang, "Customer incentives in time-based environment," *Service Enterprise Integration, Integrated Series in Information Systems* **16** (2007) 103-129. (Cited on page 212.)

[146] Chen, Jian, Nan Zhang, and Shuo Huang, "Optimal and incentive compatible pricing for heterogeneous periods," *Journal of Systems Science and Systems Engineering* **17**(1) (2008) 50-65. (Cited on page 212.)

[147] Chen, Liuxin, Youhua (Frank) Chen, and Zhan Peng, "Dynamic pricing and inventory control in a make-to-stock queue with information on the production status," *IEEE Transactions on Automation Science and Engineering* **8**(2) (2011) 361-373. (Cited on page 46.)

[148] Chen, Liuxin, Youyi Feng, Gang Hao, and Matthew F. Keblis, "Optimal pricing and sequencing in a make-to-order system with batch demand," *IEEE Transactions of Automatic Control* **60** (2015) 1455-1470. (Cited on page 48.)

[149] Chen, Liuxin, Youyi Feng, and Jihong Ou, "Joint management of finished goods inventory and demand process for a make-to-stock product: a computational approach," *IEEE Transactions on Automatic Control* **51**(2) (2006) 258-273. (Cited on page 45.)

[150] Chen, Liuxin, Youyi Feng, and Jihong Ou, "Coordinating batch production and pricing of a make-to-stock product," *IEEE Transactions on Automatic Control* **54**(7) (2009) 1674-1680. (Cited on page 46.)

[151] Chen, Liuxin, Gang Hao, and Huimin Wang, "Optimal control of a make-to-stock system with outsourced production and price-sensitive demand," *Discrete Dynamics in Nature and Society* (2014) 12 pages. (Cited on page 48.)

[152] Chen, Peishu, Wenhui Zhou, and Yongwu Zhou, "Equilibrium customer strategies in the queue with threshold policy and setup times," *Mathematical Problems in Engineering* (2015) 11 pages. (Cited on page 266.)

[153] Chen, Peishu and Yongwu Zhou, "Equilibrium balking strategies in the single server queue with setup times and breakdowns," *Operational Research: An International Journal* **15** (2015) 213-231. (Cited on page 269 and 278.)

[154] Chen, Xu, Allen H. Tai, and Yi Yang, "Optimal production and pricing policies in a combined make-to-order/make-to-stock system," *International Journal of Production Research* **52**(23) (2104) 7027-7045. (Cited on page 48.)

[155] Chen, Ying-Ju, Tingting Cui, Tianhu Deng, and Zuo-Jun Max Shen, "Revenue-maximizing pricing, scheduling, and probabilistic admission control for queueing systems under information asymmetry," manuscript (2015). (Cited on page 173.)

[156] Chen, Yizheng, Mark Holmes, and Ilze Ziedins, "Monotonicity properties of user equilibrium policies for parallel batch systems," *Queueing Systems* **70** (2012) 81-103. (Cited on page 229.)

[157] Cheng, Hsing Kenneth, "Optimal internal pricing and backup capacity of computer systems subject to breakdowns," *Decision Support Systems* **19** (1997) 93-108. (Cited on page 273.)

[158] Cheng, Hsing Kenneth, "Pricing and capacity decisions of clustered twin-computer systems subject to breakdowns," *Decision Support Systems* **25** (1999) 19-37. (Cited on page 273.)

[159] Cheng, Hsing Kenneth, Subhajyoti Bandyopadhyay, and Hong Guo, "The debate on net neutrality: a policy perspective," *Information Systems Research* **22**(1) (2011) 60-82. (Cited on page 178 and 255.)

[160] Cheng, Hsing Kenneth, Haluk Demirkan, and Gary J. Koehler, "Price and capacity competition of application services duopoly," *Information Systems and e-Business Management* **1**(3) (2003) 305-329. (Cited on page 189.)

[161] Cheng, Hsing Kenneth and Gary J. Koehler, "Optimal pricing policies of web-enabled application services," *Decision Support Systems* **35** (2003) 259-272. (Cited on page 55 and 160.)

[162] Ching, Wai-Ki, Sin-Man Choi, and Min Huang, "Optimal service capacity in a multiple-server queueing system: a game theory approach," *Journal of Industrial and Management Optimization* **6**(1) (2010) 73-102. (Cited on page 247.)

[163] Ching, Wai-Ki, Sin-Man Choi, and Min Huang, "Inducing high service capacities in outsourcing via penalty and competition," *International Journal of Production Research* **49**(17) (2011) 5169-5182. (Cited on page 247.)

[164] Ching, Wai-Ki, Sin-Man Choi, Tang Li, and Issic K.C. Leung, "A tandem queueing system with applications to pricing strategy." *Journal of Industrial and Management Optimization* **5**(1) (2009) 103-114. (Cited on page 220.)

[165] Choi, Jay Pil and Byung-Cheol Kim, "Net neutrality and investment incentives," *RAND Journal of Economics* **41**(3) (2010) 446-471. (Cited on page 255 and 256.)

[166] Choi, Sin-Man, Ximin Huang, and Wai-Ki Ching, "Minimizing equilibrium expected sojourn time via performance-based mixed threshold demand allocation in a multiple-server queueing environment," *Journal of Industrial and Management Optimization* **8**(2) (2012) 299-323. (Cited on page 246.)

[167] Choi, Sin-Man, Ximin Huang, Wai-Ki Ching, and Min Huang, "Incentive effects of multiple-server queueing networks: the principal-agent perspective," *East Asian Journal on Applied Mathematics* **1**(4) (2011) 379-402. (Cited on page 247.)

[168] Christ, Duane and Benjamin Avi-Itzhak, "Strategic equilibrium for a pair of competing servers with convex cost and balking," *Management Science* **48**(6) (2002) 813-820. (Cited on page 31, 32, 38, 74, 125, and 200.)

[169] Çil, Eren Başar, Fikri Karaesmen, and E. Lerzan Örmeci, "Dynamic pricing and scheduling in a multi-class single-server queueing system," *Queueing Systems* **67** (2011) 305-331. (Cited on page 43 and 47.)

[170] Coffman, Edward G. Jr. and Isi Mitrani, "A characterization of waiting time performance realizable by single-server queues," *Operations Research* **28**(3) (1980) 810-821. (Cited on page 15.)

[171] Cohen, Joel E. and Clark Jeffries, "Congestion resulting from increased capacity in single-server queueing networks," *IEEE/ACM Transactions on Networking* **5**(2) (1997) 305-310. (Cited on page 226.)

[172] Conforto, Paolo, Francesco Delli Priscoli, and Francisco Facchinei, "Existence and uniqueness of the Nash equilibrium in the non-cooperative QoS routing," *International Journal of Control* **83**(4) (2010) 776-788. (Cited on page 215.)

[173] Conte, Anna, M. Scarsini, and Oktay Sürücü, "Does time pressure impair performance? an experiment on queueing behavior," manuscript (2015). (Cited on page 17.)

[174] Courcoubetis, Costas and Antonis Dimakis, "Tariffs, mechanisms and equilibria at a single Internet link," *Network Control and Optimization* Lecture Notes in Computer Science, **5894** (2009) 234-248, (Cited on page 114.)

[175] Courcoubetis, Costas and Pravin Varaiya, "A game-theoretic view of two processes using a single resource," *IEEE Transactions on Automatic Control* **AC-28**(11) (1983) 1059-1061. (Cited on page 129 and 155.)

[176] Cripps, Martin W. and Caroline D. Thomas, "Strategic experimentation in queues," manuscript (2014). (Cited on page 28 and 29.)

[177] Cui, Shiliang, Xuanming Su, and Senthil Veeraraghavan, "A model of rational retrials in queues," (2014). (Cited on page 91.)

[178] Cui, Shiliang and Senthil K. Veeraraghavan, "Blind queues: the impact of consumer beliefs on revenues and congestion," *Management Science* (2015). (Cited on page 76.)

[179] Czumaj, Artur, Piotr Krysta, and Berthold Vöcking, "Selfish traffic allocation for server farms," *SIAM Journal on Computing* **39**(5) (2010) 1957-1987. (Cited on page 135.)

[180] Dai, Tinglong, Mustafa Akan, and Sridhar Tayur, "Imaging room and beyond: the underlying economics behind physicians' test-ordering behavior in outpatient services," manuscript (2012). (Cited on page 157.)

[181] Dai, Tinglong, Katia Sycara, and Michael Lewis, "A game theoretic queueing approach to self-assessment in human-robot interaction systems," *IEEE International Conference on Robotics and Automation* (2011) 58-63. (Cited on page 156.)

[182] D'Auria, Bernardo and Spyridoula Kanta, "Pure threshold strategies for a two-node tandem network under partial information," *Operations Research Letters* **43** (2015) 467-470. (Cited on page 9 and 219.)

[183] Debo, Laurens G., Refael Hassin, and Senthil K. Veeraraghavan, "Learning quality through service outcomes," manuscript (2012). (Cited on page 29.)

[184] Debo, Laurens G., Christine A. Parlour, and Uday Rajan, "Signaling quality via queues," *Management Science* **58**(5)(2012) 876-891. (Cited on page 63, 64, and 65.)

[185] Debo, Laurens G., Uday Rajan, and Senthil K. Veeraraghavan, "Signaling with prices in a congested environment," manuscript (2012). (Cited on page 64.)

[186] Debo, Laurens G., L. Beril Toktay, and Luk N. van Wassenhove, "Queueing for expert services," *Management Science* **54**(8) (2008) 1497-1512. (Cited on page 154.)

[187] Debo, Laurens G. and Senthil K. Veeraraghavan, "Models of herding behavior in operations management," *Consumer-Driven Demand and Operations Management Models*, International Series in Operations Research & Management Science **131**, S. Netessine and C.S. Tang (eds.), 2009. (Cited on page 63.)

[188] Debo, Laurens G. and Senthil K. Veeraraghavan, "Equilibrium in queues under unknown service times and service value," *Operations Research* **62**(1) (2014) 38-57. (Cited on page 2, 75, 154, and 155.)

[189] Deck, Cary, Erik O. Kimbrough, and Steeve Mongrain "Paying for express checkout: competition and price discrimination in multi-server queueing systems," *PLOS ONE* **9**(3) (2014) (2012), 13 pages. (Cited on page 32.)

[190] Dehghanian, Amin, Jeffrey P. Kharoufeh, and Mohammad Modarres, "Strategic dynamic jockeying between two parallel queues," *Probability in the Engineering and Informational Sciences* forthcoming. (Cited on page 97.)

[191] Delgado, Carlos A., Ann van Ackere, and Erik R. Larsen, "A queuing system with risk-averse customers: sensitivity analysis of performance," *IEEM* (2011). (Cited on page 205.)

[192] Dellaert, Nico P., "Due-date setting and production control," *International Journal of Production Economics* **23**(1-3) (1991) 59-67. (Cited on page 44, 265, 292, and 294.)

[193] Demirkan, Haluk, Hsing Kenneth Cheng, and Subhajyoti Bandyopadhyay, "Coordination strategies in an SaaS supply chain," *Journal of Management Information Systems* **26**(4) (2010) 119-143. (Cited on page 255.)

[194] Deng, Tianhu, Ying-Ju Chen, and Zuo-Jun Max Shen, "Optimal pricing and scheduling control of product shipping," *Naval Research Logistics* **62** (2015) 215-227. (Cited on page 172.)

[195] Deo, Sarang and Itai Gurvich, "Centralized vs. decentralized ambulance diversion: a network perspective," *Management Science* **57**(7) (2011) 1300-1319. (Cited on page 41 and 198.)

[196] Dimitrakopoulos, Yiannis and Apostolos Burnetas, "Customer equilibrium and optimal strategies in an M/M/1 queue with dynamic service control," (2011) (Cited on page 96 and 264.)

[197] DiPalantino, Dominic, Ramesh Johari, and Gabriel Y. Weintraub, "Competition and contracting in service industries," *Operations Research Letters* **39** (2011) 390-396. (Cited on page 194.)

[198] Do, Cuong T., Choong Seon Hong, and Jinpyo Hong, "Pricing control for hybrid overlay/underlay spectrum access in cognitive radio networks," *APNOMS* (2012). (Cited on page 274.)

[199] Do, Cuong T., Nguyen H. Tran, Zhu Han, Long Bao Le, Sungwon Lee, and Choong Seon Hong, "Optimal pricing for duopoly in cognitive radio networks: cooperate or not cooperate?" *IEEE Transactions on Wireless Communications* **13**(5) (2014) 2574-2587. (Cited on page 130, 192, 216, 272, and 278.)

[200] Do, Cuong T., Nguyen H. Tran, Choong Seon Hong, and Sungwon Lee, "Finding an individual optimal threshold of queue length in hybrid overlay/underlay spectrum access in cognitive radio networks," *IEICE Transactions on Communications* **E95-B**(6) (2012) 1978-1981. (Cited on page 274.)

[201] Do, Cuong T., Nguyen H. Tran, Mui Van Nguyen, Choong Seon Hong, and Sungwon Lee, "Social optimization strategy in unobserved queueing systems in cognitive radio networks," *IEEE Communications letters* **16**(12) (2012) 1944-1947. (Cited on page 274.)

[202] Do, Cuong T., Nguyen H. Tran, Dai Hoang Tran, Chuan Pham, Md. Golam Rabiul Alam, and Choong Seon Hong, "Toward service selection game in a heterogeneous market cloud computing," *IFIP* (2015). (Cited on page 192 and 216.)

[203] Dobson, Gregory and Edieal J. Pinker, "The value of sharing lead time information," *IIE Transactions* **38** (2006) 171-183. (Cited on page 69.)

[204] Dobson, Gregory and Euthemia Stavrulaki, "Simultaneous price, location, and capacity decisions on a line of time-sensitive customers," *Naval Research Logistics* **54**(1) (2007) 1-10. (Cited on page 180 and 202.)

[205] Doncel, Josu, Urtzi Ayesta, Olivier Brun, and Balakrishna J. Prabhu, "A resource-sharing game with relative priorities," *Performance Evaluation* **79** (2014) 287-305. (Cited on page 105 and 116.)

[206] Doncel, Josu, Urtzi Ayesta, Olivier Brun, and Balakrishna J. Prabhu, "Is the price of anarchy the right measure for load balancing games?" *ACM Transactions on Internet Technology* **14**(2-3) (2014) Article 18, 20 pages. (Cited on page 135.)

[207] Douligeris, Christos, and Ravi Mazumdar, "A game theoretic perspective to flow control in telecommunication networks," *Journal of the Franklin Institute* **329**(2) (1992) 383-402. (Cited on page 106 and 107.)

[208] Dube, Parijat and Rahul Jain, "Bertrand equilibria and efficiency in markets for congestible network services," *Automatica* **50** (2014) 756-767. (Cited on page 247.)

[209] Duenyas, Izak, "Single facility due date setting with multiple customer classes," *Management Science* **41**(4) (1995) 608-619. (Cited on page 295.)

[210] Duenyas, Izak and Wallace J. Hopp, "Quoting customer lead time," *Management Science* **41**(1) (1995) 43-57. (Cited on page 295.)

[211] Dutta, Debojyoti, Ashish Goel, and John Heidemann, "Oblivious AQM and Nash equilibria," *INFOCOM* (2003). (Cited on page 106.)

[212] Dziong, Zbigniew and Lorne G. Mason, "Fair-efficient call admission control policies for broadband networks: a game theoretic framework," *IEEE/ACM Transactions on Networking* **4**(1) (1996) 123-136. (Cited on page 131.)

[213] Economides, Anastasios A. and John A. Silvester, "Multi-objective routing in integrated services networks: a game theory approach," *INFOCOM* (1991). (Cited on page 208.)

[214] Economopoulos, Angelos A., Vassilis S. Kouikoglou, and Evangelos Grigoroudis, "The base stock/base backlog control policy for a make-to-stock system with impatient customers," *IEEE Transactions on Automation Science and Engineering* **8**(1) (2011) 243-249. (Cited on page 267.)

[215] Economou, Antonis, Antonio Gómez-Corral, and Spyridoula Kanta, "Optimal balking strategies in single-server queues with general service and vacation times," *Performance Evaluation* **68** (2011) 967-982. (Cited on page 26 and 270.)

[216] Economou, Antonis and Athanasia Manou, "Equilibrium balking strategies for a clearing queueing system in alternating environment, *Annals of Operations Research* **208** (2013) 489-514. (Cited on page 74 and 280.)

[217] Economou, Antonis and Athanasia Manou, "Strategic behavior in an observable fluid queue with alternating service process," *European Journal of Operational Research* forthcoming. (Cited on page 74 and 276.)

[218] Economou, Antonis and Spyridoula Kanta, "Optimal balking strategies and pricing for the single server Markovian queue with compartmented waiting space," *Queueing Systems* **59** (2008) 237-269. (Cited on page 9, 33, 70, 71, 73, and 280.)

[219] Economou, Antonis and Spyridoula Kanta, "Equilibrium balking strategies in the observable single-server queue with breakdowns and repairs," *Operations Research Letters* **36** (2008) 696-699. (Cited on page 273, 274, and 275.)

[220] Economou, Antonis and Spyridoula Kanta, "Equilibrium customer strategies and social-profit maximization in the single-server constant retrial queue," *Naval Research Logistics* **58**(2) (2011) 107-122. (Cited on page 183.)

[221] [∗ ∗ ∗] Edelson, Noel M. and David K. Hildebrand "Congestion tolls for Poisson queueing processes," *Econometrica* **43** (1975) 81-92. (Cited on page 5, 59, and 270.)

[222] Elahi, Ehsan, "Outsourcing through competition: what is the best competition parameter?" *International Journal of Production Economics* **144** (2013) 370-382. (Cited on page 249 and 290.)

[223] El Azouzi, Rachid and Eitan Altman, "Constrained traffic equilibrium in routing," *IEEE Transactions on Automatic Control* **48**(9) (2003) 1656-1660. (Cited on page 209.)

[224] El Haji, Anouar and Sander Onderstal, "Trading places: an experimental comparison of reallocation mechanisms for priority queuing," manuscript (2015). (Cited on page 138.)

[225] El-Zoghdy, Said Fathy, Hisao Kameda, and Jie Li, "Numerical studies on a paradox for non-cooperative static load balancing in distributed computer systems," *Computers & Operations Research* **33** (2006) 345-355. (Cited on page 224 and 227.)

[226] Enders, Paul, Anshul Gandhi, Varun Gupta, Laurens G. Debo, Mor Harchol-Balter, and Alan Scheller-Wolf, "Inducing optimal scheduling with selfish users," (2010). (Cited on page 65 and 130.)

[227] Engel, Roei and Refael Hassin, "Customer equilibrium in a single-server system with virtual and system queues," (2016). (Cited on page 31.)

[228] Erkoc, Murat, S. David Wu, and Haresh Gurnani, "Delivery-date and capacity management in a decentralized internal market, *Naval Research Logistics* **55**(5) (2008) 390-405. (Cited on page 49, 120, and 292.)

[229] Erlichman, Jenny, and Refael Hassin, "Strategic overtaking in a monopolistic observable M/M/1 queue," *IEEE Transactions on Automatic Control* **60**(8) (2015). (Cited on page 11, 43, 96, 102, and 139.)

[230] Ernez-Gahbiche, Ibtissem, Khaled Hadjyoussef, Abdelwaheb Dogui, and Zied Jemaï, "Stackelberg game between a customer and cooperating suppliers: stability and efficiency," *MOSIM* (2014). (Cited on page 128.)

[231] Fan, Ming, Subodha Kumar, and Andrew B. Whinston, "Short-term and long-term competition between providers of shrink-wrap software and software as a service," *European Journal of Operational Research* **196** (2009) 661-671. (Cited on page 77 and 191.)

[232] Fathian, Mohammad, M. Narenji, Ebrahim Teimoury, and Seyed Gholamreze Jalali Naini, "Applying Stackelberg game to find the best price and delivery time policies in competition between two supply chains," *International Journal of Management and Business Research* **2**(4) (2012) 313-327. (Cited on page 255.)

[233] Federgruen, Awi, and Harry Groenevelt, "Characterization and optimization of achievable performance in general queueing systems," *Operations Research* **36**(5) (1988) 733-741. (Cited on page 15.)

[234] Feng, Jiejian, Liming Liu, and Xiaoming Liu, "An optimal policy for joint dynamic price and leadtime quotation," *Operations Research* **59**(6) (2011) 1523-1527. (Cited on page 295.)

[235] Feng, Youyi, Jihong Ou, and Zhan Pang, "Optimal control of price and production in an assemble-to-order system," *Operations Research Letters* **36** (2008) 506-512. (Cited on page 46.)

[236] Filippini, Ilario, Matteo Cesana, and Ilaria Malanchini, "Competitive spectrum sharing in cognitive radio networks: a queuing theory based analysis," *BlackSeaCom* (2013).(Cited on page 213.)

[237] Filliger, Roger and Max-Olivier Hongler, "Syphon dynamics: a soluble model of multi-agents cooperative behavior," *Europhysics Letters* **70**(3) (2005) 285-291. (Cited on page 205 and 206.)

[238] Friedman, Eric J., "Genericity and congestion control in selfish routing," *CDC* (2004). (Cited on page 133, 135, and 151.)

[239] Fung, Kwok-Kwan, "It is not how long it is, but how you make it long: waiting lines in a multi-step service process," *System Dynamics Review* **17**(4) (2001) 333-340. (Cited on page 9 and 10.)

[240] Gai, Yi, Hua Liu, and Bhaskar Krishnamachari, "A packet dropping mechanism for efficient operation of M/M/1 queues with selfish users," *INFOCOM* (2011). (Cited on page 107.)

[241] Gallay, Olivier and Max-Olivier Hongler, "Market sharing dynamics between two service providers," *European Journal of Operational Research* **190** (2008) 241-254. (Cited on page 203.)

[242] Gallay, Olivier and Max-Olivier Hongler, "Circulation of autonomous agents in production and service networks," *International Journal of Production Economics* **120** (2009) 378-388. (Cited on page 9, 205, and 206.)

[243] Gallego, Guillermo, Woonghee Tim Huh, Wanmo Kang, and Robert Phillips, "Price competition with the attraction demand model: existence of unique equilibrium and its stability," *Manufacturing & Service Operations Management* **8**(4) (2006) 359-375. (Cited on page 7 and 288.)

[244] Ganesh, Ayalvadi, Sarah Lilienthal, D. Manjunath, Alexandre Proutiere, and Florian Simatos, "Load balancing via random local search in closed and open systems," *Queueing Systems* **71** (2012) 321-345. (Cited on page 97.)

[245] Gans, Noah, Ger Koole, and Avishai Mandelbaum, "Telephone call centers: tutorial, review, and research prospects," *Manufacturing & Service Operations Management* **5**(2) (2003) 79-141. (Cited on page 232.)

[246] García-Sanz, María D., Francisco R. Fernández, M. Gloria Fiestras-Janeiro, Ignacio García-Jurado, and Justo Puerto, "Cooperation in Markovian queueing models," *European Journal of Operational Research* **188** (2008) 485-495. (Cited on page 126 and 129.)

[247] Gavirneni, Srinagesh and Vidyadhar G. Kulkarni, "Self-selecting priority queues with Burr distributed waiting costs," *Production and Operations Management* forthcoming. (Cited on page 113.)

[248] Gayon, Jean-Philippe, Işilay Talay-Değirmenci, Fikri Karaesmen, and E. Lerzan Örmeci, "Optimal pricing and production policies of a make-to-stock system with fluctuating demand," *Probability in the Engineering and Informational Sciences* **23** (2009) 205-230. (Cited on page 36 and 77.)

[249] Gershkov, Alex and Paul Schweinzer, "When queueing is better than push and shove," *International Journal of Game Theory* **39** (2010) 409-430. (Cited on page 137.)

[250] Ghosh, Arka, Sarah M. Ryan, Lizhi Wang, and Ananda Weerasinghe, "Heavy traffic analysis of a simple closed-loop supply chain," *Stochastic Models* **26** (2010) 549-593. (Cited on page 151.)

[251] Giebelhausen, Michael D., Stacey G. Robinson, and J. Joseph Cronin Jr., "Worth waiting for: increasing satisfaction by making consumers wait," *Journal of the Academy of Marketing Science* **39** (2011) 889-905. (Cited on page 61.)

[252] Gilbert, Stephen M. and Z. Kevin Weng, "Incentive effects favor nonconsolidating queues in a service system: the principal-agent perspective," *Management Science* **44**(12) (1998) 1662-1669. (Cited on page 244, 246, 247, and 283.)

[253] Gilboa-Freedman, Gail and Refael Hassin "Regulation under partial information: the case of a queueing system," *Operations Research Letters* **42** (2014) 217-221. (Cited on page 35, 43, 79, and 223.)

[254] Gilboa-Freedman, Gail, Refael Hassin, and Yoav Kerner, "The price of anarchy in the Markovian single server queue," *IEEE Transactions on Automatic Control* **59**(2) (2014) 455-459. (Cited on page 27 and 28.)

[255] Gilland, Wendell G. and Donald P. Warsing, "The impact of revenue-maximizing priority pricing on customer delay costs," *Decision Sciences* **40**(1) (2009) 89-120. (Cited on page 170.)

[256] Giloni, Avi, Yaşar Levent Koçağa, and Phil Troy, "State dependent pricing policies: differentiating customers through valuations and waiting costs," *Journal of Revenue and Pricing Management* **12** (2013) 139-161. (Cited on page 53.)

[257] Glazer, Amihai and Refael Hassin, "The economics of cheating in the taxi market," *Transportation Research Part A* **17A**(1) (1983) 25-31. (Cited on page 153.)

[258] Gong, Xiting, Qiwen Wang, Honghui Deng, Nagesh N. Murthy, and Gangshu Cai, "Characterizing performance-based demand allocation policies to stimulate supplier capacities," *ICOSCM* (2010). (Cited on page 247.)

[259] González, Paula and Carmen Herrero, "Optimal sharing of surgical costs in the presence of queues," *Mathematical Methods of Operations Research* **59** (2004) 435-446. (Cited on page 125, 126, and 129.)

[260] Gopalakrishnan, Ragavendran, Sherwin Doroudi, Amy R. Ward, and Adam Wierman, "Routing and staffing when servers are strategic," manuscript (2014). (Cited on page 249.)

[261] Grossman, Thomas A. Jr. and Margaret L. Brandeau, "Optimal pricing for service facilities with self-optimizing customers," *European Journal of Operational Research* **141** (2002) 39-57. (Cited on page 225.)

[262] Grosu, Daniel and Anthony T. Chronopoulos, "Algorithmic mechanism design for load balancing in distributed systems," *IEEE Transactions on Systems, Man, and Cybernetics, Part B: Cybernetics* **34** (2004) 77-84. (Cited on page 245 and 246.)

[263] Grosu, Danial, Anthony T. Chronopoulos, and Ming-Ying Leung, "Co-operative load balancing in distributed systems," *Concurrency and Computation: Practice and Experience* **20** (2008) 1953-1976. (Cited on page 132.)

[264] Gui, Huaming and Shihua Ma, "Coordination tactic of downstream capacity expansion in supply chain," *WiCom* (2007). (Cited on page 235.)

[265] Guijarro, Luis, Vicent Pla, and Bruno Tuffin, "Entry game opportunistic access in cognitive radio networks: a priority queue model," *IFIP - WD* (2013). (Cited on page 169 and 200.)

[266] Güler, M. Güray, Taner Bilgiç, and Refik Güllü, "Joint inventory and pricing decisions when customers are delay sensitive," *International Journal of Production Economics* **157** (2014) 302-312. (Cited on page 171.)

[267] Güneş, Evrim D. and O. Zeynep Akşin, "Value creation in service delivery: relating market segmentation, incentives, and operational performance," *Manufacturing & Service Operations Management* **6**(4) (2004) 338-357. (Cited on page 242.)

[268] Guo, Chao, Haitao Xu, and Zhiyong Yao, "The optimal admission fee based congestion control scheme in space information networks," *Journal of Information & Computational Science* **12**(6) (2015) 2295-2303. (Cited on page 2.)

[269] Guo, Hong, Hsing Kenneth Cheng, and Subhajyoti Bandyopadhyay, "Net neutrality, broadband market coverage, and innovation at the edge," *Decision Sciences* **43**(1) (2012) 141-172. (Cited on page 258.)

[270] Guo, Hong, Hsing Kenneth Cheng, and Subhajyoti Bandyopadhyay, "Broadband network management and the net neutrality debate," *Production and Operations Management* **22**(5) (2013) 1287-1298. (Cited on page 258.)

[271] Guo, Hong, Hsing Kenneth Cheng, Subhajyoti Bandyopadhyay, and Yu-chen Yang, "Net neutrality and vertical integration of content and broadband services," *Journal of Management Information Systems* **27**(2) (2010) 243-275. (Cited on page 256.)

[272] Guo, Pengfei and Refael Hassin, "Strategic behavior and social optimization in Markovian vacation queues," *Operations Research* **59**(4) (2011) 986-997. (Cited on page 73, 265, 266, 267, 269, and 274.)

[273] Guo, Pengfei and Refael Hassin, "Strategic behavior and social optimization in Markovian vacation queues: the case of heterogeneous customers," *European Journal of Operational Research* **222** (2012) 278-286. (Cited on page 266.)

[274] Guo, Pengfei and Refael Hassin, "On the advantage of leadership in service pricing competition," *Operations Research Letters* **41** (2013) 397-402. (Cited on page 202.)

[275] Guo, Pengfei and Refael Hassin, "Equilibrium strategies for placing duplicate orders in a single server queue," *Operations Research Letters* **43**(3) (2015) 343-348. (Cited on page 104 and 116.)

[276] Guo, Pengfei and Refael Hassin, "Why customers place duplicate orders in queues and why firms allow such behavior," (2015). (Cited on page 105 and 187.)

[277] Guo, Pengfei, Moshe Haviv, and Yulan Wang, "To queue or not to queue? A case where the service quality is unknown to some customers," (2014). (Cited on page 65 and 68.)

[278] Guo, Pengfei and Qingying Li, "Strategic behavior and social optimization in partially-observable Markovian vacation queues," *Operations Research Letters* **41** (2013) 277-284. (Cited on page 267.)

[279] Guo, Pengfei, Robin Lindsey, and Qu Qian, "Efficiency of subsidy schemes in reducing waiting times for public health-care services," manuscript (2015). (Cited on page 123.)

[280] Guo, Pengfei, Robin Lindsey, and Zhe George Zhang, "On the Downs-Thomson paradox in a self-financing two-tier queueing system," *Manufacturing & Service Operations Management* **16**(2) (2014) 315-322. (Cited on page 124.)

[281] Guo, Pengfei, John J. Liu, and Yulan Wang, "Intertemporal service pricing with strategic customers," *Operations Research Letters* **37** (2009) 420-424. (Cited on page 149.)

[282] Guo, Pengfei, Wei Sun and Yulan Wang, "Equilibrium and optimal strategies to join a queue with partial information on service times," *European Journal of Operational Research* **214** (2011) 284-297. (Cited on page 75.)

[283] Guo, Pengfei and Zhe George Zhang, "Strategic queueing behavior and its impact on system performance in service systems with the congestion-based staffing policy" *Manufacturing & Service Operations Management* **15**(1) (2013) 118-131. (Cited on page 122 and 266.)

[284] Guo, Pengfei, and Paul Zipkin, "Analysis and comparison of queues with different levels of delay information," *Management Science* **53**(6) (2007) 962-970. (Cited on page 9, 69, and 70.)

[285] Guo, Pengfei, and Paul Zipkin, "The effects of information on a queue with balking and phase-type service time," *Naval Research Logistics* **55** (2008) 406-411. (Cited on page 69 and 70.)

[286] Guo, Pengfei, and Paul Zipkin, "The effects of the availability of waiting-time information on the balking queue," *European Journal of Operational Research* **198** (2009) 199-209. (Cited on page 69 and 70.)

[287] Guo, Pengfei, and Paul Zipkin, "The impacts of customers' delay-risk sensitivities on a queue with balking," *Probability in the Engineering and Informational Sciences* **23** (2009) 409-432. (Cited on page 69 and 70.)

[288] Guo, Xianping and Onésimo Hernández-Lerma, "Zero-sum continuous-time Markov games with unbounded transition and discounted payoff rates," *Bernoulli* **11**(6) (2005) 1009-1029. (Cited on page 260.)

[289] Gupta, Alok, Boris Jukic, Dale O. Stahl, and Andrew B. Whinston, "An analysis of incentives for network infrastructure investment under different pricing strategies," *Information Systems Research* **22**(2) (2011) 215-232. (Cited on page 212.)

[290] Gupta, Diwakar and Waressara Weerawat, "Supplier-manufacturer co-ordination in capacitated two-stage supply chain," *European Journal of Operational Research* **175** (2006) 67-89. (Cited on page 218 and 234.)

[291] Gupta, Manu K., N. Hemachandra, Bharat Singh Raghav, and J. Venkateswaran, "On a conjecture and performance of a two class delay dependent priority queue," manuscript (2014). (Cited on page 176.)

[292] Gupta, Manu K., N. Hemachandra, and J. Venkateswaran, "Optimal revenue management in two class pre-emptive delay dependent Markovian queues," manuscript (2015). (Cited on page 176.)

[293] Gupta, Manu K., N. Hemachandra, and J. Venkateswaran, "A proof of conjecture arising from joint pricing and scheduling problem," manuscript (2015). (Cited on page 176.)

[294] Gurvich, Itai, Mor Armony, and Constantinos Maglaras, "Cross-selling in a call center with a heterogeneous customer population," *Operations Research* **57**(2) (2009) 299-313. (Cited on page 243.)

[295] Haight, Frank A., "Queueing with reneging," *Metrika* **2** (1959) 186-197. (Cited on page 299.)

[296] Haji, Rasoul and Gordon F. Newell, "Optimal strategies for priority queues with nonlinear costs of delays," *Journal on Applied Mathematics* **20**(2) (1971) 224-240. (Cited on page 14.)

[297] Halfin, Shlomo and Ward Whitt, "Heavy-traffic limits for queues with many exponential servers," *Operations Research* **29**(3) (1981) 567-588. (Cited on page 15.)

[298] Hall, Joseph M., Praveen K. Kopalle, and David F. Pyke, "Static and dynamic pricing of excess capacity in a make-to-order environment," *Production and Operations Management* **18**(4) (2009) 411-425. (Cited on page 43 and 162.)

[299] Hall, Joseph M. and Evan Porteus, "Customer service competition in capacitated systems," *Manufacturing & Service Operations Management* **2**(2) (2000) 144-165. (Cited on page 44 and 205.)

[300] Harubi, Nava, Mordechai Shechter, and Abraham Subotnik, "A pricing policy for a computer facility with congestion externalities," *European Journal of Operational Research* **3** (1979) 296-307. (Cited on page 49 and 199.)

[301] Hasija, Sameer, Edieal J. Pinker, and Robert A Shumsky, "Call center outsourcing contracts under information asymmetry," *Management Science* **54**(4) (2008) 793-807. (Cited on page 239.)

[302] [∗ ∗ ∗] Hassin, Refael, "On the optimality of first-come last-served queues," *Econometrica* **53** (1985), 201-202. (Cited on page 33, 35, 36, 37, 88, 91, 140, and 168.)

[303] [∗∗∗] Hassin Refael, "Decentralized regulation of a queue," *Management Science* **41**(1) (1995) 163-173. (Cited on page 114, 137, and 177.)

[304] [∗ ∗ ∗] Hassin, Refael, "On the advantage of being the first server," *Management Science* **42**(4) (1996), 618-623. (Cited on page 9 and 206.)

[305] Hassin, Refael, "Information and uncertainty in a queueing system," *Probability in the Engineering and Informational Sciences* **21** (2007) 361-380. (Cited on page 74 and 75.)

[306] Hassin, Refael, "Equilibrium customers' choice between FCFS and random servers," *Queueing Systems* **62** (2009) 243-254. (Cited on page 13, 216, and 217.)

[307] [∗ ∗ ∗] Hassin, Refael and Moshe Haviv, "Equilibrium threshold strategies: the case of queues with priorities," *Operations Research* **45** (1997) 966-973. (Cited on page 12, 30, and 105.)

[308] [∗ ∗ ∗] Hassin, Refael and Moshe Haviv, "Nash equilibrium and subgame perfection in observable queues," *Annals of Operations Research* **113** (2002) 15-26. (Cited on page 24, 97, 228, and 296.)

[309] Hassin, Refael and Moshe Haviv, "Who should be given priority in a queue?" *Operations Research Letters* **34** (2006) 191-198. (Cited on page 16, 115, and 116.)

[310] Hassin, Refael and Yana Kleiner, "Equilibrium and optimal arrival patterns to a server with opening and closing times," *IIE Transactions* **43** (2011) 164-175. (Cited on page 11, 85, 88, and 89.)

[311] Hassin, Refael and Alexandra Koshman, "Optimal control of a queue with high-low delay announcements: the significance of a queue," *Value-tools* (2014). (Cited on page 28, 43, 72, 139, 140, 145, 152, and 162.)

[312] Hassin, Refael, Justo Puerto, and Francisco R. Fernández, "The use of relative priorities in optimizing the performance of a queueing system," *European Journal of Operational Research* **193** (2009) 476-483. (Cited on page 15 and 16.)

[313] Hassin, Refael and Ricky Roet-Green, "Equilibrium in a two dimensional queueing game: when inspecting the queue is costly," (2014). (Cited on page 13 and 67.)

[314] Hassin, Refael and Ricky Roet-Green, "The impact of inspection costs on equilibrium in a queueing system with parallel servers," (2014) (Cited on page 68 and 89.)

[315] Hassin, Refael, Yair Y. Shaki, and Uri Yovel, "Optimal service-capacity allocation in a loss system," *Naval Research Logistics* **62**(2) (2015) 81-97. (Cited on page 33, 42, and 137.)

[316] Hassin, Refael and Ran Snitkovsky, "Strategic sensing in cognitive radio networks," (2016). (Cited on page 68.)

[317] Hatoum, Karim W. and Yih-Long Chang, "Trade-off between quoted lead time and price," *Production Planning & Control* **8** (1997) 158-172. (Cited on page 292.)

[318] Haung, Yieh-Ran and Sheng-Yung Su, "An access-aware pricing strategy for the evolved packet system," *Wireless Networks* **21** (2015) 371-385. (Cited on page 142.)

[319] Haviv, Moshe, "The Aumann-Shapley price mechanism for allocating congestion costs," *Operations Research Letters* **29** (2001) 211-215. (Cited on page 125.)

[320] Haviv, Moshe, "Strategic customer behavior in a single server queue," *Wiley Encyclopedia of Operations Research and Management science*, 2011. (Cited on page 3.)

[321] Haviv, Moshe, "When to arrive at a queue with tardiness costs?" *Performance Evaluation* **70** (2013) 387-399. (Cited on page 85 and 87.)

[322] Haviv, Moshe, "Regulating an M/G/1 queue when customers know their demand," *Performance Evaluation* **77** (2014) 57-71. (Cited on page 12, 66, and 294.)

[323] Haviv, Moshe, Offer Kella, and Yoav Kerner, "Equilibrium strategies in queues based on time or index of arrival," *Probability in the Engineering and Informational Sciences* **24** (2010) 13-25. (Cited on page 89.)

[324] Haviv, Moshe and Yoav Kerner, "On balking from an empty queue," *Queueing Systems* **55** (2007) 239-249. (Cited on page 9, 26, 27, 270, and 281.)

[325] Haviv, Moshe and Binyamin Oz, "Regulating an observable M/M/1 queue," *Operations Research Letters* forthcoming. (Cited on page 35 and 140.)

[326] Haviv, Moshe and Ramandeep S. Randhawa, "Pricing in queues without demand information," *Manufacturing & Service Operations Management* **16**(3) (2014) 401-411. (Cited on page 78.)

[327] Haviv, Moshe and Liron Ravner, "Strategic timing of arrivals to a finite queue multi-server loss system" *Queueing Systems* **81** (2015) 71-96. (Cited on page 11 and 89.)

[328] Haviv, Moshe and Liron Ravner, "Strategic bidding in an accumulating priority queue: equilibrium and analysis," manuscript (2015). (Cited on page 102.)

[329] Haviv, Moshe and Ya'acov Ritov, "Externalities, tangible externalities, and queue disciplines," *Management Science* **44**(6) (1998) 850-856. (Cited on page 12.)

[330] Haviv, Moshe and Tim Roughgarden, "The price of anarchy in an exponential multi-server," *Operations Research Letters* **35** (2007) 421-426. (Cited on page 119, 133, 134, 135, and 136.)

[331] Hayel, Yezekael, Mohamed Ouarraou, and Bruno Tuffin, "Optimal measurement-based pricing for an M/M/1 queue," *Networks and Spatial Economics* **7**(2) (2007) 177-195. (Cited on page 77 and 150.)

[332] Hayel, Yezekael, Dominique Quadri, Tania Jiménez, and Luce Brotcorne, "Decentralized optimization of last-mile delivery service with non-cooperative bounded rational customers," *Annals of Operations Research* forthcoming. (Cited on page 8 and 217.)

[333] Hayel, Yezekael, David Ros, and Bruno Tuffin, "Less-than-best-effort services: pricing and scheduling," *INFOCOM* (2004). (Cited on page 169.)

[334] Hayel, Yezekael and Bruno Tuffin, "Pricing for heterogeneous services at a discriminatory processor sharing queue," *NETWORKING* 2005, *Lecture Notes in Computer Science* **3462** (2005) 816-827. (Cited on page 115.)

[335] Hayel, Yezekael and Bruno Tuffin, "An optimal congestion and cost-sharing pricing scheme for multiclass services," *Mathematical Methods of Operations Research* **64** (2006) 445-465. (Cited on page 125.)

[336] He, Qiao-Chu and Ying-Ju Chen, "Revenue-maximizing pricing and scheduling strategies in service systems with taste indifference," manuscript (2014). (Cited on page 171.)

[337] Heinhold, Michael, "An operational research approach to allocation of clients to a certain class of service institutions," *Journal of the Operational Research Society* **29**(3) (1978) 273-276. (Cited on page 224 and 225.)

[338] Hellman, Ziv, "To queue or not to queue," *The Jerusalem Report* July 6, 2009. Also, www.math.tau.ac.il/~hassin/J_Rep.pdf. (Cited on page 88.)

[339] Hemachandra, N. and Manu Gupta, "Nash equilibrium in pricing surplus server capacity model and equilibrium sets in some related models," manuscript (2015). (Cited on page 176.)

[340] Hennet, Jean-Claude and Yasemin Arda, "Supply chain coordination: a game-theory approach," *Engineering Applications of Artificial Intelligence* **21** (2008) 399-405. (Cited on page 55 and 236.)

[341] Hillier, Frederick S., "The application of waiting-line theory to industrial problems," *Journal of Industrial Engineering* **15** (1964) 3-8. (Cited on page 5.)

[342] Ho, Teck-Hua and Yu-Sheng Zheng, "Setting customer expectation in service delivery: an integrated marketing-operations perspective," *Management Science* **50**(4) (2004) 479-488. (Cited on page 188, 286, and 298.)

[343] Hong, I-Hsuan, Hsi-Mei Hsu, Yi-Mu Wu, and Chun-Shao Yeh, "Equilibrium pricing and lead time decisions in a competitive industry," *International Journal of Production Economics* **139** (2012) 586-595. (Cited on page 197.)

[344] Hong, L. Jeff, Xiaowei Xu, and Shenghao Zhang, "Capacity reservation for time-sensitive service providers: an application in seaport management" *European Journal of Operational Research* **245**(2) (2015) 470-479. (Cited on page 252.)

[345] Hong, Ki-sung and Chulung Lee, "Integrated pricing and capacity decision for a telecommunication service provider," *Multimedia Tools and Applications* **64** (2013), 389-406. (Cited on page 294.)

[346] Honnappa, Harsha and Rahul Jain, "Strategic arrivals into queueing networks: the network concert queueing game," *Operations Research* **63**(1) (2015) 247-259. (Cited on page 88.)

[347] Hsiao, Man-Tung T. and Aurel A. Lazar, "Optimal decentralized flow control of Markovian queueing networks with multiple controllers," *Performance Evaluation* **13** (1991) 181-204. (Cited on page 56 and 57.)

[348] Hsu, Vernon N., Susan H. Xu, and Boris Jukic, "Optimal scheduling and incentive compatible pricing for a service system with quality of service guarantees," *Manufacturing & Service Operations Management* **11**(3) (2009) 375-396. (Cited on page 113.)

[349] Hu, Ming, Yang Li, and Jianfu Wang, "Efficient ignorance: information heterogeneity in a queue," manuscript (2014). (Cited on page 60.)

[350] Huang, Ping, Jinting Wang, and Li Fu, "Equilibrium balking strategies in the single-server queues with Erlangian service and setup times," *LISS* (2012). (Cited on page 26 and 269.)

[351] Huang, Tingliang, Gad Allon, and Achal Bassamboo, "Bounded rationality in service systems," *Manufacturing & Service Operations Management* **15**(2) (2013) 263-279. (Cited on page 8, 18, and 287.)

[352] Huang, Tingliang and Ying-Ju Chen, "Service systems with experience-based anecdotal reasoning customers," *Production and Operations Management* **24**(5) (2015) 778-790. (Cited on page 74 and 286.)

[353] Hwang Johye, Long Gao, and Wooseung Jang, "Joint demand and capacity management in a restaurant system," *European Journal of Operational Research* **207** (2010) 465-472. (Cited on page 39.)

[354] Ilmakunnas, Pekka, "Strategic behavior in a service industry," *Managerial and Decision Economics* **23** (2002) 69-82. (Cited on page 195.)

[355] Inoie, Atsushi, Hisao Kameda, and Corinne E. Touati, "A paradox in optimal flow control of $M/M/n$ queues," *Computers & Operations Research* **33** (2006) 356-368. (Cited on page 107.)

[356] Ioannidis, Stratos, Oualid Jouini, Angelos A. Economopoulos, and Vassilis S. Kouikoglou, "Control policies for single-stage production systems with perishable inventory and customer impatience," *Annals of Operations Research* **209** (2013) 115-138. (Cited on page 38, 264, and 267.)

[357] Jagannathan, Krishna, Ishai Menache, Eytan Modiano, and Gil Zussman, "Non-cooperative spectrum access: the dedicated vs. free spectrum choice," *IEEE Journal on Selected Areas in Communications* **30**(11) (2012) 2251-2261. (Cited on page 275.)

[358] Jahnke, Hermann, Anne Chwolka, and Dirk Simons, "Coordinating service-sensitive demand and capacity by adaptive decision making: an application of case-based decision theory," *Decision Sciences* **36**(1) (2005) 1-32. (Cited on page 76 and 142.)

[359] Jain, Rahul, Sandeep Juneja, and Nahum Shimkin, "The concert queueing game: to wait or to be late," *Discrete Event Dynamic Systems* **21** (2011) 103-138. (Cited on page 86 and 88.)

[360] Janssen, Maarten C.W. and Alexei Parakhonyak, "Service refusal in regulated markets for credence goods," manuscript (2011). (Cited on page 153.)

[361] Jayaraman, Rajshri and Francis de Véricourt, "Bribe requests are associated with longer delays," manuscript (2013). (Cited on page 178.)

[362] Jayaswal, Sachin, Elizabeth M. Jewkes, and Saibal Ray, "Product differentiation and operations strategy in a capacitated environment," *European Journal of Operational Research* **210**(3) (2011) 716-728. (Cited on page 166.)

[363] Jemaï, Zied and Fikri Karaesmen, "Decentralized inventory control in a two-stage capacitated supply chain," *IIE Transactions* **39** (2007) 501-512. (Cited on page 236.)

[364] Jhang-Li, Jhih-Hua and I. Robert Chiang, "Resource allocation and revenue optimization for cloud service providers," *Decision Support Systems* **77** (2015) 55-66. (Cited on page 147 and 242.)

[365] Jiang, Bin, "The decision-making on an in-house logistic division's operation strategies," *International Journal of Production Economics* **96** (2005) 37-46. (Cited on page 143.)

[366] Jiang, Yabing and Abraham Seidmann, "Integrated capacity and marketing incentive contracting for capital-intensive service systems," *Decision Support Systems* **51** (2011) 627-637. (Cited on page 241.)

[367] Jiang, Yabing and Abraham Seidmann, "Capacity planning and performance contracting for service facilities," *Decision Support Systems* **58** (2014) 31-42. (Cited on page 241.)

[368] Jin, Youngmi and George Kesidis, "Dynamics of usage-priced communication networks: the case of a single bottleneck resource," *IEEE/ACM Transactions on Networking* **13**(5) (2005) 1041-1053. (Cited on page 106.)

[369] Jin, Yue and Jennifer K. Ryan, "Price and service competition in an outsourced supply chain," *Production and Operations Management* **21**(2) (2012) 331-344. (Cited on page 248.)

[370] Johansen, Søren Glud, "Optimal prices of a job shop with a single work station: a discrete time model," *International Journal of Production Economics* **23** (1991) 129-137. (Cited on page 49.)

[371] Johansen, Søren Glud, "Optimal prices of an M/G/1 jobshop," *Operations Research* **42**(4) (1994) 765-774. (Cited on page 49.)

[372] Johansen, Søren Glud, "Transfer pricing of a service department facing random demand," *International Journal of Production Economics* **46-47** (1996) 351-358. (Cited on page 43 and 44.)

[373] Johari, Ramesh and Sunil Kumar, "Congestible services and network effects," unpublished manuscript (2009). (Cited on page 111.)

[374] Johari, Ramesh, Shie Mannor, and John N. Tsitsiklis, "Efficiency loss in a network resource allocation game: the case of elastic supply," *IEEE Transactions on Automatic Control* **50**(11) (2005) 1712-1724. (Cited on page 133.)

[375] Johari, Ramesh, Gabriel Y. Weintraub, and Benjamin van Roy, "Investment and market structure in industries with congestion," *Operations Research* **58**(5) (2010) 1303-1317. (Cited on page 193 and 194.)

[376] Jouini, Oualid, " Analysis of a last come first served queueing system with customer abandonment," *Computers & Operations Research* **39** (2012) 3040-3045. (Cited on page 88.)

[377] Jouini, Oualid, Zeynep Akşin, and Yves Dallery, "Call centers with delay information: models and insights," *Manufacturing & Service Operations Management* **13**(4) (2011) 534-548. (Cited on page 38 and 298.)

[378] Jouini, Oualid, Yves Dallery, and O. Zeynep Akşin, "Queueing models for full-flexible multi-class call centers with real-time anticipated delays," *International Journal of Production Economics* **120** (2009) 389-399. (Cited on page 38.)

[379] Juneja, Sandeep, and Nahum Shimkin, "The concert queueing game: strategic arrivals with waiting and tardiness costs," *Queueing Systems* **74** (2013) 369-402. (Cited on page 86.)

[380] Kaman, Cumhur, Seçil Savaşaneril, and Yasemin Serin, "Production and lead time quotation under imperfect shop floor information," *International Journal of Production Economics* **144** (2013) 422-431. (Cited on page 77 and 296.)

[381] Kameda, Hisao, "Coincident cost improvement vs. degradation by adding connections to noncooperative networks and distributed systems," *Networks and Spatial Economics* **9** (2009) 269-287. (Cited on page 11.)

[382] Kameda, Hisao, "Models of paradoxical coincident cost degradation in noncooperative networks," in *Game Theory Relaunched* Hardy Hanappi (ed.), 2013. http://www.intechopen.com/books/game-theory-relaunched. (Cited on page 11 and 227.)

[383] Kameda, Hisao, Eitan Altman, Takayuki Kozawa, and Yoshihisa Hosokawa, "Braess-like paradoxes in distributed computer systems," *IEEE Transactions on Automatic Control* **45** (2000) 1687-1691. (Cited on page 226.)

[384] Kameda, Hisao and Yoshihisa Hosokawa, "A paradox in distributed optimization of performance," unpublished manuscript (2000). (Cited on page 199, 210, 224, and 227.)

[385] Kameda, Hisao, Yoshihisa Hosokawa, and Odile Pourtallier, "Effects of symmetry on Braess-like paradoxes in distributed computer systems: a numerical study," *40th IEEE Conference on Decisions and Control* (2001). (Cited on page 226.)

[386] Kameda, Hisao and Odile Pourtallier, "Paradoxes in distributed decisions on optimal load balancing for networks of homogeneous computers," *Journal of the Association for Computing Machinery* **49**(3) (2002) 407-433. (Cited on page 227.)

[387] Kaporis, Alexis C., Lefteris M. Kirousis, E.I. Politopoulou, and Paul. G. Spirakis, *WEA* 2005. (Cited on page 224.)

[388] Kaporis, Alexis C. and Paul G. Spirakis, "The price of optimum in Stackelberg games on arbitrary single commodity networks and latency functions," *Theoretical Computer Science* **410** (2009) 745-755. (Cited on page 224.)

[389] Kardeş, Erim, "Wait or share service? Customers' choice at the equilibrium," unpublished manuscript (2012). (Cited on page 13 and 217.)

[390] Kardeş, Erim, Fernando Ordóñez, and Randloph W. Hall, "Discounted robust stochastic games and an application to queueing control," *Operations Research* **59**(2) (2011) 365-382. (Cited on page 77 and 261.)

[391] Karsten, Frank, Marco Slikker, and Geert-Jan van Houtum, "Domain extensions of the Erlang loss function: their scalability and its applications to cooperative games," *Probability in the Engineering and Informational Sciences* **28** (2014) 473-488. (Cited on page 128.)

[392] Karsten, Frank, Marco Slikker, and Geert-Jan van Houtum, "Resource pooling and cost allocation among independent service providers," *Operations Research* **63**(2) (2015) 476-488. (Cited on page 128.)

[393] Katta, Akshay-Kumar and Jay Sethuraman, "Pricing strategies and service differentiation in queues: a profit maximization perspective," (2005). (Cited on page 169.)

[394] Kavurmacioglu, Emir, Murat Alanyali, and David Starobinski, "Competition in private commons: price war or market sharing?" *IEEE/ACM Transactions on Networking* forthcoming. (Cited on page 197.)

[395] Keblis, Matthew F. and Youyi Feng, "Optimal pricing and production control in an assembly system with a general stockout cost," *IEEE Transactions on Automatic Control* **57**(7) (2012) 1821-1826. (Cited on page 46.)

[396] Keon, Neil J. and G. Anandalingam, "A new pricing model for competitive telecommunications services using congestion discounts," *INFORMS Journal on Computing* **17**(2) (2005) 248-262. (Cited on page 56 and 83.)

[397] Kerner, Yoav, "Equilibrium joining probabilities for an M/G/1 queue," *Games and Economic Behavior* **71**(2) (2011) 521-526. (Cited on page 2, 27, 70, and 281.)

[398] Kesavan, Saravanan, Vinayak Deshpande, and Hyun Seok Lee, "Increasing sales by managing congestion in self-service environments: evidence from a field experiment," (2014). See also, "Sometimes self-service costs more," http://www.kenan-flagler.unc.edu/news/2015/06/ROI-selfservice (Cited on page 40.)

[399] Keskinocak, Pinar and Sridhar Tayur, "Due-Date Management Policies," in *Handbook of Quantitative Supply Chain Analysis: Modeling in the E-Business Era* D. Simchi-Levi, S.D. Wu, and Z.M. Shen (editors), Kluwer Academic Publishers, Norwell, MA, 2004, 485-553. (Cited on page 18.)

[400] Kim, Jeunghyun and Ramandeep Randhawa, "Asymptotically optimal dynamic pricing in observable queues," manuscript (2015). (Cited on page 55.)

[401] Kim, Jung Hyun, Hyun-Soo Ahn, and Rhonda Righter, "Managing queues with heterogeneous servers," *Journal of Applied Probability* **48** (2011) 435-452. (Cited on page 34 and 265.)

[402] Kim, Whan-Seon, "Price-based quality-of-service control framework for two-class network services," *Journal of Communications and Networks* **9**(3) (2007) 319-328. (Cited on page 165.)

[403] Kim, Yong J. and Michael V. Mannino, "Optimal incentive-compatible pricing for M/G/1 queues," *Operations Research Letters* **31** (2003) 459-461. (Cited on page 112.)

[404] Kim, Young Joo and Hark Hwang, "Incremental discount policy of cell-phone carrier with connection success rate constraint," *European Journal of Operational Research* **196** (2009) 682-687. (Cited on page 108.)

[405] Kittsteiner, Thomas and Benny Moldovanu, "Priority auctions and queue disciplines that depend on processing time," *Management Science* **51**(2) (2005) 236-248. (Cited on page 65 and 114.)

[406] Knight, Vincent and Paul Harper, "Selfish routing in public services," *European Journal of Operational Research* **230** (2013) 122-132. (Cited on page 136 and 224.)

[407] Koo, Minjung and Ayelet Fishbach, "A silver lining of standing in line: queuing increases value of products," *Journal of Marketing Research* **XLVII** (2010) 713-724. (Cited on page 61.)

[408] Kopalle, Praveen K. and Donald R. Lehmann, "Setting quality expectations when entering a market: what should the promise be?" *Marketing Science* **25**(1) (2006) 8-24. (Cited on page 19.)

[409] Korilis, Yannis A. and Aurel A. Lazar, "On the existence of equilibria in noncooperative optimal flow control," *Journal of the Association for Computing Machinery* **42** (1995) 584-613. (Cited on page 57.)

[410] Korilis, Yannis A., Aurel A. Lazar, and Ariel Orda, "Achieving network optima using Stackelberg routing strategies," *IEEE/ACM Transactions on Networking* **5**(1) (1997) 161-173. (Cited on page 223.)

[411] Korilis, Yannis A., Aurel A. Lazar, and Ariel Orda, "Capacity allocation under noncooperative routing," *IEEE Transactions on Automatic Control* **42**(3) (1997) 309-325. (Cited on page 214 and 223.)

[412] Korilis, Yannis A., Aurel A. Lazar, and Ariel Orda, "Avoiding the Braess paradox in non-cooperative networks," *Journal of Applied Probability* **36** (1999) 211-222. (Cited on page 215 and 226.)

[413] Kostami, Vasiliki and Sampath Rajagopalan, "Speed quality tradeoffs in a dynamic model," *Manufacturing & Service Operations Management* **16**(1) (2014) 104-118. (Cited on page 99 and 157.)

[414] Kostami, Vasiliki and Amy R. Ward, "Managing service systems with an offline waiting option and customer abandonment," *Manufacturing & Service Operations Management* **11**(4) (2009) 644-656. (Cited on page 9 and 167.)

[415] Koutsopoulos, Iordanis, Leandros Tassiulas, and Lazaros Gkatzikis, "Client-server games and their equilibria in peer-to-peer networks," *Computer Networks* **67** (2014) 201-218. (Cited on page 211.)

[416] Krämer, Jan and Lukas Wiewiorra, "Network neutrality and congestion sensitive content providers: Implications for content variety, broadband investment and regulation," *Information Systems Research* **23**(4) (2012) 1303-1321. (Cited on page 257 and 258.)

[417] Krämer, Jan, Lukas Wiewiorra, and Christof Weinhardt, "Network neutrality: a progress report," *Telecommunications Policy* **37** (2013) 794-813. (Cited on page 259.)

[418] Kremer, Mirko and Laurens Debo, "Inferring quality from wait time," *Management Science* forthcoming. (Cited on page 64.)

[419] [∗ ∗ ∗] Kulkarni, Vidyadhar, "A game theoretic model for two types of customers competing for service," *Operations Research Letters* **2**(3) (1983) 119-122. (Cited on page 90.)

[420] Kumar, Chetan, Kemal Altinkemer, and Prabuddha De, "A mechanism for pricing and resource allocation in peer-to-peer networks," *Electronic Commerce Research and Applications* **10** (2011) 26-37. (Cited on page 6.)

[421] Kumar, Piyush and Parthasarathy Krishnamurthy, "The impact of service-time uncertainty and anticipated congestion on customers' waiting time decisions," *Journal of Service Research* **10**(3) (2008) 282-292. (Cited on page 209.)

[422] Kumar, Sunil and Ramandeep S. Randhawa, "Exploiting market size in service systems," *Manufacturing & Service Operations Management* **12**(3) (2010) 511-526. (Cited on page 15, 148, 184, and 232.)

[423] Kwasnica, Anthony M. and Euthemia Stavrulaki, "Competitive location and capacity decisions for firms serving time-sensitive customers," *Naval Research Logistics* **55**(7) (2008) 704-721. (Cited on page 202.)

[424] La, Richard J., and Venkat Anantharam, "Optimal routing control: repeated game approach," *IEEE Transactions on Automatic Control* **47**(3) (2002) 437-450. (Cited on page 131.)

[425] Lachapelle, Aimé, Jean-Michel Larsy, Charles-Albert Lehalle, and Pierre-Louis Lions, "Efficiency of the price formation process in presence of high frequency participants: a mean field game analysis," *Mathematics and Financial Economics* forthcoming. (Cited on page 96.)

[426] Large, Jeremy H. and Thomas W. L. Norman, "Markov perfect Bayesian equilibrium via ergodicity," manuscript (2012). (Cited on page 60.)

[427] Lariviere, Martin A. and Jan A. Van Mieghem, "Strategically seeking service: how competition can generate Poisson arrivals," *Manufacturing & Service Operations Management* **6**(1) (2004) 23-40. (Cited on page 84.)

[428] Larsen, Christian, "Comparing two socially optimal work allocation rules when having a profit optimizing subcontractor with ample capacity," *Mathematical Methods of Operations Research* **61** (2005) 109-121. (Cited on page 122.)

[429] Le Cadre, Hélène, Mustapha Bouhtou, and Bruno Tuffin, "Competition for subscribers between mobile operators sharing a limited resource," *GameNets* (2009). (Cited on page 200.)

[430] Lee, Chihoon and Amy R. Ward, "Optimal pricing and capacity sizing for the GI/GI/1 queue," *Operations Research Letters* **42** (2014) 527-531. (Cited on page 149.)

[431] Lee, Hsiao-Hui, Edieal Pinker, and Robert Shumsky, "Outsourcing a two-level service process," *Management Science* **58**(8) (2012) 1569-1584. (Cited on page 243.)

[432] Lee, Sam C.M. and John C.S. Lui, "On the interaction and competition among Internet service providers," *IEEE Journal on Selected Areas in Communications* **26**(7) (2008) 1277-1283. (Cited on page 213.)

[433] Leeman, Wayne A., "The reduction of queues through the use of price," *Operations Research* **12**(5) (1964) 783-785. (Cited on page 56.)

[434] Leshno, Jacob, "Dynamic matching in overloaded waiting lists," manuscript (2014). (Cited on page 37.)

[435] Li, Gang, Fengfeng Huang, T.C.E. Cheng, Quan Zheng, and Ping Ji, "Make-or-buy service capacity decision in a supply chain providing after-sales service," *European Journal of Operational Research* **239**(2) (2014) 377-388. (Cited on page 238.)

[436] Li, Husheng and Zhu Han "Socially optimal queuing control in cognitive radio networks subject to service interruptions: to queue or not to queue?," *IEEE Transactions on Wireless Communications* **10**(5) (2011) 1656-1666. (Cited on page 274.)

[437] Li, Le, Jinting Wang, and Feng Zhang, "Equilibrium customer strategies in Markovian queues with partial breakdowns," *Computers & Industrial Engineering* **66**(4) (2013) 751-757. (Cited on page 275.)

[438] Li, Li, Li Jiang, and Liming Liu, "Service and price competitions when customers are naive," *Production and Operations Management* **21**(4) (2012) 747-760. (Cited on page 74 and 285.)

[439] Li, Lode, "The role of inventory in delivery-time competition," *Management Science* **38**(2) (1992) 182-197. (Cited on page 23, 103, and 187.)

[440] Li, Na and Zhibin Jiang, "Modeling and optimization of a product-service system with additional service capacity and impatient customers," *Computers & Operations Research* **40** (2013) 1923-1937. (Cited on page 38, 40, 72, and 264.)

[441] Li, Xiangyu, Jinting Wang, and Feng Zhang, "New results on equilibrium balking strategies in the single-server queue with breakdowns and repairs," *Applied Mathematics and Computation* **241** (2014) 380-388. (Cited on page 274 and 275.)

[442] Li, Xin, Pengfei Guo, and Zhaotong Lian, "Quality-speed competition in customer-intensive services with boundedly rational customers," manuscript (2015). (Cited on page 8, 155, and 290.)

[443] Libman, Lavy and Ariel Orda, "The designer's perspective to atomic noncooperative networks," *IEEE/ACM Transactions on Networking* **7**(6) (1999) 875-884. (Cited on page 214 and 226.)

[444] Libman, Lavy and Ariel Orda, "Atomic resource sharing in noncooperative networks," *Telecommunication Systems* **17**(4) (2001) 385-409. (Cited on page 209.)

[445] Libman, Lavy and Ariel Orda, "Optimal retrial and timeout strategies for accessing network resources," *IEEE/ACM Transactions on Networking* **10**(4) (2002) 551-564. (Cited on page 90 and 92.)

[446] Lin, Kyle Y., "Decentralized admission control of a queueing system: a game-theoretic model," *Naval Research Logistics* **50** (2003) 702-718. (Cited on page 14 and 118.)

[447] Liou, Cheng-Dar, "Markovian queue optimization analysis with an unreliable server subject to working breakdowns and impatient customers," *International Journal of Systems Science* **46**(12) (2015) 2165-2182. (Cited on page 265.)

[448] [∗ ∗ ∗] Littlechild, S.C., "Optimal arrival rate in a simple queueing system," *International Journal of Production Research* **12**(3) (1974) 391-397. (Cited on page 5.)

[449] Liu, Chang and Randall A. Berry, "A priority queue model for competition with shared spectrum," *IEEE 52nd Annual Allerton Conference on Communication, Control, and Computing* (2014). (Cited on page 172.)

[450] Liu, Chubo, Kenli Li, Chengzhong Xu, and Keqin Li, "Strategy configurations of multiple users competition for cloud service reservation," *IEEE Transactions on Parallel and Distributed Systems* forthcoming. (Cited on page 83 and 149.)

[451] Liu, Liming, Mahmut Parlar, and Stuart X. Zhu, "Pricing and lead time decisions in decentralized supply chains," *Management Science* **53**(5) (2007) 713-725. (Cited on page 235 and 292.)

[452] Liu, Tieming, Chinnatat Methapatara, and Laura Wynter, "Revenue management for on-demand IT services," *European Journal of Operational Research* **207** (2010) 401-408. (Cited on page 83, 149, and 286.)

[453] Liu, Weiqi, Yan Ma, and Jihong Li, "Equilibrium threshold strategies in observable queueing systems under single vacation policy," *Applied Mathematical Modelling* **36** (2012) 6186-6202. (Cited on page 272.)

[454] Liu, Yong and Marwan A. Simaan, "Non-inferior Nash strategies for routing control in parallel-link communication networks," *International Journal on Communication Systems* **18** (2005) 347-361. (Cited on page 130.)

[455] Liu, Zaiming, Yan Ma, and Zhe George Zhang, "Equilibrium mixed strategies in a discrete-time Markovian queue under multiple and single vacation policies," *Quality Technology & Quantitative Management* **12**(3) (2015) 367-380. (Cited on page 269 and 272.)

[456] Low, David W., "Optimal dynamic pricing policies for an $M/M/s$ Queue," *Operations Research* **22**(3) (1974) 545-561. (Cited on page 44 and 46.)

[457] Lozano, Macarena and Pilar Moreno, "A discrete time single-server queue with balking: economic applications," *Applied Economics* **40** (2008) 735-748. (Cited on page 39.)

[458] Lu, Lauren Xiaoyuan, Jan A. Van Mieghem, and R. Canan Savaskan, "Incentives for quality through endogenous routing," *Manufacturing & Service Operations Management* **11**(2) (2009) 254-273. (Cited on page 250.)

[459] Lu, Yina, Andrés Musalem, Marcelo Olivares, and Ariel Schilkrut, "Measuring the effect of queues on customer purchases," *Management Science* **59** (2013) 1743-1763. (Cited on page 17 and 285.)

[460] Lynn, Stephen and Kashi R. Balachandran, "Allocating costs of a shared server with stochastic service parameters and job class priorities," *European Journal of Operational Research* **180** (2007) 1155-1167. (Cited on page 120.)

[461] Ma, Yan, Wei-qi Liu, and Ji-hong Li, "Equilibrium balking behavior in the Geo/Geo/1 queueing system with multiple vacations," *Applied Mathematical Modelling* **37** 3861-3878. (Cited on page 269.)

[462] Maglaras, Constantinos, "Revenue management for a multiclass single-server queue via a fluid model analysis," *Operations Research* **54**(5) (2006) 914-932. (Cited on page 45.)

[463] Maglaras, Constantinos and Assaf Zeevi, "Pricing and capacity sizing for systems with shared resources: approximate solutions and scaling relations," *Management Science* **49**(8) (2003) 1018-1038. (Cited on page 147 and 175.)

[464] Maglaras, Constantinos and Assaf Zeevi, "Pricing and design of differentiated services: approximate analysis and structural insights," *Operations Research* **53**(2) (2005) 242-262. (Cited on page 148.)

[465] Maglaras, Costis, John Yao, and Assaf Zeevi, "Optimal price and delay differentiation in large-scale queueing systems," *Management Science* forthcoming. (Cited on page 174.)

[466] Mandelbaum, Avishai and Alexander L. Stolyar, "Scheduling flexible servers with convex delay costs: heavy-traffic optimality of the generalized $c\mu$-rule," *Operations Research* **52**(6) (2004) 836-855. (Cited on page 14.)

[467] Mandjes, Michel, "Pricing strategies under heterogeneous service requirements," *Computer Networks* **42** (2003) 231-249. Also: "Pricing strategies and service differentiation," *Netnomics* **6** (2004) 59-81. (Cited on page 168, 169, and 201.)

[468] Manjrekar, Mayank, Vinod Ramaswamy, and Srinivas Shakkottai, "A mean field game approach to scheduling in cellular systems," *INFOCOM* (2013). (Cited on page 177 and 285.)

[469] Manou, Athanasia, Antonis Economou, and Fikri Karaesmen, "Strategic customers in a transportation station: when is it optimal to wait?" *Operations Research* **62**(4) (2014) 910-925. (Cited on page 26 and 280.)

[470] Marbach, Peter, "Pricing differentiated services networks: bursty traffic," *INFOCOM* (2001). (Cited on page 101.)

[471] Marianov, Vladimir, Miguel Ríos, and Francisco Javier Barros, "Allocating servers to facilities, when demand is elastic to travel and waiting times," *RAIRO Operations Research* **39** (2005) 143-162. (Cited on page 39.)

[472] Marianov, Vladimir, Miguel Ríos, and Manuel José Icaza "Facility location for market capture when users rank facilities by shorter travel and waiting times," *European Journal of Operational Research* **191** (2008) 32-44. (Cited on page 291.)

[473] Martin, Glen E. and Lyn D. Pankoff, "Reneging in queues revisited," *Decision Sciences* **13**(2) (1982) 340-347. (Cited on page 28 and 299.)

[474] Maoui, Idriss, Hayriye Ayhan, and Robert D. Foley, "Congestion-dependent pricing in a stochastic service system," *Advances in Applied Probability* **39**(4) (2007) 898-921. (Cited on page 46 and 141.)

[475] Maoui, Idriss, Hayriye Ayhan, and Robert D. Foley, "Optimal static pricing for a service with holding costs," *European Journal of Operational Research* **197** (2009) 912-923. (Cited on page 141 and 152.)

[476] Masarani, F. and S. Sadik Gokturk, "Price setting policies for service systems in case of uncertain demand and service time," *Zeitschrift Operations Research* **31** (1987) B97-B113. (Cited on page 152 and 187.)

[477] Masuda, Yasushi and Seungjin Whang, "On the optimality of fixed-up-to tariff for telecommunications service," *Information Systems Research* **17**(3) (2006) 247-253. (Cited on page 161.)

[478] Mazalov, Vladimir V., and Julia V. Chuiko, "Nash equilibrium in optimal arrival time problem," *ISDG* (2006). (Cited on page 89.)

[479] Mazalov, Vladimir, Burkhard Monien, Florian Schoppmann, and Karsten Tiemann, *WINE* (2006) (Cited on page 134.)

[480] Mazumdar, Ravi, Lorne G. Mason, and Christos Douligeris, "Fairness in network optimal flow control: optimality of product forms," *IEEE Transactions on Communications* **39**(5) (1991) 775-782. (Cited on page 130.)

[481] McCain, Roger A., *Game Theory: A Non-Technical Introductioń to the Analysis of Strategy*, Drexel University 2010. (Cited on page 85.)

[482] Melnik, Anna V., "Equilibrium in transportation games," *Automation and Remote Control* **76**(5) (2015) 909-918. (Cited on page 223.)

[483] Melo, Emerson, "Price competition, free entry, and welfare in congested markets," *Games and Economic Behavior* **83** (2014) 53-72. (Cited on page 191 and 208.)

[484] Menache, Ishai and Nahum Shimkin, "Capacity management and equilibrium for proportional QoS," *IEEE/ACM Transactions on Networking* **16**(5) (2008) 1025-1037. (Cited on page 215.)

[485] Menasché, Daniel Sadoc, Danial Ratton Figueiredo, and Edmundo de Souza e Silva, "An evolutionary game-theoretic approach to congestion control," *Performance Evaluation* **62** (2005) 295-312. (Cited on page 106.)

[486] Mendelson, Haim and Ali K. Parlaktürk, "Product-line competition: customization vs. proliferation," *Management Science* **54**(12) (2008) 2039-2053. (Cited on page 204.)

[487] [∗ ∗ ∗] Mendelson, Haim and Seungjin Whang, "Optimal incentive-compatible priority pricing for the M/M/1 queue," *Operations Research* **38**(5) (1990) 870-883. (Cited on page 112, 113, 115, 125, and 173.)

[488] Monsef, Ehsan, Tricha Anjali, and Sanjiv Kapoor, "Price of anarchy in network routing with class based capacity guarantees," *INFOCOM* (2014). (Cited on page 117.)

[489] Musacchio, John and Shuang Wu, "The price of anarchy in computing differentiated services networks," *IEEE 46th Annual Allerton Conference on Communication, Control, and Computing* (2008) 615-622. (Cited on page 192.)

[490] Mutlu, Huseyin, Murat Alanyali, and David Starobinski, "Spot pricing of secondary spectrum access in wireless cellular networks," *IEEE/ACM Transactions on Networking* **17**(6) (2009) 1794-1804. (Cited on page 43 and 47.)

[491] Mutlu, Huseyin, Murat Alanyali, David Starobinski, and Aylin Turhan, "Online pricing of secondary spectrum access with unknown demand function," *IEEE Journal on Selected Areas in Communications* **30**(11) (2012) 2285-2294. (Cited on page 47 and 77.)

[492] Nadiminti, Raja, Tridas Mukhopadhyay, and Charles H. Kriebel, "Research report: intrafirm resource allocation with asymmetric information and negative externalities," *Information Systems Research* **13**(4) (2002) 428-434. (Cited on page 109.)

[493] Nair, Jayakrishnan, Vijay G. Subramanian, and Adam Wierman, "On competitive provisioning of cloud services," *IEEE 52nd Annual Allerton Conference on Communication, Control, and Computing* (2014). (Cited on page 112.)

[494] Nair, Jayakrishnan, Vijay G. Subramanian, and Adam Wierman, "Provisioning of large scale schemes: the interplay between network effects and strategic behavior in the user base," manuscript (2014) (Cited on page 112.)

[495] Nair, Jayakrishnan, Adam Wierman, and Bert Zwart, "Provisioning of large scale systems: the interplay between network effects and strategic behavior in the user base," *Management Science* forthcoming. (Cited on page 111 and 184.)

[496] Nam, Ick-Hyun, "Duopoly competition considering waiting cost," *Journal of Information and Operations Management* **7** (1977) 105-115. (Cited on page 192.)

[497] [∗ ∗ ∗] Naor, Pinhas (Paul), "The regulation of queue size by levying tolls," *Econometrica* **37** (1969) 15-24. (Cited on page 2, 5, 27, and 184.)

[498] Narenji, M., Mohammad Fathian, Ebrahim Teimoury, and Seyed Gholamreze Jalali Naini, "Price and delivery time analyzing in competition between an electronic and a traditional supply chain," *Mathematical Problems in Engineering* (2013), 12 pages. (Cited on page 254 and 255.)

[499] Nasrallah, Walid F., "When does management matter in a dog-eat-dog world: An Interaction Value Analysis model of organizational climate," *Computational and Mathematical Organization Theory* **12** (2006) 339-359. (Cited on page 120.)

[500] Nazerzadeh, Hamid and Ramandeep Randhawa, "Asymptotic value of customer differentiation in queueing systems: optimality of two service grades," manuscript (2015). (Cited on page 172.)

[501] Nel, Andre and Hailing Zhu, "Wireless service pricing under multiple competitive providers and congestion-sensitive users," in *Wireless Mesh Networks* N. Funabiki (ed.), 281-308, Intech, 2011. (Cited on page 169.)

[502] Ni, Guanqun, Yingfeng Xu, and Yucheng Dong, "Price and speed decisions in customer-intensive services with two classes of customers," *European Journal of Operational Research* **228**(2) (2013) 427-436. (Cited on page 156.)

[503] Nisan, Noam, Tim Roughgarden, Éva Tardos, and Vijay V. Vazirani, *Algorithmic Game Theory*, Cambridge University Press, Cambridge, UK, 2007. (Cited on page 11.)

[504] Nobel, Rein and Marije Stolwijk, "The Downs-Thomson paradox revisited," in *Computational and Mathematical Modelling*, R. Nadarajan, R.S. Lekshami, and G. Sai Sundara Krishnan (eds.), Narosa Publishing House New Delhi, 2011. (Cited on page 229.)

[505] Oh, Jaelynn and Xuanming Su, "Pricing restaurant reservations: dealing with no-shows," manuscript (2012). (Cited on page 179.)

[506] Orda, Ariel, Raphael Rom, and Nahum Shimkin, "Competitive routing in multiuser communication networks," *IEEE/ACM Transactions on*

Networking **1**(5) (1993) 510-521. (Cited on page 131, 135, 209, 211, and 214.)

[507] Owen, Susan Hesse and Mark S. Daskin, "Strategic facility location: a review," *European Journal of Operational Research* **111** (1998) 423-447. (Cited on page 225.)

[508] Oz, Binyamin, Moshe Haviv, and Martin L. Puterman, "Social and self-optimization in multi-class multi-server queueing systems with relative priorities," *NetGCoop* (2014). (Cited on page 117.)

[509] Ozdaglar, Asuman, "Price competition with elastic demand," *Networks* **52**(3) (2008) 141-155. (Cited on page 247.)

[510] Özen, Ulas, Martin I. Reiman, and Qiong Wang, "On the core of cooperative games," *Operations Research Letters* **39** (2011) 385-389. (Cited on page 126 and 128.)

[511] Paç, M. Fazil and Senthil Veeraraghavan, "False diagnosis and overtreatment in services," manuscript (2015). (Cited on page 158.)

[512] Pahlavani, Ali and Mohammad Saidi-Mehrabad, "Optimal pricing for competitive service facilities with balking and veering customers," *International Journal of Innovative Computing, Information and Control* **7**(6) (2011) 3171-3191. (Cited on page 291.)

[513] Pal, Ranjan and Pan Hui, "Economic models for cloud service markets: pricing and capacity planning," *Theoretical Computer Science* **496** (2013) 113-124. (Cited on page 196.)

[514] Palaka, Kondalarao, Steven Erlebacher, and Dean H. Kropp, "Lead-time setting, capacity utilization, and pricing decisions under lead-time dependent demand," *IIE Transactions* **30** (1998) 151-163. (Cited on page 292.)

[515] de Palma, André and Luc Leruth, "Congestion and game in capacity: a duopoly analysis in the presence of network externalities," *Annals D'Economie et de Statistique* **15/16** (1989) 389-407. (Cited on page 190.)

[516] Pangburn, Michael S. and Euthemia Stavrulaki, "Service facility location and design with pricing and waiting-time considerations," *Supply Chain Optimization* J. Geunes and P. M. Pardalos (eds), Kluwer Academic Publishers Boston/Dordrecht/London, 2005. (Cited on page 202.)

[517] Pangburn, Michael S. and Euthemia Stavrulaki, "Capacity and price setting for dispersed, time-sensitive customer segments," *European Journal of Operational Research* **184** (2008) 1100-1121. (Cited on page 165 and 180.)

[518] Park, Sangjun and Ki-sung Hong, "Optimal guaranteed service time and service level decision with time and service level sensitive demand," *Mathematical Problems in Engineering* (2014), 7 pages. (Cited on page 232 and 299.)

[519] Parkan, Celik and E. H. Warren Jr., "Optimal reneging decisions in a G/M/1 queue," *Decision Sciences* **9**(1) (1978) 107-119. (Cited on page 28, 74, and 299.)

[520] Parlaktürk, Ali K. and Sunil Kumar, "Self-interested routing in queueing networks," *Management Science* **50**(7) (2004), 949-966. See also, "Queueing theory meets your morning latte," *Stanford Business Magazine* (2003). (Cited on page 8 and 219.)

[521] Parra-Frutos, Isabel, "A queueing-based model for optimal dimension of service firms," *SERIEs* **1** (2010) 459-474. (Cited on page 143.)

[522] Paschalidis, Ioannis Ch. and Yong Liu, "Pricing in multiservice loss networks: static pricing, asymptotic optimality, and demand substitution effects, *IEEE/ACM Transactions on Networking* **10**(3) (2002) 425-438. (Cited on page 45.)

[523] Paschalidis, Ioannis Ch. and John N. Tsitsiklis, "Congestion-dependent pricing of network services," *IEEE/ACM Transactions on Networking* **8**(2) (2000) 171-184. (Cited on page 36, 43, 44, and 45.)

[524] Pazgal, Amit I. and Sonja Radas, "Comparison of customer balking and reneging behavior to queueing theory predictions: an experimental study," *Computers & Operations Research* **35** (2008) 2537-2548. (Cited on page 216.)

[525] Pekgün, Pelin, Paul M. Griffin, and Pinar Keskinocak, "Coordination of marketing and production for price and leadtime decisions," *IIE Transactions* **40**(1) (2008) 12-30. (Cited on page 121.)

[526] Pekgün, Pelin, Paul M. Griffin, and Pinar Keskinocak, "Centralized vs. decentralized competition for price and lead-time sensitive demand," manuscript (2015). (Cited on page 254.)

[527] Penmatsa, Satish and Anthony T. Chronopoulos, "Game-theoretic static load balancing for distributed systems," *Journal of Parallel and Distributed Computing* **71** (2011) 537-555. (Cited on page 125, 132, and 210.)

[528] Plambeck, Erica L., "Optimal leadtime differentiation via diffusion approximations," *Operations Research* **52**(2) (2004) 213-228. (Cited on page 15, 49, 174, and 184.)

[529] Plambeck, Erica L. and Qiong Wang, "Implications of hyperbolic discounting for optimal pricing and scheduling of unpleasant services that generate future benefits," *Management Science* **59**(8) (2013) 1927-1946. (Cited on page 164.)

[530] Plambeck, Erica L. and Qiong Wang, "Hyperbolic discounting and queue-length information management for unpleasant services that generate future benefits," manuscript (2012). (Cited on page 165.)

[531] Plambeck, Erica L. and Amy R. Ward, "Optimal control of a high-volume assemble-to-order system," *Mathematics of Operations Research* **31**(3) (2006) 453-477. (Cited on page 15, 46, and 50.)

[532] Plambeck, Erica L. and Amy R. Ward, "Optimal control of a high-volume assemble-to-order system with maximum leadtime quotation and expediting," *Queueing Systems* **60** (2008) 1-69. (Cited on page 15, 46, 50, and 184.)

[533] Plambeck, Erica L. and Stefanos A. Zenios, "Incentive efficient control of a make-to-stock production system," *Operations Research* **51**(3) (2003) 371-386. (Cited on page 15, 44, and 233.)

[534] Platz, Trine Tornøe and Lars Peter Østerdal, "The curse of the first-in-first-out queue discipline," manuscript (2015). (Cited on page 88.)

[535] Printezis, Antonios and Apostolos Burnetas, "Priority option pricing in an M/M/m queue," *Operations Research Letters* **36** (2008) 700-704. (Cited on page 31.)

[536] Printezis, Antonios and Apostolos Burnetas, "The effect of discounts on optimal pricing under limited capacity," *International Journal of Operational Research* **10**(2) (2011) 160-179 (Cited on page 141.)

[537] Printezis, Antonios, Apostolos Burnetas, and Gopalakrishnan Mohan, "Pricing and capacity allocation under asymmetric information using Paris Metro pricing," *International Journal of Operational Research* **5**(3) (2009) 265-279. (Cited on page 141 and 166.)

[538] Qiu, Lili, Yang Richard Yang, Yin Zhang, and Scott Shenker, "On selfish routing in Internet-like environment," *IEEE/ACM Transactions on Networking* **14**(4) (2006) 725-738. (Cited on page 11.)

[539] Rabieyan, Reza and Mehdi Seifbarghy, "Maximal benefit location problem for a congested system," *Journal of Industrial Engineering* **5** (2010) 73-83. (Cited on page 226.)

[540] Radhakrishnan, Suresh and Kashi R. Balachandran, "Service capacity decision and incentive compatible cost allocation for reporting usage forecasts," *European Journal of Operational Research* **157** (2004) 180-195. (Cited on page 119.)

[541] Randhawa Ramandeep S. and Sunil Kumar, "Usage restriction and sub-scription services: operational benefits with rational users," *Manufacturing & Service Operations Management* **10**(3) (2008) 429-447. (Cited on page 161.)

[542] Rapoport, Amnon, William E. Stein, Vincent Mak, Rami Zwick, and Darryl A. Seale, "Endogenous arrivals in batch queues with constant or variable capacity," *Transportation Research Part B* **44** (2010) 1166-1185. (Cited on page 74, 93, and 95.)

[543] Rapoport, Amnon, William E. Stein, James E. Parco, and Darryl A. Seale, "Equilibrium play in single-server queues with endogenously de-termined arrival times," *Journal of Economics Behavior & Organization* **55** (2004) 67-91. (Cited on page 93 and 94.)

[544] Ravner, Liron, "Equilibrium arrival times to a queue with order penal-ties," *European Journal of Operational Research* **239** (2013) 456-468. (Cited on page 87.)

[545] Ray, Saibal and Elizabeth M. Jewkes, "Customer lead time management when both demand and price are lead time sensitive," *European Journal of Operational Research* **153** (2004) 769-781. (Cited on page 143 and 235.)

[546] Reggiani, Carlo and Tommaso Valletti, "Net neutrality and innovation at the core and at the edge," manuscript (2012). (Cited on page 257.)

[547] Ren, Z. Justin and Fuqiang Zhang, "Service outsourcing: capacity, qual-ity and correlated costs," manuscript (2009). (Cited on page 240.)

[548] Ren, Z. Justin and Yong-Pin Zhou, "Call center outsourcing: coordinat-ing staffing level and service quality," *Management Science* **54**(2) (2008) 369-383. (Cited on page 232 and 239.)

[549] Richman, Oran and Nahum Shimkin, "Topological uniqueness of the Nash equilibrium for selfish routing with atomic users," *Mathematics of Operations Research* **32**(1) (2007) 215-232. (Cited on page 209.)

[550] Ros, David and Bruno Tuffin, "A mathematical model of the Paris Metro Pricing scheme for charging packet networks," *Computer Networks* **46** (2004) 73-85. (Cited on page 167.)

[551] Rubinovitch, Michael, "The Slow Server Problem: A Queue with Stalling," *Journal of Applied Probability* **22**(4) (1985) 879-892 (Cited on page 36 and 41.)

[552] Ruelas-Gonzalez, Eileen Ninette, Jorge Limon-Robles, and Neale Ri-cardo Smith-Cornejo, "Determining a checkout register opening policy to maximize profit in convenience stores chain," *Journal of Applied Research and Technology* **8**(3) (2010) 406-415. (Cited on page 265 and 284.)

[553] Sadat, Somayeh, Hossein Abouee-Mehrizi, and Michael W. Carter, "Can hospitals compete on quality?" *Health Care Management Science* **18** (2015) 376-388. (Cited on page 191.)

[554] Sahin, Ismet and Marwan A. Simaan, "A flow and routing control policy for communication networks with multiple competitive users," *Journal of the Franklin Institute* **343** (2006) 168-180. (Cited on page 105 and 209.)

[555] Sahin, Ismet and Marwan A. Simaan, "Competitive flow control in general multi-node multi-link communication networks," *International Journal of Communication Systems* **21** (2008) 167-184. (Cited on page 107.)

[556] Saidi-Mehrabad, Mohammad, Ebrahim Teimoury, and Ali Pahlavani, "Modeling customer reactions to congestion in competitive service facilities," *Journal of Service Science & Management* **3** (2010) 186-197. (Cited on page 291.)

[557] Sanders, Beverly A., "An incentive compatible flow control algorithm for rate allocation in computer networks," *IEEE Transactions on Computers* **37**(9) (1988) 1067-1072. (Cited on page 208 and 284.)

[558] Sankaranarayanan, Karthik, Erik R. Larsen, and Ann van Ackere, "Intelligent agents behavior in the queueing process: integrating cellular automata & genetic algorithms," *IEEM* (2009). (Cited on page 125 and 205.)

[559] Sankaranarayanan, Karthik, Erik R. Larsen, Ann van Ackere, and Carlos A. Delgado," Genetic algorithm based optimization of an agent based queuing system," *(IEEM)* (2010). (Cited on page 125 and 205.)

[560] Sasaki, Dan, "Affirmative priority queueing," (1997). (Cited on page 168.)

[561] Sattinger, Michael, "A queueing model of the market for access to trading partners," *International Economic Review* **43**(2) (2002) 533-547. (Cited on page 188.)

[562] Savaşaneril, Seçil, Paul M. Griffin, and Pinar Keskinocak, "Dynamic lead-time quotation for an M/M/1 base-stock inventory queue," *Operations Research* **58**(2) (2010) 383-395. (Cited on page 295.)

[563] Savaşaneril, Seçil and Ece Sayin, "Dynamic lead time quotation for a manufacturing system with responsive inventory and multiple customer classes," manuscript (2015). (Cited on page 297.)

[564] Seale, Darryl A., James E. Parco, William E. Stein, and Amnon Rapoport, "Joining a queue or staying out: effects of information structure and service time on arrival and staying out decisions," *Experimental Economics* **8** (2005) 117-144. (Cited on page 93 and 94.)

[565] Sen Sagnika, T.S. Raghu, and Ajay Vinze, "Demand heterogeneity in IT infrastructure services: modeling and evaluation of a demand approach to defining service levels," *Information Systems Research* **20**(2) (2009) 258-276. (Cited on page 56.)

[566] Serel, Doğan, A. and Erdal Erel, "Coordination of staffing and pricing decisions in a service firm," *Applied Stochastic Models Business and Industry* **24** (2008) 307-323. (Cited on page 143.)

[567] Shang, Weixin and Liming Liu, "Promised delivery time and capacity games in time-based competition," *Management Science* **57**(3) (2011) 599-610. (Cited on page 188, 286, and 298.)

[568] Shanthikumar, J. George and David D. Yao, "Multiclass queueing systems: polymatroidal structure and optimal scheduling control," *Operations Research* **40**(Supp. No. 2) (1992) S293-S299. (Cited on page 15.)

[569] Shen, Hongxia and Tamer Başar, "Optimal nonlinear pricing for a monopolistic network service provider with complete and incomplete information," *IEEE Journal on Selected Areas in Communications* **25**(6) (2007) 1216-1223. (Cited on page 150 and 160.)

[570] Shenker, Scott, "Making greed work in networks: a game-theoretic analysis of switch service disciplines," *IEEE/ACM Transactions on Networking* **3**(6) (1995) 819-831. (Cited on page 105 and 130.)

[571] Shi, Ying and Zhaotong Lian, "Optimization and strategic behavior in a passenger-taxi service system," *European Journal of Operational Research* **249**(3) (2016) 1024-1032. (Cited on page 25.)

[572] Shiang, Hsien-Po and Mihaela van der Schaar, "Queuing-based dynamic channel selection for heterogeneous multimedia applications over cognitive radio networks," *IEEE Transactions on Multimedia* **10**(5) (2008) 896-909. (Cited on page 208 and 274.)

[573] Shimkin, Nahum, and Avishai Mandelbaum, 'Rational abandonment from tele-queues: nonlinear waiting costs with heterogeneous preferences," *Queueing Systems* **47** (2004) 117-146. (Cited on page 9 and 300.)

[574] Shone, Rob, Vincent Knight, and Janet Williams, "Comparisons between observable and unobservable M/M/1 queues with respect to optimal customer behavior," *European Journal of Operational Research* **227** (2013) 133-141. (Cited on page 72.)

[575] Shone, Rob, Vincent A. Knight, Paul R. Harper, Janet E. Williams, and John Minty, "Containment of socially optimal policies in multiple-facility Markovian queueing systems," *Journal of the Operational Research Society* (2015) 1-15. (Cited on page 25.)

[576] Shrimali, Gireesh, "Competitive resource sharing by Internet service providers," *Netnomics* **11** (2010) 149-179. (Cited on page 192 and 198.)

[577] Simhon, Eran, Yezekael Hayel, David Starobinski, and Quanyan Zhu, "Optimal information disclosure policies in strategic queueing games," *Operations Research Letters* forthcoming. (Cited on page 74.)

[578] Simhon, Eran and David Starobinski, "Game-theoretic analysis of advance reservation services," *CISS* (2014). (Cited on page 179.)

[579] Singer, Gonen and Eugene Khmelnitsky, "A finite-horizon, stochastic optimal control policy for a production-inventory system with backlog-dependent lost sales," *IIE Transactions* **42** (2010) 855-864. (Cited on page 39.)

[580] Sinha, Sudhir K., Narayan Rangaraj, and N. Hemachandra, "Pricing surplus server capacity for mean waiting time sensitive customers," *European Journal of Operational Research* **205** (2010) 159-171. (Cited on page 16 and 176.)

[581] Slotnick, Susan A., "Optimal and heuristic lead-time quotation for an integrated steel mill with a minimum batch size," *European Journal of Operational Research* **210** (2011) 527-536. (Cited on page 295.)

[582] Slotnick, Susan A., "Lead-time quotation when customers are sensitive to reputations," *International Journal of Production Research* **52**(3) (2014) 713-726. (Cited on page 205 and 296.)

[583] Slotnick, Susan A. and Matthew J. Sobel, "Manufacturing lead-time rules: customer retention versus tardiness costs," *European Journal of Operational Research* **163** (2005) 825-856. (Cited on page 65 and 294.)

[584] So, Kut C., "Price and time competition for service delivery," *Manufacturing & Service Operations Management* **2**(4) (2000) 392-409. (Cited on page 285 and 287.)

[585] Sobel, J. Matthew, "Queuing processes at competing service facilities," *Management Science* **19**(9) (1973) 985-1000. (Cited on page 205.)

[586] Son, Jae-Dong, "Customer selection problem with profit from a sideline," *European Journal of Operational Research* **176** (2007) 1084-1102. (Cited on page 182 and 183.)

[587] Son, Jae-Dong, "Optimal admission and pricing control problem with deterministic service times and sideline profit," *Queueing Systems* **60** (2008) 71-85. (Cited on page 183.)

[588] Son, Jae-Dong and Yaghoub Khojasteh Ghamari, "Optimal admission and pricing control problems in service industries with multiple servers and sideline profit," *Applied Stochastic Models in Business and Industry* **24** (2008) 325-342. (Cited on page 183.)

[589] Son, Jae-Dong and Seizo Ikuta, "Customer selection problem with search cost, due date, sideline profit, and no waiting room," *Asia-Pacific Journal of Operational Research* **24**(5) (2007) 647-666. (Cited on page 182.)

[590] Stahl, Dale O., "The inefficiency of first and second price auctions in dynamic stochastic environments," *Netnomics* **4** (2002) 1-18. (Cited on page 113.)

[591] Stein, William E., Amnon Rapoport, Darryl A. Seale, Hongtao Zhang, and Rami Zwick, "Batch queue with choice of arrivals: equilibrium analysis and experimental study," *Games and Economic Behavior* **59** (2007) 345-363. (Cited on page 93, 94, and 95.)

[592] Stidham, Shaler, Jr., "Optimal control of admission to a queueing system," *IEEE Transactions on Automatic Control* **AC-42**(8) (1985) 705-713. (Cited on page 10 and 133.)

[593] Stidham, Shaler, Jr., "Pricing and congestion management in a network with heterogeneous users," *IEEE Transactions on Automatic Control* **47**(6) (2004) 976-981. (Cited on page 105 and 110.)

[594] Stidham, Shaler, Jr., *Optimal Design of Queueing Systems* CRC Press, Boca Raton, Fl., 2009. (Cited on page 1, 4, 5, 11, 15, 106, 115, 133, 169, 208, and 245.)

[595] Stidham, Shaler, Jr., "The price of anarchy for a network of queues in heavy traffic," in *Essays in Production, Project Planning and Scheduling* P. Simin Pulat, Subhash C. Sarin, and Reha Uzsoy (eds.), Springer US 2014. (Cited on page 136.)

[596] Su, Xuanming and Stefanos Zenios, "Patient choice in kidney allocation: the role of the queueing discipline," *Manufacturing & Service Operations Management* **6**(4) (2004) 280-301. (Cited on page 36 and 37.)

[597] Su, Xuanming and Stefanos Zenios, "Recipient choice can address the efficiency-equity trade-off in kidney transplantation: a mechanism design model," *Management Science* **52**(11) (2006) 1647-1660. (Cited on page 37.)

[598] Subrata, Riky, Albert Y. Zomaya, and Bjorn Landfeldt, "A cooperative game framework for QoS guided job allocation schemes in grids," *IEEE Transactions on Computers* **57**(10) (2008) 1413-1422. (Cited on page 132.)

[599] Subrata, Riky, Albert Y. Zomaya, and Bjorn Landfeldt, "Game-theoretic approach for load balancing in computational grids," *IEEE Transactions on Parallel and Distributed Systems* **19**(1) (2008) 66-76. (Cited on page 132.)

[600] Subrata, Riky, Albert Y. Zomaya, and Bjorn Landfeldt, "Cooperative power-aware scheduling in grid computing environments," *Journal of Parallel and Distributed Computing* **70** (2010) 84-91. (Cited on page 132.)

[601] Sun, Wei, Pengfei Guo, and Naishuo Tian, "Equilibrium threshold strategies in observable queueing systems with setup/closedown times," *Central European Journal of Operations Research* **18** (2010) 241-268. (Cited on page 269 and 271.)

[602] Sun, Wei and Shiyong Li, "Customer balking strategies in an observable queue with vacations," *Journal of Computational Information Systems* **8**(18) (2012) 7595-7605. (Cited on page 75 and 265.)

[603] Sun, Wei and Shiyong Li, "Customer threshold strategies in observable queues with partial information of service time," *Information Computing and Applications Communications in Computer and Information Science* **307** (2012) 456-462. (Cited on page 25.)

[604] Sun, Wei and Shiyong Li, "Effect of information, uncertainty and parameter variability on profits in a queue with various pricing strategies," *International Journal of Systems Science* **45**(8) (2014) 1781-1798. (Cited on page 75.)

[605] Sun, Wei and Shiyong Li, "Equilibrium and optimal behavior of customers in Markovian queues with multiple working vacations," *TOP* **22**(2) (2014) 694-715. (Cited on page 269.)

[606] Sun, Wei, Shiyong Li, and Cheng-Guo E, "Equilibrium and optimal balking strategies of customers in Markovian queues with multiple vacations and N-policy," *Applied Mathematical Modelling* **40**(1) (2015) 284-301. (Cited on page 266.)

[607] Sun, Wei, Shiyong Li, and Quan-Lin Li, "Equilibrium balking strategies of customers in Markovian queues with two-stage working vacations," *Applied Mathematics and Computation* **248** (2014) 195-214. (Cited on page 272.)

[608] Sun, Wei, Shiyong Li, and Naishuo Tian, "Equilibrium mixed strategies of customers in an unobservable queue with multiple vacations," *Quality Technology & Quantitative Management* **10**(4) (2013) 389-421. (Cited on page 75 and 265.)

[609] Sun, Wei, Shiyong Li, Naishuo Tian, and Hong-ke Zhang, "Equilibrium analysis in batch-arrival queues with complementary services," *Applied Mathematical Modelling* **33** (2009) 224-241. (Cited on page 221.)

[610] Sun, Wei, Naishuo Tian, and Shiyong Li, "The allocation of customers in a discrete-time multi-server queueing system," *Asia-Pacific Journal of Operational Research* **27**(6) (2010) 649-667. (Cited on page 212.)

[611] Sun, Wei, Naishuo Tian, Shiyong Li, and Hong-ke Zhang, "Priority distribution in a queue with service discrimination," *International Journal of Nonlinear Science* **4**(3) (2007) 193-200. (Cited on page 116.)

[612] Sun, Wei, Yulan Wang, and Naishuo Tian, "Pricing and setup/closedown policies in unobservable queues with strategic customers," *4OR-A Quarterly Journal of Operations Research* **10**(3) (2012) 287-311. (Cited on page 271.)

[613] Sundar, D. Krishna and K. Ravikumar, "An actor-critic algorithm for multi-agent learning in queue-based stochastic games," *Neurocomputing* **127** (2014) 258-265. (Cited on page 57.)

[614] Tan, Yong, I. Robert Chiang, and Vijay S. Mookerjee, "An economic analysis of interconnection arrangements between Internet backbone providers," *Operations Research* **54**(4) (2006) 776-788. (Cited on page 198.)

[615] Tan, Zhijia, Wan Li, Xiaoning Zhang, and Hai Yang, "Service charge and capacity selection of an inland river port with location-dependent shipping cost and service congestion," *Transportation Research Part E* **76** (2015) 13-33. (Cited on page 181.)

[616] Tandra, Rahul, N. Hemachandra, and D. Manjunath, "DiffServ node with join minimum cost queue policy and multiclass traffic," *Performance Evaluation* **55** (2004) 69-91. (Cited on page 41.)

[617] Tang, Ling and Hao Chen, "Joint pricing and capacity planning in the IaaS cloud market," *IEEE Transactions on Cloud Computing* forthcoming. (Cited on page 252.)

[618] Teimoury, Ebrahim and Mahdi Fathi, "An integrated operations-marketing perspective for making decisions about order penetration point in multi-product supply chain: a queuing approach," *International Journal of Production Research* **51**(18) (2013) 5576-5594. (Cited on page 167 and 238.)

[619] Teimoury, Ebrahim, Mohammad Modarres, A. Kazeruni Monfared, and Mahdi Fathi, "Price, delivery time, and capacity decisions in an M/M/1 make-to-order/service system with segmented market," *The International Journal of Advanced Manufacturing Technology* **57** (2011) 235-244. (Cited on page 166.)

[620] Tian, Ruiling and Dequan Yue, "Equilibrium strategies in an observable queue with single exponential vacation," *CCDC* (2011). (Cited on page 272.)

[621] Tian, Ruiling and Dequan Yue, "Optimal balking strategies in a Markovian queue with a single vacation," *Journal of Information & Computational Science* **9**(10) (2012) 2827-2841. (Cited on page 272.)

[622] Tian, Ruiling, Dequan Yue, and Wuyi Yue "Optimal balking strategies in an M/G/1 queueing system with a removal server under N-policy," *Journal of Industrial and Management Optimization* **11**(3) (2015) 715-731. (Cited on page 266.)

[623] Timmer, Judith and Werner Scheinhardt, "Cost sharing of cooperating queues in a Jackson network," *Queueing Systems* **75** (2013) 1-17. (Cited on page 127.)

[624] Toktaş-Palut, Peral and Füsun Ülengin, "Coordination in a two-stage capacitated supply chain with multiple suppliers," *European Journal of Operational Research* **212** (2011) 43-53. (Cited on page 248.)

[625] Tong, Chunyang, "Pricing schemes for congestion-prone service facilities," *Operations Research Letters* **40** (2012) 498-502. (Cited on page 108.)

[626] Tong, Chunyang and Sampath Rajagopalan, "Pricing and operational performance in discretionary services," *Production and Operations Management* **23**(4) (2014) 689-703. (Cited on page 43, 155, and 158.)

[627] Touati, Corinne, Parijat Dube, and Laura Wynter, "Performance planning, quality-of-service, and pricing under competition," *NETWORKING* (2004). (Cited on page 189.)

[628] Tran, Nguyen H., Choong Seon Hong, Zhu Han, and Sungwon Lee, "Optimal pricing effect on equilibrium behaviors of delay-sensitive users in cognitive radio networks," *IEEE Journal on Selected Areas in Communications* **31**(11) (2013) 2566-2579. (Cited on page 192, 216, and 278.)

[629] Tran, Nguyen H., Long Bao Le, Shaolei Ren, Zhu Han, and Choong Seon Hong, "Joint pricing and load balancing for cognitive spectrum access: non-cooperation vs cooperation, *IEEE Journal on Selected Areas in Communications* **33**(5) (2015) 972-985. (Cited on page 214.)

[630] Tran, Nguyen H., Dai H. Tran, Long Bao Le, Zhu Han, and Choong Seon Hong, "Load balancing and pricing for spectrum access control in cognitive radio networks," *Globecom* (2014). (Cited on page 213 and 276.)

[631] Tuffin, Bruno, Hélène Le Cadre, and Mustapha Bouhtou, "Optimal pricing strategy with compensation when QoS is not satisfied," *Annals of Telecommunications* **65** (2010) 163-170. (Cited on page 145.)

[632] Turhan, Aylin, Murat Alanyali, and David Starobinski, "Dynamic pricing of preemptive service for elastic demand," *CDC* (2012). (Cited on page 44, 48, and 300.)

[633] Veeraraghavan, Senthil K. and Laurens G. Debo, "Joining longer queues: information externalities in queue choice," *Manufacturing & Service Operations Management* **11**(4) (2009) 543-562. (Cited on page 61.)

[634] Veeraraghavan, Senthil K. and Laurens G. Debo, "Herding in queues with waiting costs: rationality and regret," *Manufacturing & Service Operations Management* **13**(3) (2011) 329-346. (Cited on page 61 and 62.)

[635] Veltman Ari and Refael Hassin, "Equilibrium in queueing systems with complementary products," *Queueing Systems* **50** (2005) 325-342. (Cited on page 221, 222, and 256.)

[636] Vorasayan, Jumpol and Sarah M. Ryan, "Optimal price and quantity of refurbished products," *Production and Operations Management* **15**(3) (2006) 369-383. (Cited on page 151.)

[637] Wang, Chia-Li, "On socially optimal queue length," *Management Science* forthcoming. (Cited on page 26, 34, and 140.)

[638] Wang, Eric Tswen-Gwo, "Information and incentives in computing services supply: the effect of limited system choices," *European Journal of Operational Research* **125** (2000) 503-518. (Cited on page 119.)

[639] Wang, Eric Tswen-Gwo and Terry Barron, "Controlling information system departments in the presence of cost information asymmetry," *Information Systems Research* **6**(1) (1995) 24-50. (Cited on page 119.)

[640] Wang, Eric Tswen-Gwo and Terry Barron, "Computing services supply management: incentives, information, and communication," *Decision Support Systems* **19** (1997) 123-148. (Cited on page 9 and 119.)

[641] Wang, Fang, Jinting Wang, and Feng Zhang, "Equilibrium customer strategies in the Geo/Geo/1 queues with single working vacations," *Discrete Dynamics in Nature and Society* (2014) 9 pages. (Cited on page 272.)

[642] Wang, Fang, Jinting Wang, and Feng Zhang, "Strategic behavior in the single-server constant retrial queue with individual removal," *Quality Technology & Quantitative Management* **12**(3) (2015) 323-340. (Cited on page 90 and 277.)

[643] Wang, Jinting and Feng Zhang, "Monopoly pricing in a retrial queue with delayed vacations for local area network applications," *IMA Journal of Management Mathematics* (2015) 20 pages. (Cited on page 90 and 277.)

[644] Wang, Jinting and Feng Zhang, "Non-cooperative and cooperative joining strategies in cognitive radio networks with random access," *IEEE Transactions on Vehicular Technology* forthcoming. (Cited on page 91 and 272.)

[645] Wang, Gang, Zhen Zeng, and Meng Liu, "Equilibrium threshold strategy in observable queueing systems in cognitive radio networks," *ICMIC* (2013). (Cited on page 274.)

[646] Wang, Jinting and Feng Zhang, "Equilibrium analysis of the observable queues with balking and delayed repairs," *Applied Mathematics and Computation* **218** (2011) 2716-2729. (Cited on page 275.)

[647] Wang, Jinting and Feng Zhang, "Strategic joining in M/M/1 retrial queues," *European Journal of Operational Research* **230**(1) (2013) 76-87. (Cited on page 90.)

[648] Wang, Jinting, Zhe George Zhang, and Zhengwu Zhang, "Performance analysis of a queue with strategic customers under a quadratic utility criterion," manuscript (2014). (Cited on page 25.)

[649] Wang, Shanshan, Junshan Zhang, and Lang Tong, "Delay analysis for cognitive radio networks with random access: a fluid queue view," *INFO-COM* (2010). (Cited on page 274.)

[650] Wang, Susheng and Lijing Zhu, "A dynamic queuing model," *The Chinese Journal of Economic Theory* **1** (2004) 14-35. (Cited on page 84 and 85.)

[651] Wang, Susheng and Lijing Zhu, "Increasing returns to scale from variable capacity utilization," *International Journal of Economic Theory* **3** (2007) 1-22. (Cited on page 85.)

[652] Wang, Xiao Jun and Shiu Hong Choi, "Optimisation of stochastic multi-item manufacturing for shareholder wealth maximisation," *Engineering Letters* **21**(3) (2013) 127-136. (Cited on page 152.)

[653] Wang, Xiao Jun and Shiu Hong Choi, "Wealth-based production planning with uncertain lot sizing," *International Journal of Applied Mathematics* **44**(2) (2014) 109-116. (Cited on page 152.)

[654] Wang, Xiaofang, Laurens G. Debo, Alan Scheller-Wolf, and Stephen F. Smith, "Design and analysis of diagnostic service centers," *Management Science* **56**(11) (2010) 1873-1890. (Cited on page 155 and 156.)

[655] Wang, Xiaofang, Laurens G. Debo, Alan Scheller-Wolf, and Stephen F. Smith, "Service design at diagnostic service centers," *Naval Research Logistics* **59**(8) (2012) 613-628. (Cited on page 155 and 156.)

[656] Wang, Xiaofang and Jun Zhuang, "Balancing congestion and security in the presence of strategic applicants with private information," *European Journal of Operational Research* **212** (2011) 100-111. (Cited on page 159.)

[657] Wee, Kwan-Eng and Ananth Iyer, "Consolidating or non-consolidating queues: a game theoretic queueing model with holding costs," *Operations Research Letters* **39** (2011) 4-10. (Cited on page 14 and 247.)

[658] Wen-sheng, Yang and Li Li, "Lead-time-contingent pricing under time and price sensitive demand in electronic markets," *WiCom* (2007). (Cited on page 143.)

[659] Weng, Kevin Z., "Manufacturing lead times, system utilization rates and lead-time-related demand," *European Journal of Operational Research* **89** (1996) 259-268. (Cited on page 292.)

[660] Whitt, Ward, "Improving service by informing customers about anticipated delays," *Management Science* **45**(2) (1999) 192-207. (Cited on page 17, 38, and 90.)

[661] Więcek, Piotr, Eitan Altman, and Arnob Ghosh, "Mean-field game approach to admission control of an M/M/∞ queue with decreasing congestion cost," *Dynamic Games and Applications* forthcoming. (Cited on page 96.)

[662] Wolisz, Adam and Volker Tschammer, "Performance aspects of trading in open distributed systems," *Computer Communications* **16**(5) (1993) 277-287. (Cited on page 36, 79, and 97.)

[663] Wu, Tao and David Starobinski, "A comparative analysis of server selection in content replication networks," *IEEE/ACM Transactions on Networking* **16**(6) (2008) 1461-1474. (Cited on page 133, 134, and 135.)

[664] Wu, Yu, Loc Bui, and Ramesh Johari, "Heavy traffic approximation of equilibria in resource sharing games," *IEEE Journal on Selected Areas in Communications* **30**(11) 2200-2209. (Cited on page 116.)

[665] Xia, Li, "Service rate control of closed Jackson networks from game theoretic perspective," *European Journal of Operational Research* **237**(2) (2014) 546-554. (Cited on page 58.)

[666] Xiao, Tiaojun and Xiangtong Qi, "A two-stage supply chain with demand sensitive to price, delivery time, and reliability of delivery," *Annals of Operations Research* forthcoming. (Cited on page 237.)

[667] Xiao, Tiaojun and Jing Shi, "Price, capacity, and lead-time decisions for a make-to-order supply chain with two production modes," *International Journal of Applied Management Science* **4**(2) (2012) 107-129. (Cited on page 165 and 237.)

[668] Xiao, Tiaojun, Danqin Yang, and Houcai Shen, "Coordinating a supply chain with a quality assurance policy via a revenue sharing contract," *International Journal of Production Research* **49**(1) (2011) 99-120. (Cited on page 151 and 237.)

[669] Xiaopan, Liu, Wang Jianjun, Zhang Binghang, and Zou Zongbao, "Price and Guaranteed delivery time competition/cooperation in a duopoly," *ICSSSM* (2014). (Cited on page 196.)

[670] Xu, Jiaming and Bruce Hajek, "The supermarket game," *Stochastic Systems* **3**(2) (2013) 405-441. (Cited on page 67.)

[671] Xu, Susan H., Long Gao, and Jihong Ou, "Service performance analysis and improvement for a ticket queue with balking customers," *Management Science* **53**(6) (2007) 971-990. (Cited on page 9, 81, and 167.)

[672] Xu, Susan H. and J. George Shanthikumar, "Optimal expulsion control: a dual approach to admission control of an ordered-entry system," *Operations Research* **41**(6) (1993) 1137-1152. (Cited on page 33 and 137.)

[673] Xu, Xiaoya, Zhaotong Lian, Xin Li, and Pengfei Guo, "A Hotelling model with probabilistic service," manuscript (2016). (Cited on page 144.)

[674] Xu, Ying, Tinglong Dai, Katia Sycara, and Michael Lewis, "A mechanism design model in robot-service-queue control with strategic operators and asymmetric information," *CDC* (2012). (Cited on page 121 and 213.)

[675] Xue, Mei and Patrick Harker, "Service co-production, customer efficiency and market competition," unpublished manuscript (2003). (Cited on page 239.)

[676] Yahalom, Tomer, J. Michael Harrison, and Sunil Kumar, "Designing and pricing incentive compatible grades of service in queueing systems," unpublished manuscript (2006). (Cited on page 174.)

[677] Yang, Bixuan, Zhenting Hou, Jinbiao Wu, and Zaiming Liu, "Analysis of the equilibrium strategies in the Geo/Geo/1 queue with multiple working vacations," manuscript (2014). (Cited on page 269.)

[678] Yang, Liu, Pengfei Guo, and Yulan Wang, "Service pricing with loss averse customers," manuscript (2014). (Cited on page 100 and 150.)

[679] Yang, Liu, Francis de Véricourt, and Peng Sun, "Time-based competition with benchmark effects," *Manufacturing & Service Operations Management* **16**(1) (2014) 119-132. (Cited on page 99 and 188.)

[680] Yang, Luyi, Laurens Debo, and Varun Gupta, "Trading time in congested environment," (2015). (Cited on page 137.)

[681] Yang, Tengteng, Jinting Wang, and Feng Zhang, "Equilibrium balking strategies in the Geo/Geo/1 queues with server breakdowns and repairs," *Quality Technology & Quantitative Management* **11**(3) (2014) 231-243. (Cited on page 274.)

[682] Yao, David D., "*S*-modular games, with queueing applications," *Queueing Systems* **21** (1995) 449-475. (Cited on page 125 and 259.)

[683] Yeh, Ruey Huei and Yi-Fang Lin, "Optimal pricing policies for services with consideration of facility maintenance costs," *International Journal of Systems Science* **43**(6) (2012) 1123-1132. (Cited on page 142 and 276.)

[684] Yildirim, Utku and John J. Hasenbein, "Admission control and pricing in a queue with batch arrivals," *Operations Research Letters* **38** (2010) 427-431. (Cited on page 26 and 53.)

[685] Yolken, Benjamin and Nicholas Bambos, "Game based capacity allocation for utility computing environments," *Telecommunication Systems* **47** (2011) 165-181. (Cited on page 151.)

[686] Yoshida, Yuichiro, "Commuter arrivals and optimal service in mass transit: does queuing behavior at transit stops matter?" *Regional Science and Urban Economics* **38** (2008) 228-251. (Cited on page 85.)

[687] Yu, Dennis Z., Xuying Zhao, and Daewon Sun, "Optimal pricing and capacity investment for delay-sensitive demand," *IEEE Transactions on Engineering Management* **60**(1) (2013) 124-136 (Cited on page 171.)

[688] Yu, Qiuping, Gad Allon, and Achal Bassamboo, "How do delay announcements shape customer behavior? An empirical study," *Management Science* forthcoming. (Cited on page 72.)

[689] Yu, Yimin, Saif Benjaafar, and Yigal Gerchak, "Capacity sharing and cost allocation among independent firms with congestion," *Production and Operations Management* (2015) **24**(8) 1285-1310. (Cited on page 129.)

[690] Yue, Dequan, Ruiling Tian, Wuyi Yue, and Yaling Qin, "Equilibrium strategies in an M/M/1 queue with setup times and a single vacation," *ISORA* (2013). (Cited on page 272.)

[691] Zarrinpoor, Naeme and Mehdi Seifbarghy, "A competitive location model to obtain a specific market share while ranking facilities by shorter travel time," *The International Journal of Advanced Manufacturing Technology* **55** (2011) 807-816. (Cited on page 291.)

[692] Zeng, Zhen and Gang Wang, "Equilibrium analysis and social optimization in cognitive radio networks," *International Journal of u- and e-Service Science and Technology* **7**(2) (2014) 189-198. (Cited on page 275.)

[693] Zhan, Dongyuan and Amy R. Ward, "Compensation and staffing to trade off speed and quality in large service systems," manuscript (2015). (Cited on page 250.)

[694] Zhang, Feng, Jinting Wang, and Bin Liu, "On the optimal and equilibrium retrial rates in an unreliable retrial queue with vacations," *Journal of Industrial and Management Optimization* **8**(4) (2012) 861-875. (Cited on page 90 and 277.)

[695] Zhang, Feng, Jinting Wang, and Bin Liu, "Equilibrium balking strategies in Markovian queues with working vacations," *Applied Mathematical
Modelling* **37** (2013) 8264-8282. (Cited on page 269.)

[696] Zhang, Feng, Jinting Wang, and Bin Liu, "Equilibrium joining probabilities in observable queues with general service and setup times," *Journal
of Industrial and Management Optimization* **9**(4) (2013) 901-917. (Cited
on page 26 and 271.)

[697] Zhang, Guanxiang, Jianhua He, Yajie Ma, Wenqing Cheng, and Zongkai
Yang, "Congestion pricing by priority auction," *SPIE* (2005). (Cited on
page 178.)

[698] Zhang, Juanjuan, "The sound of silence: observational learning in the
U.S. kidney market," *Management Science* **29**(2) (2010) 315-335. (Cited
on page 36 and 63.)

[699] Zhang, Li, Fang Wu, and Bernardo A. Huberman, "Games and Queues,"
unpublished manuscript (2008). (Cited on page 74 and 92.)

[700] Zhang, Yongbing, Hisao Kameda, and Kentaro Shimizu, "Parametric
analysis of optimal static load balancing in distributed computer systems," *Journal of Information Processing* **14**(4) (1991) 433-441. (Cited
on page 226.)

[701] Zhang, Yue, Oded Berman, Patrice Marcotte, and Vedat Verter, "A
bilevel model for preventive healthcare facility network design with congestion," *IIE Transactions* **42** (2010) 865-880. (Cited on page 225.)

[702] Zhang, Yue, Oded Berman, and Vedat Verter, "Incorporating congestion
in preventive healthcare facility network design," *European Journal of
Operational Research* **198** (2009) 922-935. (Cited on page 225.)

[703] Zhang, Zhengwu, Jinting Wang, and Feng Zhang, "Equilibrium customer strategies in the single-server constant retrial queue with breakdowns and repairs," *Mathematical Problems in Engineering* (2014), 14
pages. (Cited on page 90, 182, and 277.)

[704] Zhang, Zhongju, Debabrata Dey, and Yong Tan, "Pricing communication services with delay guarantees," *INFORMS Journal on Computing*
19(2) (2007) 248-260. (Cited on page 170.)

[705] Zhang, Zhongju, Debabrata Dey, and Yong Tan, "Price and QoS competition in data communication services," *European Journal of Operational
Research* **187** (2008) 871-886. (Cited on page 190, 197, and 288.)

[706] Zhang, Zhongju, Yong Tan, and Debabrata Dey, "Price competition
with service level guarantee in web services," *Decision Support Systems*
47 (2009) 93-104. (Cited on page 190.)

[707] Zhao, Xuying, Kathryn E. Stecke, and Ashutosh Prasad, "Lead time and price quotation mode selection: uniform or differentiated?" *Production and Operations Management* **21** (2012) 177-193. (Cited on page 166.)

[708] Zhao, Yuan, Shunfu Jin, and Wuyi Yue, "An adjustable channel bonding strategy in centralized cognitive radio networks and its performance optimization," *Quality Technology & Quantitative Management* **12**(3) (2015) 291-310. (Cited on page 276.)

[709] Zheng, Wenxiang, "Analysis of customer behavior under uncertainty on distribution of service time parameter in M/M/1 queueing system," *ICSSSM* (2015). (Cited on page 76.)

[710] Zhou, Rongrong and Dilip Soman, "Looking back: exploring the psychology of queuing and the effect of the number of people behind," *Journal of Consumer Research* **29**(4) (2003) 517-530. (Cited on page 61 and 95.)

[711] Zhou, Wenhui, Xiuli Chao, and Xiting Gong, "Optimal uniform pricing strategy of a service firm when facing two classes of customers," *Production and Operations Management* **23**(4) (2014) 676-688. (Cited on page 142.)

[712] Zhou, Wenhui, and Weixiang Huang, "Two pricing mechanisms for a service provider when customers' delay costs are value-related," *Computers & Industrial Engineering* **87** (2015) 600-610. (Cited on page 177.)

[713] Zhou, Wenhui, Zhaotong Lian, and Jinbiao Wu, "When should service firms provide free experience service?," *European Journal of Operational Research* **234** (2014) 830-838. (Cited on page 184.)

[714] Zhu, Stuart X., "Integration of capacity, pricing, and lead-time decisions in a decentralized supply chain," *International Journal of Production Economics* **164** (2015) 14-23. (Cited on page 238 and 292.)

[715] Ziani, Sofiane, Fazia Rahmoune, and Mohammed Said Radjef, "Customers' strategic behavior in batch arrivals $M^2/M/1$ queue," *European Journal of Operational Research* **247** (2015) 895-903. (Cited on page 25.)

[716] Ziedins, Ilze, "A paradox in a queueing network with state-dependent routing and loss," *Journal of Applied Mathematics and Decision Sciences* (2007) Article ID 68280, 10 pages. (Cited on page 218 and 229.)

[717] Ziya, Serhan, Haytiye Ayhan, and Robert D. Foley, "Optimal prices for finite capacity queueing systems," *Operations Research Letters* **34** (2006) 214-218. (Cited on page 141.)

[718] Ziya, Serhan, Haytiye Ayhan, and Robert D. Foley, "A note on optimal pricing for finite capacity systems with multiple customer classes," *Naval Research Logistics* **55** (2008) 412-418. (Cited on page 141.)

[719] Zohar, Ety, Avishai Mandelbaum, and Nahum Shimkin, "Adaptive behavior of impatient customers in tele-queues: theory and empirical support," *Management Science* **48**(4) (2002) 566-583. (Cited on page 17 and 95.)

Subject index

Author index

Printed and bound by CPI Group (UK) Ltd, Croydon, CR0 4YY

24/10/2024

01778283-0011